INEQUALITIES

LECTURE NOTES

IN PURE AND APPLIED MATHEMATICS

1. *N. Jacobson,* Exceptional Lie Algebras
2. *L. -Å. Lindahl and F. Poulsen,* Thin Sets in Harmonic Analysis
3. *I. Satake,* Classification Theory of Semi-Simple Algebraic Groups
4. *F. Hirzebruch, W. D. Newmann, and S. S. Koh,* Differentiable Manifolds and Quadratic Forms (out of print)
5. *I. Chavel,* Riemannian Symmetric Spaces of Rank One (out of print)
6. *R. B. Burckel,* Characterization of C(X) Among Its Subalgebras
7. *B. R. McDonald, A. R. Magid, and K. C. Smith,* Ring Theory: Proceedings of the Oklahoma Conference
8. *Y.-T. Siu,* Techniques of Extension on Analytic Objects
9. *S. R. Caradus, W. E. Pfaffenberger, and B. Yood,* Calkin Algebras and Algebras of Operators on Banach Spaces
10. *E. O. Roxin, P.-T. Liu, and R. L. Sternberg,* Differential Games and Control Theory
11. *M. Orzech and C. Small,* The Brauer Group of Commutative Rings
12. *S. Thomeier,* Topology and Its Applications
13. *J. M. Lopez and K. A. Ross,* Sidon Sets
14. *W. W. Comfort and S. Negrepontis,* Continuous Pseudometrics
15. *K. McKennon and J. M. Robertson,* Locally Convex Spaces
16. *M. Carmeli and S. Malin,* Representations of the Rotation and Lorentz Groups: An Introduction
17. *G. B. Seligman,* Rational Methods in Lie Algebras
18. *D. G. de Figueiredo,* Functional Analysis: Proceedings of the Brazilian Mathematical Society Symposium
19. *L. Cesari, R. Kannan, and J. D. Schuur,* Nonlinear Functional Analysis and Differential Equations: Proceedings of the Michigan State University Conference
20. *J. J. Schäffer,* Geometry of Spheres in Normed Spaces
21. *K. Yano and M. Kon,* Anti-Invariant Submanifolds
22. *W. V. Vasconcelos,* The Rings of Dimension Two
23. *R. E. Chandler,* Hausdorff Compactifications
24. *S. P. Franklin and B. V. S. Thomas,* Topology: Proceedings of the Memphis State University Conference
25. *S. K. Jain,* Ring Theory: Proceedings of the Ohio University Conference
26. *B. R. McDonald and R. A. Morris,* Ring Theory II: Proceedings of the Second Oklahoma Conference
27. *R. B. Mura and A. Rhemtulla,* Orderable Groups
28. *J. R. Graef,* Stability of Dynamical Systems: Theory and Applications
29. *H.-C. Wang,* Homogeneous Branch Algebras
30. *E. O. Roxin, P.-T. Liu, and R. L. Sternberg,* Differential Games and Control Theory II
31. *R. D. Porter,* Introduction to Fibre Bundles
32. *M. Altman,* Contractors and Contractor Directions Theory and Applications
33. *J. S. Golan,* Decomposition and Dimension in Module Categories
34. *G. Fairweather,* Finite Element Galerkin Methods for Differential Equations
35. *J. D. Sally,* Numbers of Generators of Ideals in Local Rings
36. *S. S. Miller,* Complex Analysis: Proceedings of the S.U.N.Y. Brockport Conference
37. *R. Gordon,* Representation Theory of Algebras: Proceedings of the Philadelphia Conference
38. *M. Goto and F. D. Grosshans,* Semisimple Lie Algebras
39. *A. I. Arruda, N. C. A. da Costa, and R. Chuaqui,* Mathematical Logic: Proceedings of the First Brazilian Conference

40. *F. Van Oystaeyen,* Ring Theory: Proceedings of the 1977 Antwerp Conference
41. *F. Van Oystaeyen and A. Verschoren,* Reflectors and Localization: Application to Sheaf Theory
42. *M. Satyanarayana,* Positively Ordered Semigroups
43. *D. L. Russell,* Mathematics of Finite-Dimensional Control Systems
44. *P.-T. Liu and E. Roxin,* Differential Games and Control Theory III: Proceedings of the Third Kingston Conference, Part A
45. *A. Geramita and J. Seberry,* Orthogonal Designs: Quadratic Forms and Hadamard Matrices
46. *J. Cigler, V. Losert, and P. Michor,* Banach Modules and Functors on Categories of Banach Spaces
47. *P.-T. Liu and J. G. Sutinen,* Control Theory in Mathematical Economics: Proceedings of the Third Kingston Conference, Part B
48. *C. Byrnes,* Partial Differential Equations and Geometry
49. *G. Klambauer,* Problems and Propositions in Analysis
50. *J. Knopfmacher,* Analytic Arithmetic of Algebraic Function Fields
51. *F. Van Oystaeyen,* Ring Theory: Proceedings of the 1978 Antwerp Conference
52. *B. Kedem,* Binary Time Series
53. *J. Barros-Neto and R. A. Artino,* Hypoelliptic Boundary-Value Problems
54. *R. L. Sternberg, A. J. Kalinowski, and J. S. Papadakis,* Nonlinear Partial Differential Equations in Engineering and Applied Science
55. *B. R. McDonald,* Ring Theory and Algebra III: Proceedings of the Third Oklahoma Conference
56. *J. S. Golan,* Structure Sheaves over a Noncommutative Ring
57. *T. V. Narayana, J. G. Williams, and R. M. Mathsen,* Combinatorics, Representation Theory and Statistical Methods in Groups: YOUNG DAY Proceedings
58. *T. A. Burton,* Modeling and Differential Equations in Biology
59. *K. H. Kim and F. W. Roush,* Introduction to Mathematical Consensus Theory
60. *J. Banas and K. Goebel,* Measures of Noncompactness in Banach Spaces
61. *O. A. Nielson,* Direct Integral Theory
62. *J. E. Smith, G. O. Kenny, and R. N. Ball,* Ordered Groups: Proceedings of the Boise State Conference
63. *J. Cronin,* Mathematics of Cell Electrophysiology
64. *J. W. Brewer,* Power Series Over Commutative Rings
65. *P. K. Kamthan and M. Gupta,* Sequence Spaces and Series
66. *T. G. McLaughlin,* Regressive Sets and the Theory of Isols
67. *T. L. Herdman, S. M. Rankin, III, and H. W. Stech,* Integral and Functional Differential Equations
68. *R. Draper,* Commutative Algebra: Analytic Methods
69. *W. G. McKay and J. Patera,* Tables of Dimensions, Indices, and Branching Rules for Representations of Simple Lie Algebras
70. *R. L. Devaney and Z. H. Nitecki,* Classical Mechanics and Dynamical Systems
71. *J. Van Geel,* Places and Valuations in Noncommutative Ring Theory
72. *C. Faith,* Injective Modules and Injective Quotient Rings
73. *A. Fiacco,* Mathematical Programming with Data Perturbations I
74. *P. Schultz, C. Praeger, and R. Sullivan,* Algebraic Structures and Applications Proceedings of the First Western Australian Conference on Algebra
75. *L. Bican, T. Kepka, and P. Nemec,* Rings, Modules, and Preradicals
76. *D. C. Kay and M. Breen,* Convexity and Related Combinatorial Geometry: Proceedings of the Second University of Oklahoma Conference
77. *P. Fletcher and W. F. Lindgren,* Quasi-Uniform Spaces
78. *C.-C. Yang,* Factorization Theory of Meromorphic Functions
79. *O. Taussky,* Ternary Quadratic Forms and Norms
80. *S. P. Singh and J. H. Burry,* Nonlinear Analysis and Applications
81. *K. B. Hannsgen, T. L. Herdman, H. W. Stech, and R. L. Wheeler,* Volterra and Functional Differential Equations

82. *N. L. Johnson, M. J. Kallaher, and C. T. Long,* Finite Geometries: Proceedings of a Conference in Honor of T. G. Ostrom
83. *G. I. Zapata,* Functional Analysis, Holomorphy, and Approximation Theory
84. *S. Greco and G. Valla,* Commutative Algebra: Proceedings of the Trento Conference
85. *A. V. Fiacco,* Mathematical Programming with Data Perturbations II
86. *J.-B. Hiriart-Urruty, W. Oettli, and J. Stoer,* Optimization: Theory and Algorithms
87. *A. Figa Talamanca and M. A. Picardello,* Harmonic Analysis on Free Groups
88. *M. Harada,* Factor Categories with Applications to Direct Decomposition of Modules
89. *V. I. Istrăţescu,* Strict Convexity and Complex Strict Convexity: Theory and Applications
90. *V. Lakshmikantham,* Trends in Theory and Practice of Nonlinear Differential Equations
91. *H. L. Manocha and J. B. Srivastava,* Algebra and Its Applications
92. *D. V. Chudnovsky and G. V. Chudnovsky,* Classical and Quantum Models and Arithmetic Problems
93. *J. W. Longley,* Least Squares Computations Using Orthogonalization Methods
94. *L. P. de Alcantara,* Mathematical Logic and Formal Systems
95. *C. E. Aull,* Rings of Continuous Functions
96. *R. Chuaqui,* Analysis, Geometry, and Probability
97. *L. Fuchs and L. Salce,* Modules Over Valuation Domains
98. *P. Fischer and W. R. Smith,* Chaos, Fractals, and Dynamics
99. *W. B. Powell and C. Tsinakis,* Ordered Algebraic Structures
100. *G. M. Rassias and T. M. Rassias,* Differential Geometry, Calculus of Variations, and Their Applications
101. *R.-E. Hoffmann and K. H. Hofmann,* Continuous Lattices and Their Applications
102. *J. H. Lightbourne, III, and S. M. Rankin, III,* Physical Mathematics and Nonlinear Partial Differential Equations
103. *C. A. Baker and L. M. Batten,* Finite Geometries
104. *J. W. Brewer, J. W. Bunce, and F. S. Van Vleck,* Linear Systems Over Commutative Rings
105. *C. McCrory and T. Shifrin,* Geometry and Topology: Manifolds, Varieties, and Knots
106. *D. W. Kueker, E. G. K. Lopez-Escobar, and C. H. Smith,* Mathematical Logic and Theoretical Computer Science
107. *B.-L. Lin and S. Simons,* Nonlinear and Convex Analysis: Proceedings in Honor of Ky Fan
108. *S. J. Lee,* Operator Methods for Optimal Control Problems
109. *V. Lakshmikantham,* Nonlinear Analysis and Applications
110. *S. F. McCormick,* Multigrid Methods: Theory, Applications, and Supercomputing
111. *M. C. Tangora,* Computers in Algebra
112. *D. V. Chudnovsky and G. V. Chudnovsky,* Search Theory: Some Recent Developments
113. *D. V. Chudnovsky and R. D. Jenks,* Computer Algebra
114. *M. C. Tangora,* Computers in Geometry and Topology
115. *P. Nelson, V. Faber, T. A. Manteuffel, D. L. Seth, and A. B. White, Jr.* Transport Theory, Invariant Imbedding, and Integral Equations: Proceedings in Honor of G. M. Wing's 65th Birthday
116. *P. Clément, S. Invernizzi, E. Mitidieri, and I. I. Vrabie,* Semigroup Theory and Applications
117. *J. Vinuesa,* Orthogonal Polynomials and Their Applications: Proceedings of the International Congress
118. *C. M. Dafermos, G. Ladas, and G. Papanicolaou,* Differential Equations: Proceedings of the EQUADIFF Conference
119. *E. O. Roxin,* Modern Optimal Control: A Conference in Honor of Solomon Lefschetz and Joseph P. Lasalle
120. *J. C. Díaz,* Mathematics for Large Scale Computing
121. *P. S. Milojević,* Nonlinear Functional Analysis

122. *C. Sadosky,* Analysis and Partial Differential Equations: A Collection of Papers Dedicated to Mischa Cotlar
123. *R. M. Shortt,* General Topology and Applications: Proceedings of the 1988 Northeast Conference
124. *R. Wong,* Asymptotic and Computational Analysis: Conference in Honor of Frank W. J. Olver's 65th Birthday
125. *D. V. Chudnovsky and R. D. Jenks,* Computers in Mathematics
126. *W. D. Wallis, H. Shen, W. Wei, and L. Zhu,* Combinatorial Designs and Applications
127. *S. Elaydi,* Differential Equations: Stability and Control
128. *G. Chen, E. B. Lee, W. Littman, and L. Markus,* Distributed Parameter Control Systems: New Trends and Applications
129. *W. N. Everitt,* Inequalities: Fifty Years On from Hardy, Littlewood and Pólya

Other Volumes in Preparation

INEQUALITIES

Fifty Years On from Hardy, Littlewood and Pólya

Proceedings of the International Conference London Mathematical Society

edited by

W. Norrie Everitt
University of Birmingham
Birmingham, England

MARCEL DEKKER, INC.   New York • Basel • Hong Kong

ISBN: 0-8247-8488-X

This book is printed on acid-free paper

MARCEL DEKKER, INC.
270 Madison Avenue, New York, New York 10016

Current printing (last digit):
10 9 8 7 6 5 4 3 2 1

PRINTED IN THE UNITED STATES OF AMERICA

To the memory of

G. H. Hardy	1877–1947
J. E. Littlewood	1885–1977
G. Pólya	1887–1985

Preface

Under the auspices of the London Mathematical Society, an International Conference on Inequalities was held July 13–17, 1987, at the University of Birmingham, England. The Conference was attended by more than 100 participants from 21 countries.

The Conference was mounted by the Society not only to encourage the continuing study of inequalities in mathematics but also to acknowledge the indebtedness of the subject to the work of G. H. Hardy, J. E. Littlewood and G. Pólya in writing the book *Inequalities*, which was first published by the Cambridge University Press in 1934. This book has to be counted as one of the outstanding achievements in mathematical scholarship in this century. Of great intrinsic interest, indeed, fascination, the book has proved an invaluable reference work for more than fifty years, and a source of lasting inspiration to workers in the vineyard of inequalities. That the book is still in print after these many years is indicative of the quality and significance of the work. Indeed, it is a stunning achievement for it still to be the main reference in an area of mathematics which has increased its bounds over the years to an extent no one could have predicted when the book first saw the light of day.

The London Mathematical Society sought to bring to the Conference some of the mathematicians who were associated in one way or another with *Inequalities* when it first appeared in 1934. We were fortunate to have Professor A. C. Offord, FRS, and Professor L. C. Young as guests of the Society during the Conference. Both enlivened the meeting with their memories and in providing a tangible link with past times. In this respect it is interesting to note that Professors A. Oppenheim, R. Rado (died 1989) and A. Zygmund are also mentioned in the book as contributors in one respect or another.

Of the 14 plenary lectures given to the Conference, 13 are represented in the contents of this volume. I thank my colleagues for the time and trouble they have taken to prepare these manuscripts as a record of their contribution to the Conference.

In addition there were more than 30 contributed lectures. However, in planning for the Conference, it had been decided that these lectures would not be represented in the conference proceedings. Nevertheless, they formed an important and significant part of the meeting.

The Organising Committee for the Conference had the following membership: W. N. Everitt, J. Gunson, J. Kyle, N. M. Queen and B. Thorpe, all of the Department of Mathematics of the University of Birmingham. The secretarial work for the Conference was placed in the ever-capable hands of the Senior Secretary of the Department, Mrs. Pat Chapman. To all these colleagues I offer my own thanks and those of the London Mathematical Society for their help, co-operation, patience and quiet efficiency. I pay special tribute to Jack Gunson and Nat Queen for their contribution to the planning and organisation of the Conference.

On behalf of the London Mathematical Society and all the participants of the Conference we offer our best thanks to the University of Birmingham for generous support in mounting the meeting, in particular to the departments of Chemistry, Mathematics and Physics for the use of lecture rooms and other accommodation. In the same vein we thank the staff of the Conference Office, the student residence High Hall, and the University Centre for their unfailing courtesy in welcoming and caring for participants of the Conference.

Throughout all the planning for the Conference, the London Mathematical Society was represented by the Meetings and Membership Secretary, Dr. Alan R. Pears. I am indebted to him for his help and guidance.

I express my gratitude to Marcel Dekker, Inc., for including these proceedings in the series Lecture Notes in Pure and Applied Mathematics. Special thanks are given to Maria Allegra for her kindness, help and advice.

Finally I record here that this volume is dedicated to the memory of G. H. Hardy, J. E. Littlewood and G. Pólya. Professor Pólya was still living at the time the Conference was planned and was informed of the meeting; he died in the early months of 1985.

W. Norrie Everitt

Contents

Contributors

Calvin D. Ahlbrandt Department of Mathematics, University of Missouri, Columbia, Missouri

J. M. Anderson Department of Mathematics, University College, London, England

William Desmond Evans School of Mathematics, University of Wales College of Cardiff, Cardiff, Wales

W. Norrie Everitt School of Mathematics and Statistics, University of Birmingham, Birmingham, England

Jack Gunson School of Mathematics and Statistics, University of Birmingham, Birmingham, England

Walter K. Hayman Department of Mathematics, University of York, Heslington, York, England

Man Kam Kwong Mathematics and Computing Science Division, Argonne National Laboratory, Argonne, Illinois

Elliott H. Lieb Departments of Mathematics and Physics, Princeton University, Princeton, New Jersey

E. Russell Love Department of Mathematics, University of Melbourne, Parkville, Victoria, Australia.

Lawrence E. Payne Department of Mathematics, Cornell University, Ithaca, New York

Johann Schröder Mathematisches Institut, Universität Köln, Cologne, Federal Republic of Germany

Giorgio G. Talenti Istituto Matematico, Università Degli Studi, Florence, Italy

Hans Triebel Sektion Mathematik, Universität Jena, Jena, German Democratic Republic

Wolfgang Walter Mathematisches Institut I, Universität Karlsruhe, Karlsruhe, Federal Republic of Germany

Anton Zettl Department of Mathematical Sciences, Northern Illinois University, DeKalb, Illinois

INEQUALITIES

1 Variational Inequalities

Calvin D. Ahlbrandt University of Missouri, Columbia, Missouri

ABSTRACT

This is a survey of integral and series inequalities from the perspective of Chapter 7 of Hardy, Littlewood, and Pólya's "Inequalities" which was entitled "some applications of the calculus of variations". Interest is focused on developments motivated by the treatment and results of that work. In particular, the work of Morse and Leighton which followed the publication of "Inequalities" is surveyed and given a different perspective as a result of work on systems due to Philip Hartman and W. T. Reid. Bliss's treatment of Mayer fields is applied to provide details of some variational arguments given in Chapter 7.

1. Introduction. On November 8,1928, G. H. Hardy gave an address to the London Mathematical Society on the occasion of his retirement from the presidency. In that address he gave reasons that he, Littlewood, and Pólya had "undertaken to contribute a tract on inequalities", (see Hardy [1929]). He gave justification for the large number of papers on inequalities which were being published in the Journal. He also pointed out the lack of a satisfactory book in the English language on the subject. At that time he sought to arrange proofs of inequalities into three classes, (see Hardy [1929,pg.66]), according to the principles which they assume, namely,

(a) "what I will call strictly elementary proofs,...";

(b) "proofs which make use of the elements of the differential calculus, or of general theorems in the theory of functions of a single variable";

(c) "proofs which depend on the theory of maxima and minima of functions of any number of variables".

He goes on to say that "No one would wish to be pedantic about method when we have a really difficult inequality to prove." "It should, however, be an axiom, when our inequality is really simple and fundamental, that we should find one proof at any rate which presupposes the absolute minimum of theory." He then speaks of difficulties in generalizing inequalities to wider classes by first giving a proof by algebraic methods and extending by a limiting process, the fundamental difficulty being in the loss of strict inequality in the limit. The lesson of his examples and discussion is given as: ***"when it is possible, avoid limiting processes"***

In a paper submitted June 13, 1929, the types of proofs a,b,c, outlined previously, had been found to be of insufficient power (see Hardy and Littlewood [1930],pg 37). They candidly admit their inadequacy in knowledge of the calculus of variations. The inequality in question had been referred to Bliss, whose paper (Bliss [1930b]) immediately follows. By 1932, (Hardy and Littlewood [1932]), the classes of proofs had evolved to types A,B,C. Proofs of type A were "strictly elementary, however the results may have been suggested, the proofs do not depend on variational theory". Proofs of type B "depend on our knowledge of the form of an extremal, i.e., of a solution of the Euler equation associated with a variational problem, but are elementary in all other respects." Proofs of type C "depend essentially on the ideas and methods proper to the calculus of variations". The interest of that paper "lies largely in the logical relations between these different types of proof". This theme is continued in Chapter 7 of Hardy, Littlewood, and Pólya's "Inequalities" [1934]. On page 182, they state "It is difficult to distinguish at all precisely between 'elementary' and 'variational' proofs, since there are many proofs of intermediate types". They give a selection of such proofs, "worked out with varying degrees of detail".

Professor Everitt has raised questions concerning certain variational proofs of integral inequalities over noncompact intervals, as presented in HLP. We shall initiate this survey by commenting on the variety of proofs given for inequality #260 of HLP [1934, pg. 188] and relevant developments that have occurred since then.

2. A Retrospective of HLP's Proof of Inequality #260. Inequality #260 asserts that for $y, y'' \in L^2(0,\infty)$, then $y' \in L^2(0,\infty)$ and

$$J[y] = \int_0^\infty \{y^2 - (y')^2 + (y'')^2\}\, dx \quad > \quad 0 \tag{2.1}$$

unless $y = AY$, for

$$Y = e^{-x/2}\sin(x\sin\gamma - \gamma), \quad \gamma = \frac{\pi}{3} \tag{2.2}$$

when there is equality. Then inequality #259, the "HLP inequality", is a consequence of a discriminate condition on an associated quadratic (see pg. 193).

Shortly after publication of "Inequalities", Morse and Leighton [1936] developed the so-called "singularity condition" in the study of quadratic variational problems with a singularity at 0. This work was continued by Leighton and Martin [1955] and Martin [1956a,1956b]. Included in Martin's study were functionals with leading term being allowed to vanish over intervals. Extensions to vector valued functions were made by E. C. Tomastik [1966]. Morse returned to variational methods in his book [1973b] and in a paper [1973a]. (Unfortunately, he did not cite Tomastik's earlier work on vector problems.) Singular quadratic functionals have been studied by means of transformation methods by Kanovsky [1983,1984]. He applies Boruvka's theory of transformations to study problems reducible to functionals with Euler equation $y'' = -y$ on (0,b). Morse and Leighton were certainly stimulated by the news of Hardy's interest in the subject of variational inequalities. They were also motivated by the work of Kemble [1933],(see also [1937,pp 125-126]), which gave a formulation of a singular point boundary condition which was not based on the powerful methods of Weyl.

The correspondence between Hardy and Bliss came at about the time of the publication by Bliss [1930a] of his readable account of the problem of Lagrange. It is quite likely that Hardy stimulated this by his questions on field theory. Bliss gives an elegant treatment there of the concept of a "Mayer field" which pertains to the details of Hardy's third proof of inequality #260. A more general treatment of field theory is given in the book of Hestenes [1966].

Another significant contribution to the understanding of these quadratic functionals of higher order is the work by Hartman [1957] and Reid [1958] on the extension of Leighton's concept of a principal solution to systems. We shall make use of those results below.

The crucial identity (HLP,pg. 189) involved in the proof of #260 is

$$\int_0^X [y^2 - (y')^2 + (y'')^2 - (y + y' + y'')^2]dx \;=\; -[(y+y')^2]\big|_0^X \tag{2.3}$$

Before giving a "slick" derivation based on work of Hartman and Reid, we will investigate the problem from a viewpoint of classical variational theory. If one considers a minimization problem associated with $J[y]$ of (2.1), questions arise as to what sort of natural boundary conditions are appropriate to impose on y at 0 and ∞. But, aside from that question, the usual necessary conditions associated with variational problems on $[a, b]$, a compact subinterval of $(0, \infty)$, must hold because a minimizing arc must also minimize over subarcs. In order to fix a suitable class of comparison curves, let us consider functions y for which

$$J_i = \int_0^\infty (y^{(i)})^2 \, dx \tag{2.4}$$

"converge", for $i = 0, 1, 2$. We will assume that y is of class $C'[0, b]$ and y' is piecewise smooth on $[0, b]$ for each positive b. (See Ewing [1969], §1.9 for terminology.) Once the identity (2.3) is recognized, the identity may be verified on a more general class of admissible arcs. The Euler equation, as well as the Jacobi equation for J is given by

$$y'''' + y'' + y = 0. \tag{2.5}$$

The "Jacobi condition" is absence of conjugate pairs (i.e., disconjugacy) for (2.5). That is, the only solution y of (2.5) which satisfies any two point boundary conditions

$$y(a) = y'(a) = 0 = y(b) = y'(b), \quad a < b, \tag{2.6}$$

is $y \equiv 0$. The computations necessary to establish disconjugacy are nontrivial. HLP evade those computations by clever methods which yield identity (2.3). Because of recent advances in computational mathematics, we can now show disconjugacy of (2.5) in a reasonably straightforward manner.

Since the Jacobi equation (2.5) has constant coefficients, disconjugacy on $(-\infty, \infty)$ is equivalent to showing that no point of $(0, \infty)$ is conjugate to 0. For that purpose, let y_1 and y_2 be the solutions determined by the initial conditions

$$\begin{gathered} y_1(0) = y_1'(0) = 0, \quad y_1''(0) = 1, \quad y_1'''(0) = 0 \\ y_2(0) = y_2'(0) = y_2''(0) = 0, \quad y_2'''(0) = 1. \end{gathered} \tag{2.7}$$

For $w(x) = \mathcal{W}[y_1(x), y_2(x)]$, the Wronskian, and

$$Y = \begin{bmatrix} y_1 & y_2 \\ y_1' & y_2' \end{bmatrix} \tag{2.8}$$

we have $w(c) = 0$ if and only if $Y(c)$ is singular if and only if c is conjugate to 0.

PROPOSITION 2.1. *For $w(x)$ the Wronskian of y_1 and y_2, we have $w(x) > 0$ for $x > 0$ and equation (2.5) is disconjugate on $(-\infty, \infty)$. Indeed, $w(x)$ is given by*

$$w(x) = \frac{\cosh x}{2} + \frac{\cos \sqrt{3}x}{6} - \frac{2}{3}, \tag{2.9}$$

which is the solution of the initial value problem

$$\begin{gathered} w'''' + 2w'' - 3w = 2 \\ w(0) = w'(0) = w''(0) = w'''(0) = 0 \end{gathered} \tag{2.10}$$

Proof: If w is given by (2.9) and there exists a positive b such that $w(b) = 0$, then Rolle's theorem implies that there exists a point c in $(0, b)$ such that

$$w'(c) = \frac{\sinh c}{2} - \frac{\sqrt{3}}{6} \sin \sqrt{3}c = 0$$

Taylor's theorem about 0 for $\sinh x$ and $(\sin\sqrt{3}x)/\sqrt{3}$ implies that for some positive α, β, we have the contradictory statement

$$1 < \cosh\alpha = \cos(\sqrt{3}\beta) \leq 1.$$

The form of $w(x)$ given by (2.9) was suggested by using the numerical package MATLAB and the symbolic manipulation package REDUCE. A direct verification of (2.9) follows from (2.10), which in turn is a consequence of the values of "bracket functions"

$$\begin{aligned} \{y_1, y_2\}_1 &= y_1'''y_2 - y_1''y_2' + y_1'y_2'' - y_1y_2''' + y_1'y_2 - y_1y_2' \\ \{y_1, y_2\}_2 &= y_1'''y_2'' - y_1''y_2''' - y_1'y_2 + y_1y_2', \end{aligned}$$

which are constant; hence,

$$\{y_1, y_2\}_1 \equiv 0, \quad \{y_1, y_2\}_2 \equiv -1.$$

Although the proof is straightforward, it is tedious. We now show how it can be avoided by use of matrix Riccati equations.

3. The Legendre-Clebsch Transformation as related to Inequality #260. As a consequence of results of Reid [1957], disconjugacy of the Jacobi equation (2.5) in some neighborhood of ∞ is equivalent to the existence of a matrix solution which is principal at ∞ for the associated canonical system

$$u' = Au + Bv, \quad v' = Cu - A^T v. \tag{3.1}$$

Furthermore, such a principal solution at ∞ exists if and only if the associated matrix Riccati equation

$$W' = C - A^T W - WA - WBW \tag{3.2}$$

has a real symmetric solution which is right extensible and minimal in the class of right extensible solutions. Furthermore, for constant coefficients, the minimal solution is constant, extends to $(-\infty, \infty)$, (hence, (3.1) is disconjugate on $(-\infty, \infty)$), and is a real symmetric solution of the algebraic Riccati equation

$$WBW + WA + A^T W - C = 0. \tag{3.3}$$

The present objective is to show that usage of this minimal solution allows a transformation of J which yields the crucial identity (2.3). First we write $J[y]$ in terms of a Lagrange problem

$$J[\eta] = \int_0^\infty \{\varsigma^T B\varsigma + \eta^T C\eta\}\, dx \tag{3.4}$$

where η is such that there exists a ς with

$$\eta' = A\eta + B\varsigma. \tag{3.5}$$

Here

$$A = \begin{bmatrix} 0 & 1 \\ 0 & 0 \end{bmatrix}, \quad B = \begin{bmatrix} 0 & 0 \\ 0 & 1 \end{bmatrix}, \quad C = \begin{bmatrix} 1 & 0 \\ 0 & -1 \end{bmatrix},$$

$$\eta^T = [\eta_1 \quad \eta_2]^T, \qquad \varsigma^T = [\varsigma_1 \quad \varsigma_2]^T.$$

Now $\eta_1 = y$, $\eta_2 = y' = \eta_1'$, $\varsigma_2 = \eta_2' = y''$ and

$$\varsigma^T B\varsigma + \eta^T C\eta = (y'')^2 - (y')^2 + y^2.$$

The "distinguished" solutions of (3.3) are

$$W_{\infty} = \begin{bmatrix} -1 & -1 \\ -1 & -1 \end{bmatrix}, \quad W_{-\infty} = \begin{bmatrix} 1 & -1 \\ -1 & 1 \end{bmatrix}$$

Here W_{∞} is the minimal symmetric solution, $W_{-\infty}$ is the maximal symmetric solution, and

$$W_{-\infty} - W_{\infty} = 2I > 0.$$

(Here, the order is that generated by the associated quadratic forms.) This condition is equivalent to the principal solutions of (3.1) at $+\infty$ and $-\infty$ being linearly independent, (see Ahlbrandt [1978], Theorem 2.1, pg. 18).

We now give the Legendre-Clebsch transformation of J.

PROPOSITION 3.1. *Suppose*

$$J_b[\eta] = \int_0^b \{\varsigma^T B\varsigma + \eta^T C\eta\}\, dx$$

on the class of piecewise smooth vectors η for which there exists a piecewise continuous ς such that (3.5) holds. If $W(x)$ is any real symmetric solution of the Riccati differential equation (3.2) on $[a,b]$, then

$$J_b[\eta] - (\eta^T W\eta)\big|_0^b = \int_0^b (\varsigma - W\eta)^T B(\varsigma - W\eta)\, dx \quad \geq 0$$

with equality if and only is $\eta' = (A + BW)\eta$.

COROLLARY 3.1. *Use $W = W_{\infty}$ to obtain identity (2.3).*

The power of these methods in suggesting identities is illustrated by application to inequality #261 of HLP (pg. 193). That is the whole line version of the HLP inequality #259, where the constant becomes 1 instead of 4. Modify the above discussion by changing C with resulting Riccati solutions

$$C = \begin{bmatrix} 1 & 0 \\ 0 & -2 \end{bmatrix}, \quad W_{\infty} = W_{-\infty} = \begin{bmatrix} 0 & -1 \\ -1 & 0 \end{bmatrix}.$$

and with this choice of W and usage of the version of Proposition 3.1 on an arbitrary interval $[a,b]$, we have the following result.

COROLLARY 3.2.

$$\int_a^b \{(y'')^2 - 2(y')^2 + y^2\}\, dx = -2yy'\big|_a^b + \int_a^b (y'' + y)^2\, dx$$

In general, replacement of $y(x)$ by $y(x/\rho)$ for $\rho > 0$, (see HLP pg. 193) transforms intervals $(0,\infty)$ and $(-\infty,\infty)$ back to themselves and give inequalities of the form

$$\rho^4 J_0 - \rho^2 kJ_1 + J_2 \geq 0, \quad k > 0,$$

which result in "discriminate" inequalities

$$J_1^2 \leq KJ_0J_2, \quad \text{for } K = \frac{4}{k^2}.$$

Here $J_i[y] = \int (y^i)^2\, dx$ over $(0,\infty)$ or $(-\infty,\infty)$. The constant K is independent of y. In #261, $k = 2$, and in #259, $k = 1$. This change of independent variable is not suitable over arbitrary intervals (a,b) and doesn't apply when one considers more general quadratic functionals with variable coefficients. In a remarkable paper on more general integral inequalities Everitt [1971] investigated the question of existence of such constants k and the question of characterization of the best such constant. That work created a fertile area of research as evidenced by the sample of citations below due to Everitt, et al; Evans and Everitt; Evans and Zettl; Franco, et al; Russell; and Bennewitz. The idea of inserting a parameter k as above was due to Everitt [1971]. He also noted that a nonpositive discriminant could arise if and only if the associated quadratic (he replaced ρ^2 by ρ) was nonnegative for all ρ. Thus, in the present notation, he started with the question of showing an inequality

$$\rho^2 J_0[y] - \rho k J_1[y] + J_2[y] \geq 0$$

for some positive constant k and all real ρ. Since the change of independent variable x to x/ρ does not generalize to series, (where x becomes n, the index of summation), it seems that Everitt's approach would be useful in the study of series inequalities. Further discussion of analogous series inequalities will be given in Section 6.

In other related work, Benson [1967] sought elementary proofs of integral inequalities by means of identities. Those methods are probably not as unified as methods based on the existing theory of the second variation. The above Legendre-Clebsch transformation, as presented in Reid [1971,pg. 325], is closely related to the Picone transformation, (see Eastham [1973] and Kratz and Peyerimhoff [1985]). It is of interest to note how much earlier this idea was employed in variational theory than in differential equations. Bolza [1904] notes on pg. 46 that Legendre used this device in 1786. A modern view of transformations of the second variation for n dependent variables was given by Bliss [1924]. The complicated transformation due to Clebsch has been greatly simplified by the development of matrix calculus. The associated Riccati theory is valuable in modern control theory.

4. The Singularity Condition and Natural Boundary Conditions. We will first back up to the study of Dirichlet integrals associated with second order linear differential equations. Morse and Leighton [1936] systematically investigated extensions of the classical calculus of variations to problems on $(0, b]$ where the functional has an integrand which is allowed to be singular at 0. In particular, they studied functionals

$$J[y]\Big|_{\epsilon}^{b} = \int_{\epsilon}^{b} f(x, y, y')\, dx, \qquad 0 < \epsilon < b < d,$$

where

$$f(x, y, y') = r(x)(y')^2 + 2q(x)yy' + p(x)y^2,$$

with continuous r, q, p on $(0, d)$. They considered a class of so-called A-admissible arcs y on $[0, b]$ with the properties

1. $y(x)$ is continuous on $[0, b]$ and $y(0) = y(b) = 0$,

2. $y(x)$ is absolutely continuous and $(y'(x))^2$ is in L on each closed subinterval of $(0, b]$.

Observe that the zero function is A-admissible and on that arc, we have $J = 0$. They sought conditions under which

$$\liminf_{\epsilon \to 0_+} \int_{\epsilon}^{b} f(x, y, y')\, dx \geq 0$$

on the class A. If this condition held on A, they said that the function $y(x) \equiv 0$ on $[0, b]$ "affords a *minimum limit* to J on the given class". They also considered several other classes of admissible arcs. A summary of these problems is given in Reid [1980], Chapter IV, Sections 3 & 4. The major contributions of that study and subsequent studies by Leighton and his students were the development of the concept of a "principal" (or recessive) solution and the so-called "singularity condition" (see Reid [1980], Theorem 4.1, pg. 212). An attempt will be made here to introduce the subject and see how it generalizes to problems of the generality needed for HLP's inequality #260. The "natural" boundary conditions which ensue must by made manageable by some hypothesis. Everitt [1971] uses the Weyl theory and a limit-point hypothesis in order to handle these end conditions.

Consider a scalar equation

$$-(r(x)u')' + p(x)u = 0 \quad \text{on } (a,b) \tag{4.1}$$

with continuous p and r. Assume r is positive on (a,b). Here $-\infty \leq a < b \leq \infty$. Initally, we make no assumptions concerning singularity or regularity at the endpoints a and b. Assume throughout that (4.1) has no conjugate pairs on (a,b), i.e., assume "disconjugacy" on (a,b). Distinct points α, β in (a,b) are called "conjugate" if the differential equation (4.1) subject to boundary conditions

$$u(\alpha) = 0 = u(\beta) \tag{4.2}$$

has only the trivial solution $u(x) \equiv 0$. Disconjugacy on (a,b) implies existence of principal (recessive) solutions u_a and u_b at a and b, respectively. (See Hartman [1964], Theorem 6.4, pg. 355.) They are unique up to constant multiples and are characterized by the properties

$$\frac{u_a(x)}{u(x)} \to 0, \quad \text{as } x \to a \tag{4.3}$$

for any solution u independent of u_a. (Replace a by b for the definition of the principal solution at b). If a is a regular point, then u_a is any nontrivial real solution with $u_a(a) = 0$. Because of disconjugacy on (a,b), neither u_a nor u_b vanish at any point of (a,b). Associated with each nontrivial solution u of (4.1) is the associated solution

$$w = \frac{ru'}{u} \tag{4.4}$$

of the Riccati equation

$$w' + \frac{w^2}{r} = p. \tag{4.5}$$

The solutions w_a and w_b associated with u_a and u_b exist and satisfy (4.5) on (a,b) and are distinguished by the following properties: (Reid [1971],pg. 349)

(i) w_b is minimal in the class of real valued right extensible solutions of (4.5);

(ii) w_a is maximal in the class of real valued left extensible solutions of (4.5).

Consequently, $w_b(x) \leq w_a(x)$ on (a,b), and any real solution extensible to (a,b) is bounded between w_b and w_a.

For regular problems on compact intervals $[a,b]$ with $r(x) > 0$, the question of

$$J[\eta] = \int_a^b \{r(\eta')^2 + pn^2\}\,dx \tag{4.6}$$

being nonnegative on the class of piecewise smooth η which vanish at a and b can by resolved in terms of conjugate point theory alone. Disconjugacy of (4.1) on (a,b) is equivalent to J being nonnegative on this class; furthermore, disconjugacy on $[a, b]$ is equivalent to positive definiteness of J. See Reid ([1971 Theorem 5.1 pg. 233].)

However, for singular functionals, disconjugacy on (a, b) is necessary for J to be nonnegative, but is not sufficient. Morse and Leighton's Example 4.1, [1936], pg. 260, is as follows. For

$$J = \int_\theta^1 \left[\frac{y'^2}{x} - \frac{y^2}{x^3}\right] dx, \quad 0 < \theta < 1,$$

we have the Euler equation

$$\left(\frac{y'}{x}\right)' + \frac{y}{x^3} = 0$$

with solutions x and $x \ln x$. Hence disconjugacy holds on $(0, 1]$, but $y(x) = x(x - 1)$ gives

$$\lim_{\theta \to 0} J[y]\Big|_\theta^1 = -\frac{1}{2}.$$

They went on to develop a new necessary condition, the "singularity condition", which asserts that in order that the portion of the axis $[0, b]$ furnish a minimum limit to J, the condition

$$\liminf_{x \to 0} [-w_b(x) y^2(x)] \geq 0, \tag{4.7}$$

holds for each A-admissible curve satisfying

$$\liminf_{\epsilon \to 0} J[y]\Big|_\epsilon^b < +\infty. \tag{4.8}$$

Condition (4.7) is the singularity condition for functionals singular at 0. For functionals singular only at the right endpoint b and regular at the left endpoint, then the singularity condition at b is (see Reid [1980,pg. 212])

$$\liminf_{x \to b} [w_a(x) y^2(x)] \geq 0 \tag{4.9}$$

for every A-admissible curve for which

$$\liminf_{d \to b} J[y]\Big|_a^d < +\infty. \tag{4.10}$$

The above cited work on singular quadratic functionals is not of sufficient generality to include the fourth order Euler equations involved in HLP's inequality #260. Tomastik and Morse extended this work to vector y, but not to "Lagrange" problems. One could avoid much of the variational theory used by Morse and Leighton by taking advantage of Proposition 3.1, restricted to this application, if only the behavior of $\eta^T W \eta$ could be ascertained. In particular, for a singularity at 0 and problems on $(0, b]$, choice of $W_b(x)$ and η a vector going to 0 as $x \to b$, one needs a l'Hospital's rule for that end condition to vanish. The study of such limiting behavior has been considered recently by Kratz and Peyerimhoff [1985]. For problems on $(0, \infty)$, the choice of W_∞ in Proposition 3.1 gives, in circumstances where $\eta^T W_\infty \eta \to 0$ at ∞,

$$J[\eta] = -(\eta^T W_\infty \eta)(0) + \int_0^\infty (\varsigma - W\eta)^T B (\varsigma - W\eta)\, dx,$$

when the limits exist. Choice of end conditions at 0 are dictated by the desire for $\eta^T(0) W_\infty(0) \eta(0)$ to vanish as in Corollary 3.1.

Furthermore, the general study of "minimum limit" problems as presented above is not sufficient to give extensions of integral inequalities like #259 and #261 of HLP to variable coefficient cases. One must study more general quadratic functionals which depend upon parameters k, ρ as in Everitt [1971]. Closely related is the concept of "strong nonoscillation" (see Reid [1980], pg. 219), in which a differential equation

$$u''(t) + \lambda q(t)u(t) = 0, \quad t \in [a, \infty)$$

is nonoscillatory for all positive λ. For example, a variational format to fourth order problems is obtained by considering

$$J[y \mid c, d, \rho, k] = \int_c^d f(x, y, y', y'', \rho, k)\, dx$$

where $[c, d] \subset (a, b)$ and

$$f(x, y, y', y'', \rho, k) = r(x)(y'')^2 - k\rho q(x)(y')^2 + \rho^2 p(x) y^2 .$$

The associated Euler equation is

$$[(r(x)y'')' + k\rho q(x)y']' + \rho^2 p(x) y = 0. \tag{4.11}$$

If

$$J_2 = \int_a^b r(x)(y'')^2, \quad J_1 = \int_a^b q(x)(y')^2\, dx, \quad J_0 = \int_a^b p(x)y^2,$$

all exist, then, for fixed positive k, the quadratic inequality

$$\rho^2 J_0 - k\rho J_1 + J_2 \geq 0$$

has associated discriminate inequality

$$J_1^2 \leq K J_0 J_2, \quad \text{for } K = \frac{4}{k^2}.$$

The crucial identity of interest results from Proposition 3.1 on $[c, d]$, i.e.,

$$J[y] = (\eta^T W \eta)\Big|_c^d + \int_c^d (\varsigma - W\eta)^T B(\varsigma - W\eta)\, dx$$

where A, B, C of Section 3 are chosen as

$$A = \begin{bmatrix} 0 & 1 \\ 0 & 0 \end{bmatrix}, \quad B = \begin{bmatrix} 0 & 0 \\ 0 & 1/r \end{bmatrix}, \quad C = \begin{bmatrix} \rho^2 p & 0 \\ 0 & -k\rho q \end{bmatrix}$$

and W is chosen as any symmetric solution of the Riccati equation which, at some point, and hence at all points of (a, b), satisfies (assuming disconjugacy of (4.11) on (a,b) for all ρ)

$$W_b(x) \leq W(x) \leq W_a(x),$$

where W_a and W_b correspond to the matrix principal solutions at a and b. For notation involving higher order formally self adjoint differential equations, see Reid [1980], Chap. V, section 14. In the special case of the above with p, q, r all 1, the largest constant k such that, for every fixed ρ, the algebraic Riccati equation has a real symmetric solution, is $k = 2$.

5. Bliss's Approach to Mayer Fields and HLP's Third Proof of #260. Professor Everitt has asked about the rigor of the third proof of #260, which uses the field theory of the calculus of variations. There are several points in regard to identity (7.8.15) which require elaboration, as they are not direct consequences of variational theory on compact intervals.

Let us start by giving the definition of a Mayer field as presented by Bliss [1930a], pg. 730. We specialize that treatment to $n = 2$ and assume a problem of Lagrange with one side condition. Consider a variational problem (here, y and z are scalar)

$$J = \int_{x_1}^{x_2} f(x, y, z, y', z')\, dx \tag{5.1}$$

on a class of arcs $y(x), z(x)$, $x_1 \leq x \leq x_2$, satisfying a differential equation

$$\phi(x, y, z, y', z') = 0 \tag{5.2}$$

and joining two given points in the space (x, y, z) for $x = x_1, x_2$. In the application of the theory used by HLP, we have

$$f(x, y, z, r, s) = y^2 - r^2 + s^2, \tag{5.3}$$
$$\phi(x, y, z, r, s) = z - r. \tag{5.4}$$

Here f and ϕ have continuous partials of all orders on $\mathbf{R}^5$.

A point (x, y, z, p, q) is called *admissible* if it lies interior to the region $\mathcal{R}$ where the continuity properties of f and ϕ have been assumed, satisfies the equation $\phi = 0$, and the matrix $[\phi_p \quad \phi_q]$ has rank 1 at that point.

A *Mayer field* is an open region $\mathcal{F}$ in (x, y, z) space having associated with it a set of functions

$$p(x, y, z), \quad q(x, y, z), \quad l(x, y, z)$$

with the following properties.

(a) p, q, l have continuous first partials in $\mathcal{F}$;

(b) the points $(x, y, z,\ p(x, y, z),\ q(x, y, z))$ defined by points (x, y, z) in $\mathcal{F}$ are admissible;

(c) the integral (the "Hilbert invariant integral")

$$J^* = \int A\, dx + B\, dy + C\, dz,$$

for A, B, C defined below, is path independent in $\mathcal{F}$. Here

$$A = F - pF_p - qF_q,$$

$$B = F_p, \quad C = F_q$$

where F and its partials are evaluated at (x, y, z, p, q, l), for p, q, l evaluated at (x, y, z), and

$$F = f(x, y, z, p, q) + l\phi(x, y, z, p, q).$$

The functions p and q are called "slope functions" and l is a "multiplier". HLP give us these functions, namely,

$$p = z, \quad q = -y - z, \quad l = 2y. \tag{5.5}$$

(We lose motivation, but gain rigor.) Then

$$F = y^2 - p^2 + q^2 + l(z-p), \tag{5.6}$$

$$F_p = -2p - l, \quad F_q = 2q, \tag{5.7}$$

and

$$A = 0, \quad B = -2y - 2z = C.$$

Hence, the integrand of J^* is an exact differential with primitive

$$\psi(x,y,z) = -(y+z)^2 \tag{5.8}$$

and p, q, l constitute a Mayer field on $\mathcal{F} = \mathbf{R}^3$. Since J^* can be written as

$$J^* = \int F\,dx + F_p(dy - p\,dx) + F_q(dz - q\,dx),$$

and J^* is path independent, then every arc joining two fixed points is a minimizing arc and every solution y, z of the system

$$\frac{dy}{dx} = p(x,y,z), \quad \frac{dz}{dx} = q(x,y,z) \tag{5.9}$$

is an extremal with multiplier $\lambda = l(x, y(x), z(x))$. In the above case, we have

$$y' = z, \quad z' = -y - z, \tag{5.10}$$

that is, the field of extremals is determined by the solutions of

$$y'' + y' + y = 0, \tag{5.11}$$

which agrees with HLP's family (7.8.10) of zero tending solutions of the Euler equation (2.5), above, i.e., (7.8.9) of HLP.

Apprehension about the field theory justification of HLP's (7.8.15) stems from the following. Bliss's field theory as presented in [1930a] only allows curves interior to the region $\mathcal{F}$ of the Mayer field, contrary to the point $(\infty, 0, 0)$ being a common endpoint of the curves. Furthermore, the second proof requires the integral identity, bottom of page 189, on $[0, X]$ in order to conclude that $y^2 + (y')^2$ has limit 0. Finally, the proof is not case free, as $y(0)$ is first assumed to be 1 and the details of the case $y(0) = 0$ are not given (see pg. 192).

What can be done is to show the identity (2.3) on $[0, b]$ by field theory. Consider an admissible arc $\eta(x)$ which is continuous, along with η', and η' is piecewise smooth on every interval $[0, b]$. Set $\alpha = \eta(0)$, $\beta = \eta'(0)$. Let Y be the solution of (5.11) determined by the initial conditions $Y(0) = \alpha$, $Y'(0) = \beta$. Then, for b fixed, and J on $[0, b]$,

$$J[Y] = J^*[Y], \tag{5.12}$$

the latter integral being path independent. In (x, y, z) space, label points $P(0, \alpha, \beta)$, $Q(b, Y(b), Y'(b))$, and $R(b, \eta(b), \eta'(b))$. Then, for RQ the straight line connecting R to Q we have by path independence

$$J^*_{PQ}[Y] = J^*_{PR}[\eta] + J^*_{RQ} \tag{5.13}$$

We now show

$$J[\eta] - J^*[\eta] = \int_0^b \mathcal{E}\,dx \tag{5.14}$$

where $\mathcal{E}$ is the *Weierstrass excess function.* In both J and J^*, we are considering the admissible curve $(x, y, z) = (x, \eta(x), \eta'(x))$. The integrand in J can be replaced by $F(B)$ for B the point

$$B = (x, y, z, y', z', l) = (x, \eta(x), \eta'(x), \eta'(x), \eta''(x), 2\eta(x))$$

Along this same curve, we have

$$J^*[\eta] = \int_0^b [F(A) + (y' - p)F_p(A) + (z' - q)F_q(A)]\, dx$$

for

$$A = (x, y, z, p, q, l) = (x, \eta(x), \eta'(x), \eta'(x), -\eta(x) - \eta'(x), 2\eta(x)).$$

Hence (5.14) holds for

$$\mathcal{E} = F(B) - F(A) - (y' - p)F_p(A) - (z' - q)F_q(A),$$

i.e., $\mathcal{E}$ is the remainder in the linear Taylor's series approximation to F in the p, q plane. But F is quadratic in p and q. Thus

$$\mathcal{E} = \frac{1}{2}[\, y' - p \quad z' - q\,] \begin{bmatrix} F_{pp} & F_{pq} \\ F_{qp} & F_{qq} \end{bmatrix} \begin{bmatrix} y' - p \\ z' - q \end{bmatrix}.$$

But

$$y' - p = \eta' - z = \eta' - \eta' = 0$$

and

$$z' - q = \eta'' + y + z = \eta'' + \eta + \eta'.$$

Thus

$$\mathcal{E} = (\eta'' + \eta' + \eta)^2. \tag{5.15}$$

Combining (5.12),(5.13),(5.14) gives

$$\begin{aligned} J[\eta] - J[Y] &= J[\eta] - J^*[\eta] - J^*_{RQ} \\ &= \int_0^b \mathcal{E}\, dx - J^*_{RQ} \end{aligned} \tag{5.16}$$

But

$$J[Y] = J^*_{PQ}[Y] = -(Y + Y')^2 \Big|_0^b$$

and

$$J^*_{RQ} = -(y + z)^2 \Big|_R^Q = (\eta(b) + \eta'(b))^2 - (Y(b) + Y'(b))^2$$

give

$$J[\eta] = \int_0^b \mathcal{E}\, dx - (\eta + \eta')^2 \Big|_0^b$$

for the crucial identity (2.3).

6. The Status of Discrete Variational Theory and Relationships with Series Inequalities. Copson [1979] presented discrete analogues of HLP's inequalities #259 and #261. His first result was that if $\{a_n\}$ was a sequence of real numbers such that

$$\sum_{n=-\infty}^{\infty} a_n^2 \quad \text{and} \quad \sum_{n=-\infty}^{\infty} (\Delta^2 a_n)^2$$

converge, then

$$\left\{\sum_{n=-\infty}^{\infty} (\Delta a_n)^2\right\}^2 \leq \left(\sum_{n=-\infty}^{\infty} a_n^2\right)\left(\sum_{n=-\infty}^{\infty} (\Delta^2 a_n)^2\right)$$

with equality if and only if $a_n = 0$ for all n. The companion result was

$$\left\{\sum_{n=0}^{\infty} (\Delta a_n)^2\right\}^2 \leq 4\left(\sum_{n=0}^{\infty} a_n^2\right)\left(\sum_{n=0}^{\infty} (\Delta^2 a_n)^2\right)$$

To relate the first result to Everitt's observation, introduce the notation

$$S_0 = \sum_{n=-\infty}^{\infty} (a_n)^2, \quad S_j = \sum_{n=-\infty}^{\infty} (\Delta^j a_n)^2, \quad j = 1, 2.$$

Consider a quadratic

$$Q(\rho, k) = \rho^2 S_0 - k\rho S_1 + S_2.$$

Then, for fixed positive k, Q is nonnegative for all ρ if and only if the discriminant condition

$$S_1^2 \leq K S_0 S_2, \quad K = \frac{4}{k^2}$$

holds. Hence, Copson's first result implies that $Q(\rho, 2) \geq 0$ for all ρ. Hence, a "variational" problem is suggested. Consider

$$J[a_n, \rho, k] = \sum_{n=-\infty}^{\infty} (\Delta^2 a_{n-2})^2 - k\rho(\Delta a_{n-1})^2 + \rho^2 a_n^2.$$

If J is to be nonnegative for all ρ and sequences a_n for which S_0, S_1, S_2 converge, then restriction of a_n to be 0 for all n which satisfy, for some $M < N, a_n = 0, n < M, a_n = 0, n > N$, gives a finite sum J which must be nonnegative for all ρ. The status of finite variational theory is not, so far as the author knows, developed to the stage where one can read Copson's inequalities off from discrete Riccati equations in a manner analogous to Corollaries 3.1 and 3.2 above. Of course, one cannot make variable changes n to n/ρ here, but the theory is also inadequate in regard to being able to write even order symmetric equations in terms of an analogue of first order Hamiltonian systems. However, there are recent developments in discrete theory which are worthy of mention.

In Ahlbrandt and Hooker [1985], the details of discrete variational theory were presented for the problem of minimizing

$$J[Y] = \sum_{n=M}^{N} f(n, y_n, \Delta y_{n-1}) \tag{6.1}$$

on the class of vectors Y with real components $y_{M-1}, \ldots, y_N$ which satisfy fixed end conditions

$$y_{M-1} = c, \quad y_N = d \tag{6.2}$$

Here Δ is the forward difference operator and M, N are fixed integers with $M < N$. Suppose that a vector Y' provides a local minimum for J. Let a vector H with components $\eta_{M-1}, \ldots, \eta_N$ be an admissible variation, i.e.,

$$\eta_{M-1} = 0 = \eta_N \tag{6.3}$$

Then the function $\phi(t) = J[Y' + tH]$ has a relative minimum at the interior point $t = 0$. The necessary conditions

$$I: \quad \phi'(0) = 0$$

and

$$II: \quad \phi''(0) \geq 0$$

become, under an assumption of continuous second partials on f, and for

$$\begin{aligned} P_n &= f_{yy}(n, y'_n, \Delta y'_{n-1}) \\ Q_n &= f_{yr}(n, y'_n \Delta y'_{n-1}) \\ R_n &= f_{rr}(n, y'_n, \Delta y'_{n-1}), \end{aligned} \tag{6.4}$$

the conditions

$$I: \quad J_1[H] \equiv \sum_M^N f_y(n, y'_n, \Delta y'_{n-1})\eta_n + f_r(n, y'_n, \Delta y'_{n-1})\Delta\eta_{n-1} = 0$$

and

$$II: \quad J_2[H] \equiv \sum_M^N P_n \eta_n^2 + 2Q_n \eta_n \Delta\eta_{n-1} + R_n(\Delta\eta_{n-1})^2 \geq 0,$$

for all admissible variations H. Condition I becomes the *Euler equation*

I: for f_y and f_r evaluated at $(n, y'_n, \Delta y'_{n-1})$ we have

$$f_y = \Delta f_r, \quad n = M, \ldots, N-1. \tag{6.5}$$

If the left or right end is free, then we have the respective *transversality conditions*

$$f_r(M, y'_M, \Delta y'_{M-1}) = 0 \tag{6.6}$$

and

$$f_y(N, y'_N, \Delta y'_{N-1}) + f_r(N, y'_N, \Delta y'_{N-1}) = 0. \tag{6.7}$$

The second variation can be rewritten in the form

$$J_2[H] = \sum_M^N c_{n-1}(\Delta\eta_{n-1})^2 + a_n\eta_n^2, \tag{6.8}$$

for $a_N = P_N + Q_N$ and

$$a_n = P_n - \Delta Q_n, \quad c_{n-1} = R_n + Q_n, \qquad M \leq n \leq N-1 \tag{6.9}$$

Alternatively, using (6.3) we can also write J_2 as a symmetric tridiagonal quadratic form

$$J_2[H] = \sum_M^N b_n\eta_n^2 - 2c_{n-1}\eta_n\eta_{n-1} \tag{6.10}$$

where $b_n = c_n + c_{n-1} + a_n$, $M \leq n \leq N-1$.

The Jacobi equation is the Euler equation for J_2, namely

$$L[y]_n = -\Delta(c_{n-1}\Delta y_{n-1}) + a_n y_n = 0 \tag{6.11}$$

At this point, we have a variance from the continuous theory where the Legendre condition is $R = f_{rr} \geq 0$ along a minimizing arc. Here, the Legendre necessary condition is $b_n \geq 0$ for $n = M, \ldots, N-1$. Some authors made the assumption $c_n > 0$ in the study of (6.11) because of the parallel with the nonsingular continuous theory. However, the parallel condition becomes $b_n \geq 0$ and the nonsingularity condition $c_n \neq 0$. The continuous case where the lead coefficient is allowed to change sign is a topic of considerable interest, see for example, Everitt and Knowles [1986]. Criteria for positive definiteness of the second variation are presented in Ahlbrandt and Hooker [1985]. Those criteria have been extended to matrix versions, where y_n is a vector and (6.7) has matrix coefficients. (Ahlbrandt and Hooker [1987a],[1987b]). Also, concepts of matrix recessive (principal) solutions have been obtained. Much of the theory parallels that of Reid in terms of Legendre-Clebsch transformations, Riccati equations and disconjugacy. However, the theory is not yet general enough to use matrix methods on the variational analogues of Copson's inequalities. Another possibility is to write out the theory of variational problems with f involving second differences and pattern the transformation theory after Eastham [1971].

7. Miscellaneous Developments based on HLP's Chapter 7. Several generalizations and new proofs based on Hardy's Chapter 7 are worth noting. We have thus far emphasized developments related to inequalities 259, 260, and 261. In particular, it has been shown above how certain results extend to vector problems, i.e., for problems in n dependent variables. For results pertaining to multiple integrals, i.e., n independent variables, the reader is referred to Lewis [1984], where extensions of inequalities related to #253 are given. Inequality 257 and the identity (7.7.3) concern a case where the principal solutions at 0 and π coincide. Inequality (7.7.1),

$$\int_0^{\pi/2} y^2\,dx \leq \int_0^{\pi/2} (y')^2\,dx$$

for functions which vanish at 0, with equality only when y is a multiple of $\sin x$ is an example where $w(x)$ the solution of the Riccati equation may be chosen as any solution with

$$-\tan x \leq w(x) \leq -\tan(x + \pi/2),$$

the latter choice being used by Hardy. If $w(x)$ were chosen as $-\tan x$, then $y(0) = 0$ would be replaced by $y(\pi/2) = 0$. Beesack [1958] presents generalizations of those of Section 7.7 of Hardy. Included in that paper are extensions of Wirtinger's inequality, #258 of HLP. Further references are contained in the book of Beckenbach and Bellman [1965]. Of special interest is the paper by Reid [1959] in which the Jacobi form of a Legendre transformation was expressed in polar coordinates. Reid [1972], pg, 465, unifies results of Leighton [1970] concerning Wirtinger-type inequalities for a fourth order equation. Reid's work, (as well as that of Reid [1956]), was based to some extent upon the paper of Birkhoff and Hestenes [1935] which developed the theory of the second variation for extremals which were not necessarily minimizing arcs. They showed that all extremals may "be obtained as a solution of a properly formulated minimum problem". They do this by showing that "by adding a suitable set of 'natural isoperimetric conditions', *which are automatically satisfied by every possible extremal fulfilling the given conditions*, there is obtained a related isoperimetric problem for which the extremal arc in question is a minimizing arc".

Reid [1973, 1977] used variational methods to study Liapunov inequalities.

Discrete analogues of Hardy's results in Section 7.7 and extensions to certain higher order inequalities have been given by Fan, Taussky, and Todd [1955].

REFERENCES

[1978] Ahlbrandt, C. D., Linear independence of the principal solutions at ∞ and $-\infty$ for formally self-adjoint differential systems, J. Differential Eqs. **29** (1978), pp. 15–27.

[1985] Ahlbrandt, C. D., and Hooker, J. W. A variational view of nonoscillation theory for linear difference equations, *in* "Proceedings of the Thirteenth Midwest Differential Equations Conference, J. L. Henderson, ed., Univ. of Missouri-Rolla, 1985, pp.1–21.

[1987a] Ahlbrandt, C. D., and Hooker, J. W., Riccati matrix difference equations and disconjugacy of discrete linear systems, SIAM J. Math. Analysis, (to appear).

[1987b] Ahlbrandt, C. D., and Hooker, J. W., Recessive solutions of symmetric three term recurrence relations, *in* CMS-AMS Conference Proceedings of 1986 Toronto Conference, A. Mingarelli, editor, American Math Soc., Providence, (to appear).

[1965] Beckenbach, E. F., and Bellman, R., "Inequalities", Springer-Verlag, New York, 1965.

[1958] Beesack, P. R., Integral inequalities of the Wirtinger type, Duke Math. J. **25** (1958), pp. 477–498.

[1984] Bennewitz, C., A general version of the Hardy-Littlewood-Pólya-Everitt (HELP) inequality, Proc. Royal Soc. Edinburgh **97A** (1984), pp. 9–20.

[1967] Benson, D. C., Inequalities involving integrals of functions and their derivatives, J. Math. Anal. Appl. **17**(1967),pp. 292–308.

[1935] Birkhoff, G. D., and Hestenes, M. R., Natural isoperimetric conditions in the calculus of variations, Duke Math. J. **1** (1935), pp. 198–286.

[1924] Bliss, G. A., The transformation of Clebsch in the calculus of variations, Proc. Int. Congress held in Toronto, 1922, **1**(1924), pp. 589–603.

[1930a] Bliss, G. A., The problem of Lagrange in the calculus of variations, American J. Math. **52** (1930), pp. 673–744.

[1930b] Bliss, G. A., An integral inequality, J. London Math. Soc. **5**(1930), pp. 40–46.

[1904] Bolza, O., Lectures on the Calculus of Variations, Chelsea reprint of 1904 University of Chicago edition, Chelsea Publishing Co., New York.

[1979] Copson, E. T., Two series inequalities, Proc. Royal Soc. Edinburgh **83A** (1979), pp. 109–114.

[1973] Eastham, M. S. P., The Picone identity for self-adjoint differential equations of even order, Mathematika **20** (1973), pp. 197–200.

[1982] Evans, W. D. and Everitt, W. N., A return to the Hardy-Littlewood inequality, Proc. Royal Soc. London **A 380** (1982), pp. 447–486.

[1985] Evans, W. D., and Everitt, W. N., On an integral inequality of Hardy-Littlewood type: I

Proc. Royal Soc. Edinburgh **A 101** (1985), pp. 131–140.

[1986] Evans, W. D., and Everitt, W. N., Hayman, W. K., and Ruscheweyh, S., On a class of integral inequalities of Hardy-Littlewood type, J. d' Analyse Mathematique, (to appear).

[1978] Evans, W. D., and Zettl, A., Norm inequalities involving derivatives, Proc. Royal Soc. Edinburgh **A 82** (1978), pp. 51–70.

[1971] Everitt, W. N., On an extension to an integro-differential inequality of Hardy, Littlewood, and Polya, Proc. Royal Soc. Edinburgh **A 69** (1971), pp. 295–333.

[1974] Everitt, W. N., Integral inequalities and the Liouville transformation, *in* "Lecture Notes in Mathematics", vol. **415**, pp. 338–352. Berlin, Springer-Verlag.

[1985] Everitt, W. N., A note on linear ordinary quasi-differential equations, Proc. Royal Soc. Edinburgh **101A** (1985), pp. 1–14.

[1986] Everitt, W. N., and Knowles, I. W., Limit-point and limit circle criteria for Sturm-Liouville equations with intermittently negative principal coefficients, Proc. Royal Soc. Edinburgh **103A** (1986), pp. 215–218.

[1983] Everitt, W. N., and Wray, S. D., On quadratic integral inequalities associated with second-order symmetric differential expressions, *in* "Lecture Notes in Mathematics", vol. **1032**, (1983), pp. 170–223.

[1969] Ewing, G. M., "Calculus of Variations with Applications", W. W. Norton, New York 1969. (reprinted by Dover, 1985).

[1955] Fan, K., Taussky, O., and Todd, J., Discrete analogs of inequalities of Wirtinger, Monatshefte für Mathematik **59** (1955), pp. 73–90.

[1985] Franco, Z. M., Kaper, H. G., Kwong, M. K., and Zettl, A., Best constants in norm inequalities for derivatives on a half line, Proc. Royal Soc. Edinburgh **100A** (1985), pp.67–84.

[1929] Hardy, G. H., Prolegomena to a chapter on inequalities, J. London Math. Soc. **4** (1929), pp. 61–78, (addenda, ibid., **5**, p. 80).

[1930] Hardy, G. H. and Littlewood, J. E., Notes on the theory of series (XII): On certain inequalities connected with the calculus of variations, J. London Math. Soc. **5**(1930), pp. 34–46.

[1932] Hardy, G. H. and Littlewood, J. E., Some integral inequalities connected with the calculus of variations, Quarterly J. Math. **3** (1932), pp. 241–252.

[1934] Hardy, G. H., Littlewood, J. E., and Pólya, G., "Inequalities", Cambridge University Press, 1934.

[1957] Hartman, P., Self-adjoint, non-oscillatory systems of ordinary, second order, linear differential equations, Duke Math. J. **24** (1957), pp. 25–35.

[1966] Hestenes, M. R., "Calculus of Variations and Optimal Control Theory", Wiley, New York,

1966.

[1983] Kanovsky, V., Global transformations of linear differential equations and quadratic functionals, I, Arch. Math **3**, Scripta Fac. Sci. Nat. Ujep Brunensis **XIX** (1983), pp.161–172.

[1984] Kanovsky, V., Global transformations of linear differential equations and quadratic fumctionals, II, Arch. Math. **3**, Scripta Fac. Sci. Nat. Ujep Brunensis **XX** (1984), pp. 149–156.

[1933] Kemble, E. C., Note on the Sturm-Liouville eigenvalue-eigenfunction problem with singular end-points, Proceedings National Academy of Sciences **19** (1933), pp. 710–714.

[1937] Kemble, E. C., "The Fundamental Principles of Quantum Mechanics", McGraw-Hill, New York, 1937.

[1985] Kratz, W., and Peyerimhoff, A., A treatment of Sturm-Liouville eigenvalue problems via Picone's identity, Analysis **5** (1985), pp. 97–152.

[1949] Leighton, W., Principal quadratic functionals, Trans. Amer. Math. Soc. **67** (1949), pp. 253–274.

[1969] Leighton, W., Regular singular points in the nonanalytic case; singular functionals, J. Math. Anal. Appl. **28** (1969), pp. 59–76.

[1970] Leighton, W., Quadratic functionals of second order, Trans. Amer. Math. Soc. **151** (1970), pp. 309–322.

[1955] Leighton, W., and Martin, A. D., Quadratic functionals with a singular end point, Trans. Amer. Math. Soc. **79** (1955), pp. 98–128.

[1984] Lewis, R. T., A Friedrichs inequality and an application, Proc. Royal Soc. Edinburgh **97** (1984), pp. 185–191.

[1956a] Martin, A. D., A regular singular functional, Canadian J. Math **8** (1956), pp. 53–68.

[1956b] Martin, A. D., A singular functional, Proc. Amer Math. Soc. **7** (1956), pp. 1031–1035.

[1973a] Morse, M., Singular quadratic functionals, Math. Ann. **201** (1973), pp. 315–340.

[1973b] Morse, M., "Variational Analysis: Critical Extremals and Sturmian Extensions", Wiley-Interscience, New York, 1973.

[1936] Morse, M., and Leighton, W., Singular quadratic functionals, Trans. Amer. Math. Soc. **40** (1936), pp. 252–286.

[1956] Reid, W. T., Oscillation criteria for linear differential systems with complex coefficients, Pacific J. Math. **6** (1956), pp. 733–751.

[1958] Reid, W. T., Principal solutions of non-oscillatory self-adjoint linear differential systems, Pacific J. Math. **8** (1958), pp. 147–169.

[1959] Reid, W. T., The isoperimetric inequality and associated boundary problems, J. Math. and

Mech. **8** (1956), pp. 897–906.

[1971] Reid, W. T., "Ordinary Differential Equations", Wiley, New York, 1971.

[1972] Reid, W. T., Variational aspects of oscillation phenomena for higher order differential equations, J. Math. Anal. Appl., **40** (1972), pp. 446–470.

[1973] Reid, W. T., A generalized Liapunov inequality. J. Diff. Eqs. **13** (1973), pp. 182–196.

[1977] Reid, W. T., Interrelations between a trace formula and Liapunov type inequalities, J. Differential Equ., **23** (1977), pp. 448–458.

[1980] Reid, W. T., "Sturmian Theory for Ordinary Differential Equations", Applied Math. Sciences **31**, Springer-Verlag, New York, 1980.

[1978] Russell, A., On a fourth order singular integral inequality, Proc. Royal Soc. Edinburgh, Sec. A **80** (1978), pp. 249–260.

[1966] Tomastik, E. C., Singular quadratic functionals of n dependent variables, Trans. Amer. Math. Soc., **124** (1966), pp. 60–76.

2 The Grunsky Inequalities

J. M. Anderson University College, London, England

1. Introduction

Let Σ denote the class of functions

$$g(z) = z + \sum_{n=1}^{\infty} b_n z^{-n} \tag{1}$$

which are univalent in $\Delta = \{z: |z| > 1\}$. We may then form the expression

$$\log \frac{g(z) - g(\zeta)}{z - \zeta} = - \sum_{k=1}^{\infty} \sum_{\ell=1}^{\infty} b_{k\ell}\, z^{-k}\zeta^{-\ell} \tag{2}$$

for z and ζ in Δ. The numbers $b_{k\ell}$ so obtained are called the Grunsky coefficients of g. The Grunsky inequalities, as formulated in [6], see also [16], Chapter 3, state that $g \in \Sigma$ if and only if

$$\left| \sum_{k=1}^{\infty} \sum_{\ell=1}^{\infty} b_{k\ell}\, \lambda_k \lambda_\ell \right| \leq \sum_{k=1}^{\infty} \frac{|\lambda_k|^2}{k},$$

for all sequences $\{\lambda_k\} \subset \mathbb{C}$. In more modern terminology we set $B_{k\ell} = \surd(k\ell) b_{k\ell}$ and let $B = B(g)$ be the (infinite) matrix with entries $B_{k\ell}$. Then $g \in \Sigma$ if and only if B is a bounded operator on ℓ^2 with $||B|| \leq 1$, where $||B||$ denotes the usual operator norm of $B(g)$.

Associated with (2) we have the quantity

$$U(z,\zeta,g) = \frac{\partial^2}{\partial z \partial \zeta} \log \frac{g(z)-g(\zeta)}{z - \zeta} = \frac{g'(z)g'(\zeta)}{(g(z)-g(\zeta))^2} - \frac{1}{(z-\zeta)^2} . \tag{3}$$

The Schwarzian derivative, $S(g,z)$, of g is defined to be 6 $U(z,z,g)$, with the usual limiting procedure, and it is easy to establish that

$$S(g,z) = \left[\frac{g''(z)}{g'(z)}\right]' - \frac{1}{2} \left[\frac{g''(z)}{g'(z)}\right]^2 . \tag{4}$$

Much work has been done recently on the case when $||B|| < 1$, giving rise to questions concerning quasi conformal mappings and Fredholm eigenvalues. To explain this some further definitions are required.

As usual we set $D = \{z: |z| < 1\}$ and write

$$\frac{\partial g}{\partial z} = \frac{1}{2}\left(\frac{\partial g}{\partial x} - i\frac{\partial g}{\partial y}\right), \quad \frac{\partial g}{\partial \bar{z}} = \frac{1}{2}\left(\frac{\partial g}{\partial x} + i\frac{\partial g}{\partial y}\right).$$

If $0 < k < 1$, the function $g \in \Sigma$ is said to have a k-quasi conformal extension to to $\mathbb{C}$ if it has a homeomorphic extension $\breve{g}(z)$ to $\mathbb{C}$ such that

(i) $\breve{g}(z) = g(z), \quad z \in \Delta,$

(ii) $\mu(z) = \dfrac{\partial \breve{g}}{\partial \bar{z}} \Big/ \dfrac{\partial \breve{g}}{\partial z}$ satisfies $|\mu(z)| \leq k$ for a.e. $z \in D$.

We call $\mu(z)$, defined almost everywhere, the complex dilation of g and write (ii) as

$$||\mu|| \leq k . \tag{5}$$

We exclude from our considerations the trivial case $k = 0$ when $g(z) \equiv z$.

If g has such a k-quasi conformal extension, the boundary of $g(\Delta)$, denoted by Γ is called a k-quasi circle. In particular, Γ will be a simple closed Jordan curve, possibly non-rectifiable. The nicest characterization of quasi-circles is due to Ahlfors[2]. If Γ is a k-quasi circle then there is a constant $c = c(k)$ such that, for any two points z_1, z_2 on Γ, dividing Γ into two arcs Γ_1 and Γ_2,

$$\min(\text{diam } \Gamma_1, \text{ diam } \Gamma_2) \leq c|z_1 - z_2|. \tag{6}$$

Conversely, if (6) holds for some c and all possible choices of z_1 and z_2 then there is a $k = k(c) < 1$ as that Γ is a k-quasi circle. This elegant criterion is very useful for proving that a curve Γ is a k-quasi circle for some $k < 1$ but it is frequently difficult to determine the best (i.e. smallest) such value k.

For a closed Jordan curve Γ let $H(\Gamma)$ denote the set of all functions $h(z)$, continuous in $\mathbb{C}$ and harmonic in both the interior and exterior of Γ. For such an h, let $D(h)$ and $D_1(h)$ denote the Dirichlet integral of h with respect to the exterior (interior) of Γ. The number $\lambda = \lambda(\Gamma)$ defined by

$$\frac{1}{\lambda} = \sup \left\{\frac{D(h) - D_1(h)}{D(h) + D_1(h)}, \ h \in H(\Gamma)\right\}$$

is called the Fredholm eigenvalue associated with Γ. Clearly $\lambda \geq 1$ and it is known that $\lambda > 1$ if and only if Γ is a quasi circle. The trivial case $\lambda = \infty$

again corresponds to the function $g(z) = z$. In the case when Γ is sufficiently smooth, λ corresponds to the Fredholm eigevalue of the classical Neuman-Poincaré integral equation in conformal mapping theory; for further informaion on this see [1].

We now make the following definition.

Definition 1. Let $g(z)$, given by (1), be univalent in Δ and let $\Gamma = \partial g(\Delta)$. Suppose that k is given with $0 < k < 1$. Then

(i) If $||B|| \leq k$ we say that $g \in \Sigma_1(k)$

(ii) If Γ is a k-quasi circle we say that $g \in \Sigma_2(k)$

(iii) If $\lambda(\Gamma) \geq \frac{1}{k}$ we say that $g \in \Sigma_3(k)$

2. The Classes $\Sigma_1(k)$, $\Sigma_2(k)$, $\Sigma_3(k)$.

The relation between the above three classes is now resolved, thanks to the following two theorems.

Theorem 1. For $0 < k < 1$,

(i) $\Sigma_2(k) \subset \Sigma_1(k)$.

(ii) $\Sigma_2(k) \not\subset \Sigma_1(k_1)$ if $k_1 < k$.

(iii) $\Sigma_1(k) \subset \Sigma_2(k_2)$ for some k_2 with $k < k_2 < 1$.

Theorem 2. For $0 < k < 1$,

(i) $\Sigma_1(k) = \Sigma_3(k)$ and $\Sigma_2(k) \subsetneq \Sigma_1(k)$

(ii) $\Sigma_3(k) \not\subset \Sigma_1(k_1)$ if $k_1 < k$.

Theorem 1 is due to many people, see e.g. [1], [8], [13], [17] [11] but Theorem 2 is due to Kühnau [9], [10], who first realised the importance of the class $\Sigma_3(k)$.

The example to show that $\Sigma_2(k) \neq \Sigma_1(k)$ shows in fact, by an ingenious argument, that $\Sigma_2(k) \neq \Sigma_3(k)$. We set $0 < r < 1$ and define $g(z)$ by the relations

$$g(w) = w^{\frac{2}{3}},\quad w(\zeta) = \zeta + r^2\zeta^{-1},\quad |\zeta| \geq 1$$
$$= \zeta + r^2\bar{\zeta},\quad |\zeta| \geq 1$$

and $\zeta(z) = z^{\frac{3}{2}}$. The resulting boundary curve Γ exhibits 3-fold symmetry. It is easy to show that $g \in \Sigma_2(r^2)$ but $g \notin \Sigma_2(k)$ for any $k < r^2$. Potential theory considerations, making essential use of the 3-fold symmetry then show that $\lambda > r^{-2}$.

Thus a whole range of examples is obtained for $\lambda^{-1} < k < r^2$; see [9] for details.

3. Löwner k-chains.

If $0 < k \leqslant 1$, a function

$$f(z,t) = e^t z + a_2(t)z^2 + \ldots\ldots\ldots,$$

analytic in D for each $t \in [0,\infty)$ is called a Löwner k-chain if it satisfies the Löwner differential equation

$$\frac{\partial}{\partial t} f(z,t) = z\, f'(z,t)\, p(z,t),$$

where

$$p(z,t) = \frac{1 + \psi(z,t)}{1 - \psi(z,t)} \quad \text{and} \quad |\psi(z,t)| \leqslant k,\ 0 \leqslant t < \infty.$$

It is more convenient in this case to consider such questions in D rather than in Δ. If $g(z)$ is of the form (1) in Δ and $g(z) \neq 0$ there then

$$f(z) = (g(^1/z)\}^{-1} \tag{7}$$

is analytic and univalent in D. Also f will have a k-quasi conformal extension to $\mathbb{C}$ if and only if $g \in \Sigma_2(k)$.

Theorem 3. Let f and g be related by (7) and suppose that, for $0 < k < 1$, there is a k-chain with $f(z,0) = f(z)$ Then $g \in \Sigma_2(k)$. In fact, $f(z,t)$ is continuous in $\bar{D}$ for $t \geqslant 0$ and the function

$$\begin{aligned} \breve{f}(z) &= f(z,0), \quad z \in \bar{D}, \\ &= f(e^{i\theta}, \log r), \quad z = re^{i\theta} \in \Delta \end{aligned}$$

provides the necessary k-quasi conformal extension for f.

This theorem of Becker [5] is derived from the celebrated Löwner univalence criterion, [14] or [16] Theorem 6.2, in the case $k = 1$.

Definition 2. If $0 < k < 1$ and a Löwner k-chain exists for f, then we say that $g \in \Sigma_4(k)$.

Thus Theorem 3 states that $\Sigma_4(k) \subset \Sigma_2(k)$ and there are relatively simple examples to show that the inclusion is strict for each k. It is an open question, asked in slightly more general terms by Pommerenke [15], whether a function $g \in Z_2(k)$ need belong to $\Sigma_4(k_1)$ for some k_1 with $k \leqslant k_1 < 1$. The more ambitious problem of characterising functions in $\Sigma_2(k)$ or $\Sigma_4(k)$ in terms of their Grunsky coefficients, seems very difficult indeed.

4. A Theorem of Schur.

One obvious way to start a study of the condition $||B|| \leq k$ is to consider the $n \times n$ submatrices B_n of B, whose entries are $B_{k\ell}$ for $1 \leq k, \ell \leq n$. Clearly $||B_n|| \leq k$ and B_n is symmetric, though not real symmetric and so not Hermitian.

Theorem 4. *If B_n is a symmetric $n \times n$ matrix, then there exists a unitary matrix U and numbers $d_r \geq 0$, $1 \leq r \leq n$, such that*

$$B_n = U^T D U, \tag{8}$$

where D is a diagonal matrix with entries d_r, $1 \leq r \leq n$.

This theorem is due to Schur, [18] or see [16] p.89. Since in general $U^T \neq U^{-1}$ the numbers d_r are not the eigenvalues of B_n; but they are related to the eigenvalues. If $||B_n|| \leq k$ then $0 \leq d_r \leq k$ for $1 \leq r \leq n$.

We define the numbers λ_n by

$$(\lambda_n)^{-1} = \max \{ | \sum_{k,\ell=1}^{n} B_{k\ell}\, x_k x_\ell | \ ; \ \sum_{k=1}^{n} |x_k|^2 \leq 1 \},$$

or, what is the same thing, λ_n^{-1} is the largest of the numbers d_r in the diagonal matrix appearing in (8). Clearly the sequence $\{\lambda_n^{-1}\}$ is non-decreasing and if $||B|| \leq k$ then $\lambda_n^{-1} \to k$ as $n \to \infty$. Thus, from Theorem 2, $\lambda_n \to \lambda$ as $n \to \infty$, providing an effective method of calculating the Fredholm eigenvalue λ.

Further important conclusions about the Fredholm eigenvalue have been drawn by Kühnau [9] p.38. For $\rho > 1$ we let Γ_ρ denote the level line of the function $g(z)$ corresponing to $|z| = \rho$ and let $\lambda(\rho)$ be the correpsonding Fredholm eigenvalue.

Theorem 5. *If Γ is a Jordan curve then $\rho^{-2}\lambda(\rho)$ is monotonic increasing for $\rho \geq 1$. In particular $\lambda(\rho) \to \lambda = \lambda(1)$ as $\rho \to 1+$.*

This result of Kuhnau, [9], Folgerung 2, is substantially better than that of [3], Corollary 1, obtained in ignorance of [9] and by a method involving the Schwarzian derivative.

5. The Diagonal Representation.

Using the diagonal representation (8) Hinkkanen has recently given a nice unified treatment of the Grunsky inequalities and their ensuing results [7]. It is too much to expect that a representation like (8) will hold for the infinite matrix B, but Hinkkanen shows that (8) does hold if B is compact ([7], Theorem 1). It is, however, not clear for which functions $g(z)$ the operator $B(g)$ is compact. But for fixed $\rho > 1$ we may set $g_\rho(z) = \rho^{-1}g(\rho z)$ and assert that $B(g_\rho)$ is compact; it is actually in the Hilbert-Schmidt class. Thus, for such a $\rho > 1$ there is an infinite sequence $\{d_k\}_1^\infty = \{d_k(\rho)\}_1^\infty$ with $d_k \geqslant 0$ for all k such that

$$B(g_\rho) = U^T D U, \tag{9}$$

where D is a diagonal matrix with entries $\{d_k\}$ as before and U is a unitary operator acting on ℓ^2.

Now let $B^2(\Delta)$ denote the Bergman space of Δ, i.e. the Hilbert space of functions analytic in Δ and belonging to $L^2(\Delta)$ with respect to the area measure dxdy. The functions $\psi_k(z) = \sqrt{\left[\frac{k}{\pi}\right]} z^{-k-1}$, $k = 1,2,\ldots$ form an orthonormal basis for $B^2(\Delta)$ and hence so do the functions

$$\phi_k(z) = \sum_{\ell=1}^{\infty} u_{k\ell}\psi_\ell(z)$$

where the $u_{k\ell}$ are the entries of the unitary matrix U appearing in (9). Note that the numbers $u_{k\ell}$ depend on ρ, though this is not important in what follows.

Theorem 5. With U and S defined by (3) and (4) we have, for each $\rho > 1$

$$U(z,\zeta,g_\rho) = -\pi \sum_{k=1}^{\infty} d_k\, \phi_k(z)\phi_k(\zeta)$$

$$S(g_\rho, z) = -6\pi \sum_{k=1}^{\infty} d_k\, (\phi_k(z))^2$$

for $z, \zeta, \in \Delta$, where $\phi_k(z)$ is defined by (10).

We remark also that

$$\sum_{k=1}^{\infty} \phi_k(z)\overline{\phi_k(\zeta)} = K(z,\zeta) = \frac{1}{\pi}(1-z\bar{\zeta})^{-2},$$

which is the kernal function of $B^2\Delta)$. Thus

$$K(z,z) = \frac{1}{\pi}(1-|z|^2)^{-2} = \frac{1}{\pi}(\rho(z))^2.$$

where ρ is the denisty of the Poincaré metric of Δ.

Using the representation of U and S afforded by Theorem 5, Hinkkanen [7] gave quick proofs of several well-known results concerning $\Sigma_1(k)$. We mention three of those, for which we need the following notation:-

$$||S(g,z)|| = \sup\{\rho(z)^{-2}|S(g,z)| : z \in \Delta\}$$

$$d(U,g) = d(U(z,\zeta,g)) = \sup\left\{\rho(\zeta)^{-1}\left[\iint_{\Delta}|U(z,\zeta,g)|^2 dx\, dy\right]^{1/2} : \zeta \in \Delta\right\}.$$

Theorem 6. The following estimate holds.

$$||S(g)|| \leq 6\, d(U,g) \leq 6\, ||B(g)||.$$

In particular if $g \in \Sigma_1(k)$ then $||S(g)|| \leq 6k$ and $d(U,g) \leq k$.

The first inequality is due to several persons [8], [12], [13] while the second is due to Bazilevitch and Zuravlev [19]. This method of proof of Hinkkanen is very promising and deserves further study. In paticular it would be interesting to see if the result of [4] Theorem 1, that if $g \in \Sigma_2(k)$ then $d(U,g) \leq K||S(g)||$ for some suitable $K = K(k)$, could be devised by these means.

In conclusion I thank my friends D. Bshouty and A. Hinkkanen for several interesting conversations on the subject of this report.

References

1. L.V. Ahlfors, "Remarks on the Neumann-Poincaré integral equation", Pacific J. Math. 2, (1952), 271-280.

2. L.V. Ahlfors, "Quasiconformal reflections", Acta Math. 109, (1963), 291-301.

3. J.M. Anderson, "The Fredholm eigenvalue problem for plane domains", Complex Variables, 5, (1986), 95-100.

4. J.M. Anderson and A. Hinkkanen, "Univalent Functions and domains bounded by quasi circles", J.London Math. Soc. (2), 35, (1982), 253-260.

5. J. Becker, "Löwnersche Differentialgleichung and quasikonform fortsetzbare schlichte Funktionen", J.Reine Angew. Math. 255, (1972),23-43.

6. H. Grunsky, "Koeffizientenbedingungen für schlicht abbildende meromorphe Funktionen", Math. Zeitschr. 45, (1939), 29-61.

7. A. Hinkkanen, "The diagonal representation of the Grunsky operator", to appear in Complex Variables.

8. R. Kühnau, "Verzerrungssätze und Koeffizientenbedingungen vom Grunskyschen Typ fur quasi konforme Abbildungen, Math. Nach. 48, (1971), 77-105.

9. R. Kühnau, "Zuden Grunskyschen Koeffizientenbedingungen", Ann. Acad. Sci. Fennicae, Ser A.1. Math. 6, (1981), 125-130.

10. R.Kühnau, "Quasikonforme Fortsetzbarkeit, Fredholmsche Eigenwerte und Grunskysche Koeffizienten bedingungen", Ann. Acad. Sci. Fennicae, Ser A.1. Math. 7, (1982), 383-391.

11. R.Kühnau, "Zum Koeffizientenproblem bei den quasi konform fortsetzbaren schlichten konformen Abbildungen", Math. Nach., 55, (1973), 225-231.

12. O.Letho,"Schlicht Functions with a quasi-conformal extension", Ann. Acad. Sci. Fennicae, Ser A.1., 500, (1971).

13. O.Lehto, "Quasi conformal mappings in the plane", Lecture Note 14, University of Maryland, (1975).

14. K. Löwner, "Untersuchungen uber schlichte konforme Abbildungen des Einheitskreises" Math. Ann., 89, (1923), 103-121.

15. C.Pommerenke, Problem, Ann. Polonici Math., 20, (1968),321-322.

16. C.Pommerenke, Univalent Functions, Vandenhoeck and Rupprecht, 1975.

17. G.Schober, Univalent Functions -Selected topics, Lecture notes in Math. 478, Springer, 1975.

18. J.Schur, "Bemerkung zur Theorie der beschrankten Bilinearforrnen mit unendlich vielen Veranderlichen", J. Reine Angew. Math., 140,(1911), 1-28.

19. I.V. Zuravlev, "Some sufficient conditions for the quasiconformal extension of analytic functions", Dokl. Akad. Nauk. S.S.S.R, 243, (1978): English Translation in Soviet Math. Doklady, 19, (1978), 1549-1552.

3 Hardy-Littlewood Integral Inequalities

William Desmond Evans University of Wales College of Cardiff, Cardiff, Wales
W. Norrie Everitt University of Birmingham, Birmingham, England

1. Introduction

In 1932 Hardy and Littlewood obtained the following result:

Theorem 1.1 If y and y'' are in $L^2(0,\infty)$ then

$$\left[\int_0^\infty |y'|^2dx\right]^2 < 4 \int_0^\infty |y|^2dx \int_0^\infty |y''|^2dx \tag{1.1}$$

unless $y(x) = AY(Bx)$, where $A \in \mathbb{C}$, B is a positive number and

$$Y(x) = e^{-\frac{1}{2}x} \sin(x \sin \tfrac{\pi}{3} - \tfrac{\pi}{3}) \tag{1.2}$$

when there is equality in (1.1).

This appeared in the paper [20] which subsequently formed the basis of Chapter VII of the book [21] Inequalities by Hardy, Littlewood and Polya, published in 1934. In addition to the three proofs of Theorem 1.1 in [21] a number of others have appeared since then: Kato [24] proved it by Hilbert-space methods, Copson [8] gave an elementary proof, while others have obtained the result as a special case of more general integral inequalities (see [13], [14], [26]).

Of the various extensions of Theorem 1.1, the one that is our prime concern in this paper is that considered by Everitt in [14] in which a connection is established between the Hardy-Littlewood inequality (1.1) and the Titchmarsh-Weyl theory for the second-order linear differential equation

$$-(py')' + qy = \lambda wy \quad \text{on } [a,b) \tag{1.3}$$

where $-\infty < a < b \leqslant \infty$, $y' = \frac{dy}{dx}$, $\lambda \in \mathbb{C}$ is the spectral parameter and p, q, w are real-valued functions which satisfy appropriate conditions (see §2). In the special case $a = 0$, $b = \infty$, $w = 1$, Everitt [14] investigated the inequality

$$\left[\int_a^b (p|f'|^2 + q|f|^2)dx\right]^2 \leqslant K \int_a^b w|f|^2dx \int_a^b w|w^{-1}M[f]|^2dx \tag{1.4}$$

where

$$M[f] := -(pf')' + qf \tag{1.5}$$

for all f in the maximal subspace Δ of the weighted Hilbert space $L^2_w(a,b)$ on which the right-hand side of (1.4) is finite. The Hardy-Littlewood inequality (1.1) is, of course, the special case $a = 0$, $b = \infty$, $p = 1$, $q = 0$ and $w = 1$ of (1.4). Everitt proved that the validity of (1.4) and the existence of extremal functions which give equality in (1.4) when K is the optimal constant, depends on the properties of the Titchmarsh-Weyl m-function associated with (1.3). His proof is modelled on one of the three proofs of (1.1) given in [21] which is based on the methods of the calculus of variations. An alternative proof of Everitt's result was given by Evans and Zettl in [13] using the von Neumann-Stone theory of extensions of symmetric operators in Hilbert space. In [10] Evans and Everitt surveyed the progress made up to that time in the study of (1.4) and also proved (1.4) under minimal conditions on the coefficients of (1.3) in the case when a is a regular end-point and b a singular end-point of (1.3) at which the strong limit-point condition prevails (see §2 for these terms). Since then further light has been shed, particularly for the regular problem, by Bennewitz in [4] and [5] and also new examples have been analysed which serve to illustrate the complex nature of the problem (see§4).

Our objective in this article is to give an up-to-date account of the work that has been done on (1.4), paying particular attention to the many varied and interesting examples now known.

We use the following notation throughout: $AC_{\ell oc}(I)$ denotes the set of function which are locally absolutely continuous on the interval I; if $w \in L_{\ell oc}[a,b)$ with $w(x) > 0$ a.e in $[a,b)$ then $L^2_w(a,b)$, stands for both the Lebesgue w-weighted integration space

$$\{f : [a,b) \mathbb{C} : \int_a^b w|f|^2 \equiv \int_a^b w(x)|f(x)|^2\, dx < \infty\}$$

and the corresponding Hilbert space of equivalence classes; if $f \in L_{loc}[a,b)$ and $\lim_{x\to b-} \int_a^x f(t)dt$ exists then the limit is written as $\int_a^{\to b} f$; $\mathbb{C}_+$ and $\mathbb{C}_-$ denote the open upper and lower half-planes of the complex plane $\mathbb{C}$ respectively.

2. The Titchmarsh-Weyl theory for (1.3)

The coefficients p, q of M in (1.5) are assumed to be real-valued and to satisfy

$$\left.\begin{aligned} &p(x) \neq 0 \quad \text{for a.e} \quad x \in [a,b), \\ &\frac{1}{p},\ q \in L_{loc}[a,b). \end{aligned}\right\} \tag{2.1}$$

For any f in the set

$$D(M) := \{f: [a,b) \to \mathbb{C} : f,\ pf' \in AC_{loc}[a,b)\} \tag{2.2}$$

$M[f] = -(pf')' + qf$ is defined as an element of $L_{loc}[a,b)$ and M is formally symmetric in the sense that, for all f, g $\in$ D(M), the Green's formula

$$\int_\alpha^\beta (\bar{g}Mf - f\,\overline{Mg}) = [fg](\beta) - [fg](\alpha) \tag{2.3}$$

holds for every compact subinterval $[\alpha,\beta]$ of $[a,b)$; here $[fg]$ denotes the skew-symmetric sesquinlinear form

$$[fg](x) = f(x)\,(\overline{pg'})(x) - (pf')(x)\,\bar{g}(x). \tag{2.4}$$

For $f \in D(M)$, M[f] is a quasi-differential expression; this refers to the fact that, in general, neither p nor f' is differentiable (even a.e.) and the term $(pf')'$ is to be taken as the derivative of the so-called quasi-derivative pf' of f.

In (1.3) w is assumed to be real-valued and to satisfy

$$w(x) > 0 \quad \text{for a.e. } x \in [a,b), \quad w \in L_{loc}[a,b). \tag{2.5}$$

Under the conditions (2.1) and (2.5) th end-point a if the interval [a,b) is a regular point of (1.3). The end-point b is regular if $b < \infty$ and $\frac{1}{p}$, q, w $\in$ L(a,b); otherwise b is singular. In this section we shall be concerned with a singular end-point b. The fact that a is a regular end-point of (1.3) implies that the initial-value problem for (1.3) with initial conditions at a has a unique solution; in particular there are unique solutions $\theta(\cdot,\lambda)$, $\phi(\cdot,\lambda)$ of (1.3) which satisfy

$$\left.\begin{aligned} \theta(a,\lambda) &= 0, \quad (p\theta')(a,\lambda) = 1, \\ \phi(a,\lambda) &= -1, \quad (p\phi')(a,\lambda) = 0. \end{aligned}\right\} \tag{2.6}$$

In 1910 Weyl proved that for $\operatorname{Im}\lambda \neq 0$ there is at least one non-trivial solution of (1.3) in $L^2_w(a,b)$. If there is precisely one (up to constant multiples), (1.3) is said to be in the limit-point case (LP) at b, while if all the solutions are in $L^2_w(a,b)$ then (1.3) is said to be in the limit-circle case (LC) at b. The limit-point classification of (1.3) is equivalent to the criterion that

$$\lim_{x\to b-} [fg](x) = 0 \tag{2.7}$$

for all f,g in

$$\Delta = \Delta(p,q,w) := \{f: f \in D(M),\ f \text{ and } \tfrac{1}{w} M[f] \in L^2_w(a,b)\}. \tag{2.8}$$

It is worth noting that Δ is the largest subspace of $L^2_w(a,b)$ on which the right-hand side of (1.4) is meaningful.

We shall also need the following notions: the equation (1.3) (or M) is

(i) strong limit-point (SLP) at b if

$$\lim_{x\to b-} f(x)(pg')(x) = 0 \qquad (f,g \in \Delta)$$

(ii) Dirichlet (D) at b if

$$|p|^{1/2}f' \text{ and } |q|^{1/2}f \in L^2(a,b) \qquad (f \in \Delta)$$

(iii) conditional Dirchlet (CD) at b if

$$|p|^{1/2}f' \in L^2(a,b) \text{ and } \int_a^{\to b} qf\bar{g} \text{ exists} \qquad (f,g \in \Delta)$$

(iv) weak Dirchlet (WD) at b if

$$\int_\alpha^{\to b} (pf'\bar{g}' + qf\bar{g}) \text{ exists} \qquad (f,g \in \Delta). \tag{2.9}$$

The motivation for, and connection between, these definitions come from the Dirichlet formula for (1.3):

$$\int_\alpha^\beta (pf'\bar{g}' + qf\bar{g}) = \bar{g}(pf')\Big|_\alpha^\beta + \int_\alpha^\beta w(\tfrac{1}{w} M[f])\bar{g} \tag{2.10}$$

which is valued for all f, $g \in \Delta$ and $[\alpha,\beta] \subset [a,b]$. It is clear that the following implications hold:

$$\text{LP} \Leftarrow \text{SLP} \Rightarrow \text{WD} \Leftarrow \text{CD} \Leftarrow \text{D}. \tag{2.11}$$

In general, these implications are known to be strict, except possibly for SLP $\Rightarrow$ WD which is undecided; in special circumstances D and even CD implies SLP (see [15], [23] and [25]). We shall be assuming in §3 that (1.3) is SLP at b and, by (2.11), this has the important consequence for us that (1.3) is WD at b and hence that both sides of (1.4) are finite for all $f \in \Delta$.

A discussion of the above concepts and abundant references may be found in [10,§3]; see also [1] and [27] for the general theory.

If (1.3) is LP at b there exist a pair of functions m_+, m_- defined on $\mathbb{C}_+, \mathbb{C}_-$ respectivly, which have the following properties:

$$\left.\begin{array}{ll}
\text{(i) } m_\pm : \mathbb{C}_\pm \to \mathbb{C}_\pm, & \\
\text{(ii) } m_\pm \text{ is analytic on } \mathbb{C}_\pm, & \\
\text{(iii) } m_\pm(\lambda) = \overline{m_\mp(\bar{\lambda})} & (\lambda \in \mathbb{C}_\pm), \\
\text{(iv) } \psi_\pm(\cdot,\lambda) := \theta(\cdot,\lambda) + m_\pm(\lambda)\phi(\cdot,\lambda) \in L^2_W(a,b) & (\lambda \in \mathbb{C}_\pm);
\end{array}\right\} \tag{2.12}$$

in fact

$$\int_a^b |\psi_\pm(\cdot,\lambda)|^2 = \frac{\operatorname{Im}[m_\pm(\lambda)]}{\operatorname{Im}[\lambda]} \qquad (\lambda \in \mathbb{C}_\pm).$$

In (2.11) and (2.12) read the upper signs or the lower signs together. The Titchmarsh-Weyl m-function is defined by

$$m(\lambda) := m_\pm(\lambda) \qquad (\lambda \in \mathbb{C}_\pm).$$

Its behaviour near the real-axis is linked with the spectral properties of the self-adjoint operator generated in $L^2_W(a,b)$ by M and Neumann boundary conditions at a. Indeed, the spectral measure of this operator is determined by m through the Titchmarsh-Kodaira formula.

Since M is LP at b, it follows from (2.12)(iv) that for $\operatorname{Im}\lambda \neq 0$, any $L^2_W(a,b)$ solution $\Psi(\cdot,\lambda)$ of (1.3) must be a constant (depending on λ) multiple of $\psi(\cdot,\lambda) := \theta(\cdot,\lambda) + m(\lambda)\,\phi(\cdot,\lambda)$. Thus there exists $k(\lambda)$ such that $\psi(x,\lambda) = k(\lambda)\,\Psi(x,\lambda)$, whence, on using (2.6),

$$m(\lambda) = -\Psi(a,\lambda)/(p\Psi')(a,\lambda) \tag{2.13}$$

and

$$\psi(x,\lambda) = \Psi(x,\lambda)/(p\Psi')(a,\lambda). \tag{2.14}$$

The identity (2.13) provides a handle for determining the m-functions; the choice of Ψ will be dictated by the properties of the solutions of (1.3).

A comprehensive survey of the properties of the m-function may be found in [6] and references therein; results concerning the behaviour of m near the real axis which are important to the inequalities discussed here, were given in [1],[2] and [7].

3. Everitt's inequality

Throughout this section the following conditions are assumed:

$$\left.\begin{array}{ll}\text{(i)} & p,q,w \text{ satisfy (2.1) and (2.5),} \\ \text{(ii)} & M \text{ is SLP at } b.\end{array}\right\} \tag{3.1}$$

In order to state the main result from [10] we need some extra notation. We shall write $\lambda = r\,e^{i\theta}$, where $r \in (0,\infty)$ and $\theta \in [0,2\pi)$, and define the line segments

$$L_+(\theta) = \{re^{i\theta} : r \in (0,\infty)\}, \quad L_-(\theta) = \{re^{i(\theta+\pi)} : r \in (0,\infty)\}. \tag{3.2}$$

for $\theta \in (0,\frac{\pi}{2}]$. Set

$$\theta_\pm := \inf\{\theta \in (0,\tfrac{\pi}{2}] : \text{for all } \phi \in [\theta,\tfrac{\pi}{2}],\ \mp\,\mathrm{Im}[\lambda^2 m_\pm(\lambda)] \geqslant 0 \quad (\lambda \in L_\pm(\phi))\}, \tag{3.3}$$

$$\theta_o := \max(\theta_+,\theta_-), \tag{3.4}$$

$$E_\pm := \{r \in (0,\infty) : \lambda \in L_\pm(\theta_o) \text{ and } \mathrm{Im}\,[\lambda^2 m_\pm(\lambda)] = 0\}, \tag{3.5}$$

$$Y_\pm(x;r) := \mathrm{Im}[\lambda\,\psi_\pm(x,\lambda)] \qquad (\lambda \in L_\pm(\phi)), \tag{3.6}$$

where $\psi_\pm$ are defined in (2.12)(iv). Note that the set in (3.3) is non-empty since it contains $\theta = \frac{\pi}{2}$, by (2.12)(iv).

Theorem 3.1 If (3.1) is satisfied we have

(a) $\theta_o \neq 0$;

(b) there is an inequality

$$\left\{\int_a^{\to b}(p|f'|^2 + q|f|^2)\right\}^2 \leqslant K \int_a^b w|f|^2 \int_a^b w|\tfrac{1}{w} M[f]|^2 \qquad (f \in \Delta) \tag{3.7}$$

for some positive constant K if and only if

$$\theta_0 \neq \frac{\pi}{2}\ ; \tag{3.8}$$

(c) if (3.8) is satisfied the best constant K in (3.7) is

$$K = \sec^2\theta_0\ ; \tag{3.9}$$

(d) if (3.8) is satisfied and K is given by (3.9) then all the $f \in \Delta$ that give equality in (3.7) are determined by the following three mutually exclusive conditions:

(i) $f = 0$,

(ii) there exists $f \in \Delta$ with $f \neq 0$, $M[f] = 0$ and either $f(a) = 0$ or $(pf')(a) = 0$, in which case both sides of (3.7) are equal to zero,

(iii) $E_+ \cup E_- \neq \emptyset$ and $f(x) = A\, Y_\pm(x;r)$ $(r \in E_\pm)$, with $A \in \mathbb{C}$, $A \neq 0$, in which case the two sides of (3.7) are equal but not zero.

Proof (sketch). In view of (2.12)(i) and (ii), m is a Nevanlinna function and the $m_\pm$ can be represented in the form

$$m_\pm(\lambda) = \alpha + \beta\lambda + \int_{-\infty}^{\infty}\left[\frac{1}{t-\lambda} - \frac{t}{t^2+1}\right]d\rho(t) \qquad (\lambda \in \mathbb{C}_\pm), \tag{3.10}$$

where ρ is a uniquely determined function satisfying

(i) $\rho : \mathbb{R} \to \mathbb{R}$ is monotonic increasing,

(ii) $\rho(t) = \frac{1}{2}[\rho(t-0)+\rho(t+0)]$ and $\rho(0) = 0$,

(iii) $\int_{-\infty}^{\infty}(1+t^2)^{-1}\, d\rho(t) < \infty$,

and α,β are uniquely determined real numbers with $\beta \geqslant 0$; for this representation of $m_\pm$ see [1,§69],[2,§3] and [6,§4]. Using (3.10) it is shown in [10,§4] that $\theta_0 = 0$ implies that m is analytic in $\mathbb{C} \setminus \{0\}$ with a simple pole at the origin.

This in turn implies that the spectrum of the self-adjoint operator T defined in $L^2_w(a,b)$ by $Tf = \frac{1}{w} M[f]$ on the domain $D(T) = \{f \in \Delta : (pf')(a) = 0\}$ is either empty or consists of $\{0\}$. Since T is unbounded its spectrum is an unbounded subset of $\mathbb{R}$ and consequently the asumption $\theta_0 = 0$ is contradicted.

The next step in the proof follows [21,§7.8] in observing that (3.7) is valid (i.e. there is an inequality for some $K \in (0,\infty)$) if and only if

$$J_{\rho,k}(f) = \int_a^{\to b} \{\rho^2 w|f|^2 - \rho k\,(p|f'|^2+q|f|^2) + w|w^{-1}M[f]|^2\} \tag{3.11}$$

satisfies

$$J_{\rho,k}(f) \geq 0 \qquad (f \in \Delta, \rho \in \mathbb{R}) \tag{3.12}$$

for some $k \in (0,\infty)$; if K,k are the best constants in (3.7) and (3.12) they are related by

$$K = 4k^{-2}. \tag{3.13}$$

Hereafter it is the inequality (3.12) which is analysed and, following the proof in [10], two different methods are employed. The first proceeds along the lines in [14]. Setting

$$\lambda_{\rho,k}(x) := \tfrac{1}{2}\rho k + \tfrac{1}{2} ik\surd(4-k^2)$$

for $\rho \in \mathbb{R}$ and $k \in [0,2)$, we define

$$\Psi_{\rho,k}(x) := 2\rho^{-1}(4-k^2)^{-\frac{1}{2}} \operatorname{Im}[\lambda_{\rho,k}\, \psi_+(x,\lambda_{\rho,k})] \qquad (\rho > 0) \tag{3.14}$$

with ψ_+ replaced by ψ_- for $\rho < 0$. Arguing, without loss of generality, with real $f \in \Delta$, it can be shown that with

$$g(x) := f(x) - (pf')(a)\, \Psi_{\rho,k}(x), \tag{3.15}$$

we have

$$\begin{aligned} J_{\rho,k}(f) &= J_{\rho,k}(g) + [(pf')(a)]^2 J_{\rho,k}(\Psi_{\rho,k}) \\ &\geq [pf'(a)]^2\, J_{\rho,k}(\Psi_{\rho,k}) \qquad (\rho \in \mathbb{R} \setminus \{0\}) \end{aligned} \tag{3.16}$$

with equality if and only if $g = 0$ and hence f is a constant-multiple of $\Psi_{\rho,k}$. Also, for $\rho > 0$

$$J_{\rho,k}(\Psi_{\rho,k}) = -2\rho^{-1}(4-k^2)^{-\frac{1}{2}}\,\mathrm{Im}[\lambda_{\rho,k}^2\, m_+(\lambda_{\rho,k})] \qquad (3.17)$$

with m_+ being replaced by m_- on the right-hand side if $\rho < 0$. It follows from (3.26) and (3.17) that

$$J_{\rho,k}(f) \geqslant 0 \qquad (f \in \Delta,\ \rho \in \mathbb{R} \setminus \{0\}) \qquad (3.18)$$

if and only if

$$\mp\, \mathrm{Im}[\lambda_{\rho,k}^2\, m_\pm\,(\lambda_{\rho,k})] \geqslant 0 \qquad (\rho \in \mathbb{R} \setminus \{0\}). \qquad (3.19)$$

The remainder of the theorem is then shown in [10] to be a consequence of (3.19).

The second proof in [10] gives an alternative way of establishing the equivalence of (3.18) and (3.19). It is based on the operator-theoretic argument in [13] and depends on the fact that Δ is the domain of the adjoint $T_0{}^*$ of the minimal operator T_0 generated by $\frac{1}{w}M$ in $L_w^2(a,b)$; T_0 is the closed symmetric operator defined in $L_w^2(a,b)$ by

$$\left.\begin{aligned} D(T_0) &= \Delta_0 = \{f \in \Delta : f(a) = (pf')(a) = 0\}, \\ T_0 f &= \frac{1}{w}M[f] \qquad (f \in D(T_0)). \end{aligned}\right\} \qquad (3.20)$$

From the von Neumann formula for closed symmetric operators (see [1,§102] and [27,§14.5]) we have the direct sum representation

$$\Delta = \Delta_0 \oplus N_\lambda \oplus N_{\bar\lambda} \qquad (\mathrm{Im}\lambda \neq 0), \qquad (3.21)$$

where $N_\lambda, N_{\bar\lambda}$ are the deficiency subspaces of T_0, i.e.

$$N_\lambda = \mathrm{kernel}\ (T_0{}^* - \lambda I) = \{\mathrm{range}\ (T_0 - \bar\lambda I)\}^\perp,$$

I being the identity operator on $L_w^2(a,b)$. A crucial fact for this method of proof is the observation that (3.21) is an orthogonal sum decomposition with respect to the inner product $(\cdot,\cdot)_\lambda$ defined for $\lambda = \mu + i\nu \in \mathbb{C}_+ \cup \mathbb{C}_-$ by

$$(f,g)_\lambda = (\tfrac{1}{w}M[f] - \mu f,\ \tfrac{1}{w}M[g] - \mu g) + \nu^2(f,g), \qquad (3.22)$$

where $(\cdot,\cdot)$ is the $L_w^2(a,b)$ inner product. Since M is LP at b, the deficiency subspaces N_λ, $N_{\bar\lambda}$ are spanned by $\psi(\cdot,\lambda)$ and $\psi(\cdot,\bar\lambda) = \bar\psi(\cdot,\lambda)$ respectively, and it follows from (2.6) and (3.21) that any real $f \in \Delta$ can be written as

$$f = f_0 + 2\mathrm{Re}[A_\lambda \psi_\pm(\cdot,\lambda)] \qquad (\lambda \in \mathbb{C}_\pm), \qquad (3.23)$$

where $f_0 \in \Delta_0$ and $A_\lambda = \alpha_\lambda + i\beta_\lambda \in \mathbb{C}$. A calculation then shows that

$$J_{\rho,k}(f) = ||f_0||_\lambda{}^2 + 4\nu(\beta_\lambda + \mu\nu^{-1}\alpha_\lambda)^2 \mathrm{Im}[m_\pm(\lambda)] - 4\nu^{-1}\alpha_\lambda{}^2 \mathrm{Im}[\lambda^2 m\pm(\lambda)] \tag{3.24}$$

with $\lambda = \lambda_{\rho,k}$. The equivalence of (3.18) and (3.19) is then established from (3.24) on using the fact that $\nu\mathrm{Im}[m\pm(\lambda)] > 0$ for all $\lambda \in C_+ \cup \mathbb{C}_-$, this being a consequence of (2.12)(iv).

An immediate consequence of Theorem 3.1 is that the constant K in (3.7) satisfies $K > 1$. We shall see in §4, Example 3 that any value in $(1,\infty)$ can be realised by the best possible constant K by an appropriate choice of the interval $[a,b)$ and the coefficients p,q and w.

4. Examples

In this section we give the examples known to us of the ienquality (3.7) when the conditions of Theorem 3.1 are satisfied. Other examples will be discussed later, in §§5,6. In all except Examples 5 and 6, the fact that M is SLP (and also D) at b will follow from the following lemma, communicated to us by C.Bennewitz.

Lemma 4.1 Let the interval be $[a,b)$ and suppose that

(i) $p(x) > 0$ a.e on $[a,b)$,

(ii) $q(x) \geqslant -kw$ on $[a,b)$, for some non-negative number k.

(iii) $w \notin L(a,b)$.

Then M is D and SLP at b.

Proof: Suppose $M[u] = f$ with u and $f \in L^2_w(a,b)$ and let $x \in [a,b)$. Integratation by parts gives

$$\int_a^x \{p|u'|^2 + (q + kw)|u|^2\} = pu'\bar{u}(x) - pu'\bar{u}(a) + \int_a^x f\bar{u}w + k\int_a^x |u|^2 w.$$

The two last terms have finite limits as $x \to b$ so the real part of the first term tends to a finite limit or $= \infty$. The limit can not be ∞ because that would imply that $u'\bar{u} + \bar{u}'u$ is eventually positive so that $|u|^2$ increases. Therefore $u \in L^2_w(a,b)$ would force $w \in L^1(a,b)$. Thus M is D. It also follows that $pu'\bar{u}(x)$ has a finite limit since all other terms have this property. Now we must have $pu'\bar{u}(x) \to 0$ as $x \to b$ because otherwise $1/(p\bar{u}'u(x))$ would be bounded in a left neighbourhood of b. Since $p|u'|^2 \in L^1(a,b)$, multiplication would then show that

$u'/u \in L^1(a,b)$ so that u has a non-zero limit at b. This again contradicts $w \notin L^1(a,b)$. It follows that M is SLP at b so the proof is complete.

The identity (2.13) is the tool used for determining $m_\pm$ and hence the other quantities in Theorem 3.1. Thus the first step is always to find the unique (up to constant multiples) solution $\Psi(\cdot,\lambda)$ of $M[f] = \lambda wf$ for $\mathrm{Im}\lambda \neq 0$, which lies in $L^2_w(a,b)$. Apart from the Examples 5 and 6 the functions m_+, m_- will be analytic continuations of one another; in the first four examples m is analytic in $\mathbb{C}\backslash[0,\infty]$, while in Example 7 it is meromorphic in $\mathbb{C}$ with simple poles on the real axis. Example 7 is the only one in which there is equality in (3.7) under the circumstances of d(ii) in Theorem 3.1, i.e. when both sides of (3.7) are equal to zero.

In all the examples given below the function $\lambda \to \lambda^\nu$ for $\nu \in (0,1)$ is defined by $\lambda = re^{i\theta}(r \in [0,\infty), \theta \in [0,2\pi))$ and $\lambda^\nu = r^\nu e^{i\nu\theta}$, so that $\lambda \to \lambda^\nu$ is analytic in $\mathbb{C} \setminus [0,\infty)$. Also, we use the notation $K(\epsilon)$, $\theta_\pm(\epsilon)$ etc. to indicate dependence of the constants in Theorem 3.1 on any parameter ϵ that appears in the example.

Example 1 $\quad a = 0,\ b = \infty,\ p = w = 1,\ q = 0.$ In this case (3.7) is the Hardy-Littlewood inequality (1.1) We have

$$\Psi(x,\lambda) = \exp(ix\sqrt{\lambda})$$

$$m(\lambda) = i\lambda^{-\frac{1}{2}}$$

$$\theta_+ = \frac{\pi}{3},\quad \theta_- = 0,\qquad \boxed{K = 4}$$

$$E_+ = (0,\infty),\quad E_- = \emptyset$$

$$Y(x;r) = \exp(-\tfrac{1}{2}rx)\sin(\tfrac{1}{2}rx\sqrt{3} - \tfrac{1}{3}\pi)$$

which is in accord with Theorem 1.1.

Example 2 $\quad a = 0,\ b = \infty,\ p = w = 1,\ q(x) = -\mu \in \mathbb{R}$, see ([3],[7],[9],[28]).

We have

$$\Psi(x,\lambda) = \exp(ix\sqrt{[\lambda+\mu]})$$

$$m(\lambda) = i(\lambda+\mu)^{-\frac{1}{2}}$$

$$\theta_+(\mu) = \begin{cases} \tfrac{1}{2}\pi & \text{if } \mu \in (-\infty,0), \\ \tfrac{1}{3}\pi & \text{if } \mu \in (0,\infty). \end{cases} \qquad \theta_-(\mu) = \begin{cases} 0 & \text{if } \mu \in (-\infty,0], \\ \pi/4 & \text{if } \mu \in (0,\infty), \end{cases}$$

$$K(\mu) = \begin{cases} \infty & \text{if } \mu \in (-\infty,0), \\ 4 & \text{if } \mu \in [0,\infty). \end{cases}$$

$$E_+(\mu) = E_-(\mu) = \emptyset \text{ if } \mu \in (0,\infty).$$

Hence there is a valid inequality if and only if $\mu \in [0,\infty)$. There are no non-trivial cases of equality when $\mu \in (0,\infty)$; the case $\mu = 0$ is given in Example 1.

Example 3 $a = 0$, $b = \infty$, $p(x) = x^\beta (\beta < 1)$, $q(x) = 0$, $w(x) = x^\alpha$ $(\alpha > -1)$ on $(0,\infty)$; see [19]. We now have

$$\Psi(x,\lambda) = x^{\frac{1}{2}(1-\beta)} H_\nu^{(1)}(k^{-1}x^k \sqrt{\lambda}),$$

$$m(\lambda) = \left[(2k)^{2\nu}\Gamma(1+\nu)/(1-\beta)\Gamma(1-\nu)\right]e^{\nu\pi i}\lambda^{-\nu},$$

where $k = \frac{1}{2}(\alpha-\beta+2)$, $\nu = (1-\beta)/(\alpha-\beta+2)$ and $H_\nu^{(1)}$ is the Hankel-Bessel function of order ν and type 1;

$$\theta_+(\alpha,\beta) = \pi(1-\nu)/2-\nu), \quad \theta_-(\alpha,\beta) = 0,$$

$$K(\alpha,\beta) = \left\{\cos\left[\frac{(1-\nu)\pi}{2-\nu}\right]\right\}^{-2}$$

$$E_+(\alpha,\beta) = (0,\infty), \quad E_-(\alpha,\beta) = \phi,$$

$$Y(x,r) = (xr)^{\frac{1}{2}} \operatorname{Im}\left[\exp(\tfrac{1}{2} i\pi[2+\nu])H_\nu^{(1)}\{k^{-1}(xr)^k \exp[i\pi(1-\nu)(4-2\nu)^{-1}]\}\right].$$

It is easily shown that with $\beta = 0$, $K(\alpha,0)$ is a continuous, monotonic increasing function of α on $(-1,\infty)$ and satisfies

$$\lim_{\alpha\to-1} K(\alpha,0) = 1, \qquad \lim_{\alpha\to\infty} K(\alpha,0) = \infty .$$

Hence, this example shows that the best constant K in (3.7) can attain every value in the permitted range $1 < K < \infty$.

Example 4 $a = 1$, $b = \infty$, $p(x) = x^\tau$ for some $\tau \in \mathbb{R}$, $q = 0$, $w = 1$; see [18] and [19],§4].

For $\tau > \frac{3}{2}$ the equation $M[f](x) \equiv -(x^\tau f'(x))' = 0$ on $[1,\infty)$ has a solution $f(x) = x^{1-\tau}$ in $L^2(1,\infty)$ and hence, since this $f \in \Delta$ the inequality (3.7) is not valid. If $\tau < \frac{3}{2}$ (in fact $\tau < 2$) we have

$$\Psi(x,\lambda) = x^{\frac{1}{2}(1-\tau)} H_{\nu}^{(1)}\left[2[2-\tau]^{-1}x^{\frac{1}{2}(1-\tau)}\sqrt{\lambda}\right]$$

and

$$m(\lambda) = - H_{\nu}^{(1)}(2[2-\tau]^{-1}\sqrt{\lambda})/\lambda^{\frac{1}{2}} H_{\nu-1}^{(1)}(2[2-\tau]^{-1}\sqrt{\lambda}).$$

It follows that

$$\theta_+(\tau) = \begin{cases} \frac{1}{3}\pi \text{ if } \tau \in (-\infty,0], \\ (3-\tau)^{-1}\pi \quad \text{if } \tau \in [0,1), \\ \frac{1}{2}\pi \quad \text{if } \tau \in [0,\infty), \end{cases}$$

$$\theta_-(\tau) = 0 \quad (\tau \in (-\infty,\infty)),$$

$$\boxed{K(\tau) = \begin{cases} 4 & \text{if } (\tau \in (-\infty,0], \\ \{\cos[(3-\tau)^{-1}\pi]\}^{-2} & \text{if } \tau \in (0,1), \\ \infty & \text{if } \tau \in [1,\infty). \end{cases}}$$

$$E_+(\tau) = E_-(\tau) = \emptyset \quad \text{if } \tau \neq 0,$$

$$E_+(0) = (0,\infty),\ E_-(0) = \emptyset$$

Apart from the Hardy-Littlewood case $\tau = 0$, there are no non-trivial cases of equality.

Example 5 $a = 0$, $b = \infty$, $p = w = 1$, $q(x) = - x$; see [11]. This time $M[f](x) = -f''(x) - xf(x)$ is SLP (and hence WD) at ∞ but is not D, nor even CD, at ∞; see [10,§3] and [16,§4]. Another interesting feature of this example is that m_+, m_- are not analytic continuations of one another. It can be shown that

$$\Psi(x,\lambda) = \begin{cases} (x+\lambda)^{\frac{1}{2}} H_{\frac{1}{3}}^{(1)}[\frac{2}{3}(x+\lambda)^{\frac{3}{2}}] & (\lambda \in \mathbb{C}_+), \\ (x+\lambda)^{\frac{1}{2}} H_{\frac{1}{3}}^{(2)}[\frac{2}{3}(x+\lambda)^{\frac{3}{2}}] & (\lambda \in \mathbb{C}_-), \end{cases}$$

$$m_+(\lambda) = - H_{\frac{1}{3}}^{(1)}(\tfrac{2}{3}\lambda^{\frac{3}{2}})/\lambda^{\frac{1}{2}} H_{-\frac{2}{3}}^{(1)}(\tfrac{2}{3}\lambda^{\frac{3}{2}})$$

$$m_-(\lambda) = - H_{\frac{1}{3}}^{(2)}(\tfrac{2}{3}\lambda^{\frac{3}{2}})/\lambda^{\frac{1}{2}} H_{-\frac{2}{3}}^{(2)}(\tfrac{2}{3}\lambda^{\frac{3}{2}})$$

$$\theta_+ = \tfrac{1}{3}\pi, \quad \theta_- = \tfrac{1}{6}\pi, \qquad \boxed{K = 4}$$

$$E_+ = (0,\infty),\ E_- = \emptyset,$$

$$Y(x,r) = \operatorname{Re}\left[e^{-\frac{1}{3}\pi i}(x+\lambda)^{\frac{1}{2}} H_{\frac{1}{3}}^{(1)}(\tfrac{2}{3}(x+\lambda)^{\frac{3}{2}})\right]$$

with $\lambda = re^{\frac{1}{3}\pi i}$ and $r \in E_+ = (0,\infty)$.

Example 6 $a = 0$, $b = \infty$, $p = w = 1$, $q(x) = -x^2$; see [22]. As in Example 5, M is SLP and WD at ∞; the proof that it is not CD at ∞ is similar to that for $q(x) = -x$. We also have that m_+, m_- are not analytic continuations of one another.

We have

$$\Psi(x,\lambda) = \begin{cases} D_{\frac{1}{2}i\lambda-\frac{1}{2}}([1-i]x) & \text{if } \lambda \in \mathbb{C}_+ , \\ D_{-\frac{1}{2}i\lambda-\frac{1}{2}}([1+i]x) & \text{if } \lambda \in \mathbb{C}_- , \end{cases}$$

where the $D_\nu(\cdot)$ are parabolic cylinder functions. It can be proved that

$$m_+(\lambda) = e^{i\pi/4}\,\Gamma\left[\tfrac{1}{4} - \tfrac{i\lambda}{4}\right] / 2\Gamma(\tfrac{3}{4} - i\tfrac{\lambda}{4}) ,$$

$$m_-(\lambda) = e^{-i\pi/4}\,\Gamma\left[\tfrac{1}{4} + \tfrac{i\lambda}{4}\right] / 2\Gamma(\tfrac{3}{4} + i\tfrac{\lambda}{4}),$$

$$\theta_+ = 3\tfrac{\pi}{8}, \quad \theta_- = \tfrac{\pi}{8} \qquad \boxed{K = 4+2\sqrt{2}}$$

$$E_+ = E_- = \emptyset.$$

There are no non-trivial cases of equality.

Example 7 $a = 0$, $b = \infty$, $p = w = 1$, $q(x) = x^2 - \tau$ for $\tau \in \mathbb{R}$; see [14,§17] and [12].

From [29,§4.2]

$$\Psi(x,\lambda) = e^{-\frac{1}{2}x^2} \int_\infty^{(0+)} \exp[-xz - \tfrac{1}{4}z^2]\, z^{-\mu/2} dz$$

and, denoting the Titchmarsh-Weyl function by m_τ we obtain

$$m_\tau(\lambda) = m_0(\lambda+\tau),$$

where

$$m_0(\mu) = \Gamma(\tfrac{1}{4} - \tfrac{1}{4}\mu)/2\Gamma(\tfrac{3}{4} - \tfrac{1}{4}\mu)$$

Therefore m_0 is meromorphic in $\mathbb{C}$ with simple poles at $\mu = 4k-3$, $k = 1,2,\ldots$ and simple zeros at $\mu = 4k-1$, $k = 1,2,\ldots$. The poles are in fact the eigenvalues of the boundary-value problem determined by the Hermite equation $M_0[f](x) \equiv -f''(x) + f(x) = \lambda f(x)$ on $[0,\infty)$ with the Neumann boundary condition $f'(0) = 0$, while the zeros are the eigenvalues for the Dirchlet boundary condition $f(0) = 0$. Collecting these poles and zeros of m_0 together in the set $\{2n + 1 : n = 0,1,\ldots\}$ the corresponding eigenfunctions $\{\Phi_n : n = 0,1,\ldots\}$ in the respective problems are given by

$$\Phi_n(x) = e^{-\frac{1}{2}x^2} H_n(x) \tag{4.1}$$

where H_n is the Hermite polynomial. The functions m_+, m_- are analytic continuations of one another.

It was proved in [14,§17, Example 4] that (3.7) is valid if and only if $\tau = 2n+1$, $n = 0,1,2\ldots$. In [12] it was proved that when $\tau = 2n+1$, the problem depends significantly on whether n is even or odd.

When n is <u>even</u>

$$\theta_+(2n+1) = \tfrac{1}{3}\pi,$$

$$\theta_-(2n+1) \begin{cases} \in (0,\tfrac{1}{3}\pi) & \text{if } n \neq 0, \\ = 0 & \text{if } n = 0, \end{cases}$$

$$\boxed{K(2n+1) = 4}$$

$$E_+(2n+1) = E_-(2n+1) = \emptyset.$$

There is equality if and only if, for some $A \in \mathbb{C}$, $f = A\,\Phi_n$, where Φ_n is defined in (4.1), when both sides of (3.7) are zero.

When n is <u>odd</u>

$$\theta_+(2n+1) = \pi - \theta_n, \text{ where } \theta_n \in (\tfrac{1}{2}\pi, \tfrac{2}{3}\pi)$$

$$\theta_-(2n+1) = \tfrac{1}{3}\pi,$$

$$\boxed{K(2n+1) = \sec^2\theta_n}$$

As n increases through odd values the θ_n are strictly increasing and as $n \to \infty$

$$\theta_n = \tfrac{2}{3}\pi - \frac{\sqrt{3}}{8\pi^2 n^2} + \frac{O(1)}{n^3},$$

$$K(2n+1) = 4 + \frac{3}{\pi^2 n^2} + \frac{O(1)}{n^3}.$$

Also, we have

$$E(2n+1) = \{r_n\}, \quad E_-(2n+1) = \emptyset,$$

where $r_n \in (0,4)$ and satisfies

$$r_n = \frac{6}{\pi^2 n} + \frac{O(1)}{n^2}$$

as $n \to \infty$. There are the following two non-trivial cases of equality in (3.7):

(i) $f = A\Phi_n$, $A \in \mathbb{C}$, when both sides are zero,

(ii) $f(x) = A \operatorname{Im} [r_n e^{-i\theta_n} j_n \Psi(x, 2n+1- r_n e^{-i\theta_n})]$, where

$$\frac{1}{j_n} = \Psi'(0,2n+1 - r_n e^{-i\theta_n})$$

$$= i(-1)^{(n-1)/2} 2^{1-n+r_n e^{-\frac{1}{2}i\theta_n}} \left\{\cos(\pi r_n e^{-i\theta_n}) - \sin(\pi r_n e^{-i\theta_n})\right\} \Gamma\left(\tfrac{3}{4} - \tfrac{n}{2} + r_n e^{-i\theta_n}\right) .$$

5. Validity of (3.7) in other cases

Theorem 3.1 determines the validity of (3.7) in the case when a is a regular end-point, b is singular and M is SLP at b. It is natural to ask if (3.7) can still be valid if any one of these assumptions is dropped. There are four cases to consider; the case of M being SLP at a and b a regular end-point is analogous to that in Theorem 3.1 and is omitted. If a and b are assumed to be singular end-points, it is to be understood that the open interval (a,b) replaces $[a,b)$ in (2.1), (2.2) and (2.5).

Case 1: a a regular end-point and M LP but not SLP at b.

It was conjectured in [10,§12] that (3.7) is not valid in this case and this was confirmed by Bennewitz in [4.§2]. His argument is based on the observation that (3.7) is valid only if the map defined for $f \in \Delta$ by

$$\{f,T_0^*f\} \to \int_a^{\to b} (p|f'|^2 + q|f|^2) =: D[f] \tag{5.1}$$

is bounded in $L^2_w(a,b) \times L^2_w(a,b)$; recall that T_0^* is the adjoint of the closed symmetric operator T_0 defined in (3.20) and $D(T_0^*) = \Delta$. If M is LP at b then it is well-known that the set $\Delta' \subset \Delta$, whose members have compact support in (a,b), is a core of T_0. Furthermore, if (3.7) is valid and hence (5.1) is bounded on $L^2_w(a,b) \times L^2_w(a,b)$ then it follows from the Dirichlet formula (2.10) that for all $f \in D(T_0)$

$$(f, T_0 f) = D[f]$$

and consequently $\lim_{x\to 0-} \bar{f}(pf')(x) = 0$. But, if $f \in \Delta$ there exists $\phi \in \Delta$ which vanishes in a neighbourhood of b such that

$$\phi(a) = f(a), \ (p\phi')(a) = (pf')(a)$$

and hence $f - \phi \in D(T_0)$.

We conclude that $\lim_{x\to 0-} \bar{f}(pf')(x) = 0$ for all $f \in \Delta$. On applying this result to $f + g$ and $f + ig$ it follows that M is SLP at b.

Case 2: a and b regular end-points

In [10,§12] it was conjectured that there is no valid inequality in this case. However, Bennewitz showed in [4,§5] that $p = 1$, $q = -1$, $w = 1$ gives a counter-example to this conjecture if $b-a$ is an integer multiple n of π. In this example the best constant K is approximately $(2.48)^2$ when $n = 1$ and is strictly decreasing with n to the limit 4, which is the value for the problem $[a,\infty)$. Bennewitz goes on to show in [5] that a necessary condition for (3.7) to be valid in the regular problem is that the origin is a double eigenvalue for (1.3) with a boundary condition of the form

$$u(a) = Cu(b), \ (pu')(b) = C(pu')(a) \tag{5.2}$$

for some non-zero real number C; this can be shown (see[5,§2]) to be equivalent to the necessary condition obtained by Everitt that the origin must be an eigenvalue for (1.3) with both the Dirchlet conditions $u(a) = u(b) = 0$ and the Neumann conditions $(pu')(a) = (pu')(b) = 0$.

In [5] Bennewitz obtains a necessary and sufficient condition for the inequality to be valid in the regular problem under conditions which govern the oscillatory behaviour of p near a and b. To be precise, he assumes that with

$$P_a(x) := \sup_{a \leqslant s \leqslant t < x} \left| \int_s^t {}^1/p \right| \quad \text{or} \quad P_b(x) := \sup_{x \leqslant s \leqslant t \leqslant b} \left| \int_s^t {}^1/p \right| \quad \text{then}$$

$$\left.\begin{aligned} P_a(x) &\backsim \int_a^x {}^1/p \quad \text{or} \quad P_a(x) \backsim -\int_a^x {}^1/p \quad \text{as} \quad x \to a+, \\ P_b(x) &\backsim \int_x^b {}^1/p \quad \text{or} \quad P_b(x) \backsim -\int_x^b {}^1/p \quad \text{as} \quad x \to b-, \end{aligned}\right\} \tag{5.3}$$

Let $W_a(x) := \int_a^x w$, $W_b(x) := \int_x^b w$ and denote the inverse functions by W_a^{-1}, W_b^{-1}; Bennewitz allows w to vanish on non-null subsets of (a,b) and defines W_a^{-1}, W_b^{-1} to be the generalised inverses

$$W_a^{-1}(x) := \inf\{y > a : W_a(y) > x\}, \quad W_b^{-1}(x) := \sup\{y < b : W_b(y) > x\}.$$

Define

$$\left.\begin{aligned} S_a(x) &:= \overline{\lim_{u \to 0+}}\ P_a \circ W_a^{-1}(xu) / P_a \circ W_a^{-1}(u), \\ S_b(x) &:= \overline{\lim_{u \to 0+}}\ P_b \circ W_b^{-1}(xu) / P_b \circ W_b^{-1}(u). \end{aligned}\right\} \tag{5.4}$$

The function S_a, S_b are increasing for $x \geqslant 0$ and in $(0,1)$ either $S_a \equiv 1$ or $\lim_{x \to 0+} S_a(x) = 0$; similarly for S_b. The result obtained by Bennewitz in [5] is

<u>Theorem 5.1</u> A necessary condition for (3.7) to be valid in the regular problem is that the origin is a double eigenvalue for (1.3) and non-separated boundary conditions (5.2). Furthermore, if p satisfies (5.3) then a necessary and sufficient condition for (3.7) to be valid is that in addition to the double eigenvalue property we have that neither S_a nor S_b is identically 1 in the interval $(0,1)$.

<u>Case 3:</u> <u>a a regular end-point and M LC at b</u>

It has been pointed out to us by C.Bennewitz that the change of variable $x \to {}^1/x$ in his example in [4] for the interval $[0,\pi]$, gives rise to a valid inequality in a case when M is LC at b. The inequality is

$$\left\{\int_{1/\pi}^{\infty}(x^2|f'(x)|^2 - x^{-2}|f(x)|^2)\right\}^2 \leqslant K\int_{1/\pi}^{\infty} x^{-2}|f(x)|^2 \int_{1/\pi}^{\infty}|x(x^2f'(x)'+ x^{-1}f(x)|^2,$$

where the best cosntant K is the same as that in Bennewitz's example for $[0,\pi]$. Note that the weight w in this example is $w(x) = x^{-2}$, and $1/p$, q, w all lie in $L^1(1/\pi,\infty)$.

Bennewitz has conjectured that there is no valid inequality in the LC case unless $1/p$, q, $w \in L^1(a,b)$.

Case 4: M SLP at a and b

The inequality is always valid in this case with best possible constant $K = 1$. For, on allowing $\alpha \to a+$, $\beta \to b-$ in the Dirichlet formula (2,10) we obtain for all $f \in \Delta$

$$\int_a^b (p|f'|^2 + q|f|^2) = \int_a^b w\{\frac{1}{w} M[f]\}\bar{f} .$$

Hence, by the Cauchy-Schwarz inequality, $K = 1$. For real f, there is equality if and only if $M[f] = \lambda wf$ for some $\lambda \in \mathbb{R}$. In the Hardy-Littlewood case there is clearly no $L^2(-\infty,\infty)$ solutions of $-f'' = \lambda f$ for $\lambda \in \mathbb{R}$ and hence there are no non-trivial cases of equality.

6. A Hardy-Littlewood type integral inequality with a monotonic weight function.

In this final section we report briefly ona remarkable extension of the orginal Hardy-Littlewood integral inequality, as given by (1.1) above. This new inequality was first noted by Kwong and Zettl [30] in 1979, and given in a more detailed account in [31] in 1981. There is an alternative discussion of this inequality given in 1987 by Everitt and Guinand.

Suppose that w is any positive, non-decreasing function on $(0,\infty)$ or $(-\infty,\infty)$; let $L_w^2(0,\infty)$ respectively $L_w^2(-\infty,\infty)$ denote the corresponding integrable-square weighted spaces, i.e. the collection of all Lebesgue measurable complex-valued functions such that

$$\int_0^\infty w(x)|f(x)|^2dx < \infty \quad \text{or} \quad \int_{-\infty}^\infty w(x)|f(x)|^2dx < \infty \;. \tag{6.1}$$

Let the linear manifold $D(w)$ in these spaces be determined by

$$D(w):= \{f:(0,\infty) \to \mathbb{R} \mid \quad \text{(i)} \quad f \text{ and } f' \in AC_{loc}(0,\infty) \qquad \text{(ii)} \quad f \text{ and } f'' \in L_w^2(0,\infty)\} \tag{6.2}$$

with a corresponding definition when the interval is $(-\infty,\infty)$.

Theorem 6.1 Let the weight w be given on the interval $(0,\infty)$ as above; let $D(w)$ be defined as in (6.2); then

(i) $f' \in L_w^2(0,\infty)$

(ii) the following inequality holds

$$\left[\int_0^\infty w(x)\; f'(x)^2dx\right]^2 \leqslant 4 \int_0^\infty w(x)f(x)^2dx \int_0^\infty w(x)\; f''(x)^2dx \tag{6.3}$$

for all $f \in D(w)$.

There are entirely similar results on the interval $(-\infty,\infty)$.

Proof For the proof of (6.3) see Zettl and Kwong [31] and Everitt and Guinand [17].

We make the following remarks on the inequality (6.3):

(i) the number 4 in (6.3) applies to all weights w of the kind prescribed above; in this sense it is a global number for the whole class of inequalities

(ii) if $\lim_{x\to 0+} w(x) > 0$ then the number 4 in (6.3) is best possible and all cases of equality can be determined

(iii) if $w(x) = 1$ $(x \in (0,\infty))$ then (6.3) is valid with the number 4 best possible; the general inequality then reduces to the Hardy-Littlewood inequality (1.1); the general theory also yields up all cases of equality given by (1.2)

(iv) if w is chosen so that $\lim_{x\to 0+} w(x) = 0$ then the best possible number in the inequality (6.3) may be strictly less than 4

(v) in the case of the interval $(-\infty,\infty)$ the best possible number in the inequality can be less than 4; there are no known cases when the best possible number is equal to 4.

The proof of these statements concerning the inequality (6.3) may be found in [31] and [17]. The methods used in [31] are operator theoretic, whilst those used in [17] are essentially based on classical analysis.

There is no equivalent of the Titchmarsh Weyl m-coefficient methods for the analysis of the integral inequality (6.3).

Finally we mention two examples which relate to the statements made above:

(i) $w(x) = 1$ $(x \in (-\infty,\infty)$; this example is mentioned in Case 4 in section 5 above; the best possible number in the inequality is 1, and there are no non-trivial cases of equality

(ii) $w(x) = x$ $(x \in (0,\infty))$; this example is considered in detail in [17, section 6]; the best possible number K_0 in the inequality

$$\left[\int_0^\infty xf'(x)^2dx\right]^2 \leqslant K_0 \int_0^\infty xf(x)^2dx \int_0^\infty xf''(x)^2dx$$

can be shown to satisfy the bounds

$$2.35070 < K_0 < 2.35075$$

and there is a continuum of non-trivial, distinct cases of equality.

References

1. Akhiezer, N.I. and J.M. Glazman. Theory of linear operators in Hilbert Space I. (Translated from the third Russian edition) Pitman and Scottish Academic Press, London and Edinburgh 1981.

2. Aronszajn, N. On a problem of Weyl in the theory of singular Sturm-Liouville equations. Amer. J. Math. 79 (1957), 597-610.

3. Beesack, P.R. A simpler proof of two inequalitites of Brodlie and Everitt. Proc, Roy. Soc, Edinb. A84(1979), 259-261.

4. Bennewitz, C. A general revision of the Hardy-Littlewood-Polya-Everitt (HELP) inequality. Proc. Roy. Soc. Edinb 97A (1984), 9-20.

5. Bennewitz, C. The HELP inequality in the regular case. General Inequalities 5, Oberwolfach (1987), Birkhäuser, Basel.

6. Bennewitz, C. and W.N. Everitt. Some remarks on the Titchmarsh-Weyl m-coefficient. Tribute to Åke Pleijel (Proceedings of the Pleijel Conference, Uppsala, 1979), pp49-108. Uppsala, Sweden: Department of Mathematics, University of Uppsala.

7. Brodlie, K.W. and W.N. Everitt. On an inequality of Hardy and Littlewood. Proc.Roy.Soc. Edinb. A72 (1973/4), 179-186.

8. Copson, E.T. On two integral inequalities. Proc.Roy.Soc.Edinb. A77 (1977), 325-328.

9. Copson, E.T. On two inequalities of Brodlie and Everitt. Proc.Roy.Soc.Edinb. A77 (1977), 329-333.

10. Evans, W.D. and W.N. Everitt. A return to the Hardy-Littlewood integral inequality. Proc.Roy.Soc.Lond. A380 (1982), 447-486.

11. Evans, W.D. and W.N. Everitt. On an inequality of Hardy-Littlewood type: I. Proc.Roy.Soc.Edinb. 101A (1985), 131-140.

12. Evans, W.D., Everitt, W.N., Hayman, W.K. and S. Ruscheweyh. On a class of integral inequalities of Hardy-Littlewood type. J.d'Anal. Math. 46 (1986), 118-147.

13. Evans, W.D. and A.Zettl. Norm inequalities involving derivatives. Proc. Roy.Soc.Edinb. A82 (1978), 51-70.

14. Everitt, W.N. On an extension to an integro-differential inequality of Hardy, Littlewood and Polya. Proc.Roy.Soc.Edinb. A69 (1971/2), 295-333.

15. Everitt, W.N. A note on the Dirichlet condition for second-order differential expressions. Can.J.Math. 28 (1976), 312-320.

16. Everitt, W.N., Giertz, M. and J.B.McLeod. On the strong and weak limit-point classification of second-order differential expressions. Proc.Lond.Math.Soc. (3) 29 (1974), 142--158.

17. Everitt, W.N. and A.P. Guinand. On a Hardy-Littlwood integral inequality with a monotonic weight function. General Inequalities 5, Oberwolfach (1987), Birkhäuser, Basel.

18. Everitt, W.N. and D.S. Jones. On an integral inequality. Proc.Roy.Soc.Lond. A357 (1977), 271-288.

19. Everitt, W.N. and A.Zettl. On a class of integral inequalities. J.Lond.Math.Soc. (2) 17(1978), 291-303.

20. Hardy, G.H. and J.E. Littlewood. Some integral inequalities connected with the calculus of variations. Quart J.Math(Oxford)(2) 3 (1932), 241-252.

21. Hardy, G.H., Littlewood, J.E. and G.Polya. Inequalities. Cambridge University Press 1934.

22. Hayman, W.K. Private communication, 1986.

23. Kalf, H. Remarks on some Dirichlet type results for semi-bounded Sturm-Liouville operators. Math. Ann. 210 (1974), 192-205.

24. Kato, T. On an inequality of Hardy, Littlewood and Polya. Adv.Math. 7 (1971), 217-218.

25. Kwong, M.K. Note on the strong limit-point condition of second-order differential expressions. Quart. J.Math.(Oxford) 28 (1977), 201-208.

26. Ljubic, Ju. I. On inequalities between the powers of a linear operator. Amer. Math. Soc.Transl. (2) 40 (1964), 39-84.

27. Naimark, M.A. Linear differential operators II. Ungar: New York 1968.

28. Phong, Vu Quôc. On inequalities for powers of linear operators and for quadratic forms. Proc.Roy.Soc.Edinb. A89 (1981), 25-50.

29. Titchmarsh, E.C. Eigenfunction expansions I. Oxford University Press 1962.

30. A. Zettl and M.K.Kwong. An extension of the Hardy-Littlewood inequality. Proc. Amer. Math. Soc. 77 (1979), 117-118.

31. A. Zettl and M.K. Kwong. Norm inequalities of product form in weighted L^p spaces. Proc. Royal Soc. Edinb. A 89 (1981), 293-307.

4 Inequalities in Mathematical Physics

Jack Gunson University of Birmingham, Birmingham, England

1. Introduction.

This review covers applications and developments of the classical inequalities in the study of stationary states in quantum mechanics and of equilibrium states in quantum statistical mechanics. The last two decades have seen many remarkable and elegant applications of inequalities in these areas, particularly in the treatment of fundamental issues and in qualitative studies of very complex systems. Most of the non-trivial problems are quite intractable for analytic solution and we often have to settle for bounds as close as we can get on the quantities of physical interest. One good example in the case of quantum hamiltonians is the construction of spectral inequalities which have increased our understanding of the stability of atoms, molecules and bulk matter and of the existence and qualititative properties of thermodynamic functions such as entropy. Many of the large variety of approximation methods used in quantum theory, such as those associated with the names of Rayleigh, Hartree and Thomas-Fermi, can be interpreted as providing applications of inequalities. In selecting material however, I have attempted to stress the inequalities which possess a more general mathematical interest and which indeed could be candidates for inclusion in some future successor to the classic work of Hardy, Littlewood and Pólya (1934). These have been roughly classified under the headings: Sobolev and spectral inequalities, majorization inequalities, convexity inequalities and rearrangement inequalities. Unfortunately, limitations on space and time have not permitted any coverage of some other important classes: eigenfunction inequalities, scattering inequalities and correlation inequalities.

Much of the discussion in later sections centres on the N-body non-relativistic Schrödinger equation. By a one-body Schrödinger operator we mean the partial differential operator of the form

$$H = H_0 + V = -\Delta + V \tag{1.1}$$

where Δ is the ν-dimensional Laplacian and $V = V(x)$ is a fixed one-body potential. If we introduce pair potentials $V_{ij} = V_{ij}(x_i - x_j)$ depending on the coordinates x_i and x_j of a pair of particles, we can write a typical N-body Schrödinger operator as

$$H = -\sum_{i=0}^{N} (\Delta_i - V_i) + \sum\sum_{i<j} V_{ij} \tag{1.2}$$

All Schrödinger operators considered are self-adjoint on a suitable domain of definition in the Hilbert space $L^2(\mathbb{R}^\nu)$ of quantum mechanical states in ν dimensions. In studying eigenvalue bounds, it is convenient to impose separate conditions of the positive (repulsive) and negative (attractive) parts of the potential, which are defined by

$$V_+(x) = \begin{cases} V(x) & V(x)>0 \\ 0 & V(x)\le 0 \end{cases} \qquad V_- = V_+ - V \tag{1.3}$$

In particular, we will usually impose stronger restrictions on the attractive part V_-, in order to ensure that the hamiltonian is lower semi-bounded. We use the notation $L^p(\mathbb{R}^\nu)$ to denote for $1 \le p < \infty$, the Banach space of pth power complex Lebesgue integrable functions on $\mathbb{R}^\nu$ under the usual norm

$$\|f\|_p = \left(\int |f(x)|^p \, d^\nu x \right)^{1/p} \tag{1.4}$$

Proofs will not in general be given unless they are very short or new or a suitable reference cannot be found.

2. Sobolev type inequalities and spectral inequalities.

For suitably restricted potentials and suitable $p,q \in (1,\infty)$, the resolvent of the Hamiltonian H defines a bounded map from L^p to L^q. Any inequality expressing the boundedness of such a map we call a Sobolev type inequality. For the one-body Schrödinger operator, Simon (1982) has proved a fairly general result, for potentials lying in the Kato class K_ν consisting of real-valued measurable functions on $\mathbb{R}^\nu$ for which

$$\text{if } \nu \ge 3 \text{ then } \lim_{\alpha\downarrow 0} \left[\sup_x \int_{|x-y|\le\alpha} |x-y|^{-(\nu-2)} |V(y)| d^\nu y \right] = 0$$

$$\text{if } \nu = 2 \text{ then } \lim_{\alpha\downarrow 0} \left[\sup_x \int_{|x-y|\le\alpha} \log\{|x-y|^{-1}\} |V(y)| d^2 y \right] = 0$$

$$\text{if } \nu = 1 \text{ then} \quad \sup_x \int_{|x-y|\le 1} |V(y)|dy < \infty \tag{2.1}$$

The kernel functions used in (2.1) are formed from the fundamental solutions for the differential operator Δ. In addition we denote by K_ν^{loc} the class of functions which are only locally in K_ν.

Theorem 2.1. *Let* $H = -\Delta + V$ *satisfy*

(i) $V_+ \in K_\nu^{loc}$ *and* $V_- \in K_\nu$,

(ii) $\alpha > 0$,

(iii) $1 \le p \le r \le \infty$ *and* $p^{-1} - r^{-1} < (2\alpha/\nu)$.

Then for any complex number obeying $\mathrm{Re}(z) < \inf(\mathrm{spec}(H))$, *the operator* $(H - z)^{-\alpha}$ *is bounded from* L^p *to* L^r.

If we take the case $V = 0$, $\alpha = 1/2$, $z = -1$, this gives the inequality

$$\|(-\Delta+1)^{-1/2}f\|_r \le C_{\nu pr}\|f\|_p \qquad p^{-1} - r^{-1} < \nu^{-1} \qquad f \in L^p \tag{2.2}$$

which is equivalent to the more usual form

$$\|g\|_{1,p} \ge K_{\nu pr}\|g\|_r \qquad g \in W^{1,p} \tag{2.3}$$

where $W^{1,p}(\mathbb{R}^\nu)$ is the Sobolev space with norm (see Adams (1975))

$$\|f\|_{1,p} = \left(\int (|\underline{\nabla} f|^p + |f|^p) d^\nu x\right)^{1/p} \tag{2.4}$$

This is a standard Sobolev inequality which holds also for $p^{-1} - r^{-1} = \nu^{-1}$ where $1 < p < r < \infty$. It is not known whether the latter remains true for all the potentials V in theorem 2.1. The best constants appearing in (2.3) can be expressed in terms of the three-norm Sobolev constants $C_{\nu,p,q,\vartheta}$, defined by

$$C_{\nu,p,q,\vartheta} = \inf_{f \in \mathcal{D}^{1,p} \cap L^q} \left(\frac{\|\underline{\nabla} f\|_p^{\vartheta} \|f\|_q^{1-\vartheta}}{\|f\|_r} \right) \tag{2.5}$$

where $0 \le \vartheta \le 1$, $\frac{1}{r} = \frac{\vartheta}{p} - \frac{\vartheta}{\nu} + \frac{1-\vartheta}{q}$, $1 \le p < \nu$, and $\mathcal{D}^{1,p}(\mathbb{R}^\nu)$ is the completion of $\mathcal{D}(\mathbb{R}^\nu)$ in the norm $f \longmapsto \left(\int \|\nabla f\|^p d^\nu x\right)^{1/p}$ (cf P.L. Lions (1985)). The relations between the parameters p, q, r, ϑ, ν ensure that the ratio of norms in (2.5) is homogeneous in f and scale invariant under dilations $x \longrightarrow \alpha x$, $\alpha > 0$. This excludes the trivial cases where the infimum is zero. In terms of these constants, a simple scaling argument gives the required Sobolev inequality

$$\|f\|_{1,p} \ge C_{\nu,p,p,\vartheta}\, \vartheta^{-\vartheta/p}\, (1-\vartheta)^{-(1-\vartheta)/p}\, \|f\|_r$$

where $\vartheta = \nu/p - \nu/r$, $0 \le \vartheta \le 1$.

Setting $\vartheta = 1$ in (2.5), we get the two-norm Sobolev constants

$$C_{\nu,p} = \inf_{f \in \mathfrak{D}^{1,p}} \left(\frac{\|\nabla f\|_p}{\|f\|_r} \right); \qquad \frac{1}{r} = \frac{1}{p} - \frac{1}{\nu}; \qquad 1 \le p \le \nu \tag{2.6}$$

which were evaluated by Aubin (1976) and Talenti (1976). Explicitly,

$$C_{\nu,p} = \begin{cases} \pi^{1/2}\nu^{1/p}\left(\dfrac{\nu-p}{p-1}\right)^{1-1/p}\left(\dfrac{\Gamma(\nu/p)\,\Gamma(\nu+1-\nu/p)}{\Gamma(1+\nu/2)\,\Gamma(\nu)}\right)^{1/\nu} & 1 < p < \nu \\ \pi^{1/2}\nu\,(\Gamma(1+\nu/2))^{-1/\nu} & p = 1,\ \nu \ge 2 \end{cases} \tag{2.7}$$

The only other cases where the constants are known explicitly occur in the trivial case $\vartheta = 0$ (when $C_{\nu,p,q,0} = 1$), and the one-parameter family

$$C_{\nu,p,q,\vartheta} = \left[\frac{\pi^{\nu/2}\,\chi^{-\xi(p-1)/p}\,\Gamma(\xi(p-1)/p)\,\Gamma(1+\nu(p-1)/p)}{\left(\frac{q-p}{p}\right)^{\nu(p-1)/p}\left(\frac{q}{\nu}\right)^{\nu/p}\Gamma\left(\frac{q(p-1)}{q-p}\right)\Gamma\left(\frac{\nu}{2}-1\right)} \right]^{\vartheta/\nu}$$

where $\xi = \left(\dfrac{pq - \nu(q-p)}{q-p}\right)$, $\chi = \left(\dfrac{pq - \nu(q-p)}{qp}\right)$,

$$\vartheta = \frac{\nu(q-p)}{(pq - \nu(q-p))(q-1)}, \qquad p \le q \le \frac{p(\nu-1)}{\nu-p} \tag{2.8}$$

A proof of the above results follows the pattern (cf. e.g. Weinstein (1983)):

(a) use a Schwarz symmetrization argument (for these and other rearrangement inequalities see section 5) to reduce the problem to the class S of functions which are positive, spherically symmetric and monotonic decreasing radially outwards,

(b) use an argument originally due to Strauss (1977) (see also P.L. Lions (1982)) to show that the restriction to the subspace of spherically symmetric functions of the natural injection of $\mathfrak{D}^{1,p} \cap L^q$ into L^r is compact,

(c) from (b) show that the infimum is actually attained by some minimising function in the class S and show that there is a unique solution to the Euler equation for the variational problem (2.5), which then gives the desired minimum. In the explicitly known cases this solution is always of the form $(a + b|x|^c)^{-d}$ for some positive values of the constants a,b,d and $c = p/(p-1)$.

For particular values of the parameters, the Euler equation is readily integrated using a shooting method (Levine (1980)). Veling (1985) has shown that, for fixed ν,p and q, that $C_{\nu,p,q,\vartheta}$ is logarithmically convex in ϑ. This is a simple consequence of Hölder's inequality applied to (2.5). Hölder's inequality can also be used to show that

$$C_{\nu,p,q,\vartheta} \geq (\vartheta C_{\nu,s})^{\vartheta}; \quad \frac{\vartheta}{s} = \frac{\vartheta}{p} + \frac{(1-\vartheta)}{q}; \quad (1+q(1-1/p))^{-1} \leq \vartheta \leq 1 \quad (2.9)$$

As shown by Dhesi and Gunson (1987), the same best constants $C_{\nu,p,q,\vartheta}$ are still valid for the corresponding function spaces on $\mathbb{R}^{\nu}$ taking vector or matrix values.

The case p = 2 is of most relevance for quantum mechanics. The important work of Lieb and Thirring (1976a) was one of the first applications using these inequalities to provide estimates on the discrete spectra of Schrödinger hamiltonians. The connection is simply illustrated by taking the case $\nu = 3$, $p = q = 2$ and $\vartheta = 3/5$ in (2.5). We can then write, using the notation of Lieb and Thirring (1976a),

$$\int |\underline{\nabla} f|^2 d^3x \geq K_3^{\,1} \int \rho^{5/3} d^3x \Big/ \left(\int \rho d^3x\right)^{2/3} \quad (2.10)$$

in which $\rho = |f|^2$ and $K_3^{\,1} = (C_{3,2,2,3/5})^2 = 3.880\ldots$. The ground state energy E_0 of a one-body hamiltonian can be expressed through the variational principle

$$E_0 = \inf_{\|f\|=1} \left(\int |\underline{\nabla} f|^2 d^3x + \int V|f|^2 d^3x\right)$$

The chain of inequalities

$$E_0 \geq \inf_{\|f\|=1} \left(\int |\underline{\nabla} f|^2 d^3x - \int V_-|f|^2 \, d^3x\right) \geq$$

$$\inf\left(K_3^{\,1}\lambda^{5/3} - \|V_-\|_{5/2}\lambda\right) \geq \left(\frac{3}{2K_3^{\,1}}\right)^{3/2}\left(\frac{5}{2}\right)^{-5/2}\int V_-^{\,5/2}(x)d^3x \quad (2.11)$$

provides a lower bound to E_0, in terms of the potential V. Conversely, if we set $V_- = \alpha\rho^{2/3}$ in the E_0 bound and maximise with respect to α, we recover the Sobolev inequality. The equivalence of the spectral bound to a Sobolev inequality is thus established.

More detailed spectral information is contained in the moment inequalities of Lieb and Thirring (1976a)

$$\sum_j |e_j|^\gamma \le L_{\gamma,\nu} \int (V_-(x))^{\gamma+\nu/2} d^\nu x \tag{2.12}$$

where $\gamma \ge 0$ and the $\{e_j\}$ are the eigenvalues in the negative spectrum of H. For a recent summary of what is known about these inequalities see Lieb (1984a). We have inequalities of the type (2.12) in the cases $\gamma > 1/2$ for $\nu = 1$, $\gamma > 0$ for $\nu = 2$ and $\gamma \ge 0$ for $\nu \ge 3$.

Analagously to the relation between the Sobolev inequality (2.10) and the eigenvalue bound (2.11), we can relate (2.12) to a generalised Sobolev type inequality, called by Lieb (1986a) a <u>Sobolev inequality for orthonormal functions</u>. Let $\{\phi_i : i = 1,2,..,n\}$ be an orthonormal set of functions in $L^2(\mathbb{R}^\nu)$, with the density function ρ defined by

$$\rho(x) = \sum_{i=1}^n \|\phi_i(x)\|^2 \tag{2.13}$$

then we can show

$$\sum_{i=1}^n \int |\underline{\nabla}\phi_i|^2 d^\nu x \ge K_\nu \int \rho^{1+2/\nu} d^\nu x \tag{2.14}$$

where K_ν is a positive constant which can be chosen so that (2.14) holds for all n. Orthogonality is essential for this latter property. In the standard vector-valued Sobolev inequality corresponding to (2.10) (cf. Dhesi and Gunson (1987))

$$\sum_{i=1}^n \int |\underline{\nabla} f_i|^2 d^\nu x \ge K_\nu^1 \int \rho^{1+2/\nu} d^\nu x \Big/ \left(\int \rho d^\nu x\right)^{2/\nu} \tag{2.15}$$

where $K_\nu^1 = (C_{\nu,2,2,\nu/(\nu+2)})^2$, $\rho = \sum_i \|f_1\|^2$ and no orthogonality is assumed. By normalising each component f_i in (2.15), we get (2.14) except that K_ν is replaced by $K_\nu^1 n^{-2/\nu}$. This tends to zero as $n \longmapsto \infty$ and so the latter bound is weaker for n large enough. In the presence of orthogonality the $n^{2/\nu}$ dependence can be replaced by a positive constant. In applications, particularly to many-particle Fermi systems, we need to extend (2.14) to the case of antisymmetric wave functions, i.e. functions $\psi(x_1,x_2,..,x_n) \in L^2(\mathbb{R}^{\nu n})$ with arguments $x_i \in \mathbb{R}^\nu$, $i=1,2,..,n$ and completely antisymmetric under interchanges of the x_i. For this we define the single-particle density as

$$\rho(x) = N \int |\psi(x_1,x_2,..,x_n)|^2 d^\nu x_2 ... d^\nu x_n \tag{2.16}$$

then (2.14) still holds if the left hand side is replaced by

$$N \int |\nabla_1 \psi|^2 \, d^\nu x.$$

Inequalities of the type (2.14) may be regarded as mathematical formulations of the quantum mechanical uncertainty principle which are qualitatively stronger than the classical Heisenberg uncertainty principle inequality: the more localised the one-particle density ρ, the larger the kinetic energy expression on the left hand side of (2.14). A discussion of these connections is given by Faris (1978) and an extension to the relativistic case has been carried out by Daubechies (1983a). The principal application of these inequalities has been to establish the stability and the existence of thermodynamic functions for bulk Coulomb matter. For the simplest system of this type we take a Hamiltonian of the type (1.2) for a system of N electrons interacting through Coulomb forces with a number of fixed charged nuclei. Stability means the existence of a finite energy per particle, i.e. an inequality of the form $H \geq -cN$ for all N and some positive constant c. For details of the history of this problem see the review by Lieb (1976c) and the final chapter of Thirring (1983). Another method to attack this problem has been developed by Fefferman (1983, 1985, 1986), based on a very detailed local uncertainty principle analysis. The corresponding stability problem with relativistic kinetic energy expression $-(\Delta + m^2)^{1/2}$ replacing $-\Delta$ has been treated by Lieb (1983a), Daubechies and Lieb (1983b), Conlon (1984) and Fefferman (1986). Another recent application by Lieb and Loss (1986b) is to the stability of atoms in external magnetic fields. The above results depend essentially on the fermi nature of the electrons. For a Bose gas with Coulomb interactions, the corresponding lower bounds have at best an N dependence given by $H \geq -cN^{7/5}$, where c is another universal constant. This has been discussed by Conlon (1985,1987a) and proved in general by Conlon, Lieb and Yau (1987b).

Equivalent forms of the two-norm inequality (2.6) are given by

$$\int |\nabla \phi|^p d^\nu x \; - \; \alpha \log \int |\phi|^{p^*} d^\nu x \;\; \geq \;\; \beta_{\nu,p,\alpha} \qquad \phi \in \mathcal{D}^{1,p} \qquad (2.17)$$

where $p^* = \nu p/(\nu - p)$, $\alpha > 0$ and

$$\beta_{\nu,p,\alpha} \;=\; \frac{\alpha\nu}{\nu-p}\left(1 \;-\; \log\left(\frac{\alpha\nu}{\nu-p}\right) \;+\; p \,\log\left(C_{\nu,p}\right)\right) \qquad (2.18)$$

If we denote the left hand side expression in (2.17) by $S(\phi)$, then the inequality (2.17) is readily verified by minimising $S(\lambda\phi)$ first with respect to the positive multiplier λ and then over $\phi \in \mathcal{D}^{1,p}$. The case with p = 2 of (2.17) has been used by Breen (1983), Magnen and Rivasseau (1985) and Magnen, Nicolò, Rivasseau and Sénéor (1987) in estimating the Lipatov bounds in perturbative expansions occurring in various quantum field theories. The Lipatov bound determines the radius of convergence of the Borel transformed series obtained by a Borel resummation procedure applied to the divergent perturbation series. For a theory with a $g\phi^4$ interaction in four space-time dimensions, Magnen et al (1987) obtain the Lipatov bound C_1 for the ground state energy perturbation series (modulo some renormalisation counterterms) as

$$C_1 = \exp\left(2 - \inf_{\phi}\left(\frac{1}{2}\int(\underline{\nabla}\phi)^2 d^4x - \log\left(\oint \phi^4 d^4x\right)\right)\right) \qquad (2.19)$$

from which, using (2.18) and (2.7), we get $C_1 = 3/2\pi^2$.

Another closely related inequality, the logarithmic Sobolev inequality, is readily obtained by applying Jensen's inequality to the logarithmic function in equation (2.17). Setting $\|\phi\|_p = 1$ and $\alpha = (\nu-p)/p^2$, we get

$$\int|\underline{\nabla}\phi|^p d^\nu x \geq \int|\phi|^p \log(|\phi|)d^\nu x + \gamma_{\nu,p} \qquad \|\phi\|_p = 1 \qquad (2.20)$$

for which the expression in (2.18) no longer necessarily gives the best value for $\gamma_{\nu,p}$. We can recalculate the best constant by following the standard procedure described earlier in this section or the alternative method given for p = 2 by Adams and Clarke (1979). Using optimising functions of the form $\lambda\exp(-\rho|x|^c)$, where $c = p/(p-1)$ and $\lambda,\rho > 0$, we get, for $1 < p < \infty$,

$$\gamma_{\nu,p} = \frac{\nu}{p} + \frac{\nu\log\pi}{2p} + \frac{\nu\log p}{p^2} - \frac{\nu(p-1)\log(p-1)}{p^2} - \frac{1}{p}\log\left(\frac{\Gamma(1+\nu/2)}{\Gamma(1+\nu(p-1)/p)}\right)$$

For p = 2, we get

$$\int|\underline{\nabla}\phi|^2 d^\nu x \geq \int|\phi|^2 \log(|\phi|)d^\nu x + \nu\left(\frac{1}{2} + \frac{1}{4}\log(2\pi)\right) \qquad \|\phi\|_2 = 1 \qquad (2.21)$$

which is the Lebesgue measure form of the logarithmic Sobolev inequality first obtained by L. Gross (1976) in terms of a Gaussian measure. The Gaussian function $G_\nu(x) = (2\pi)^{-\nu/4}\exp(-|x|^2/4)$ is optimising for (2.21). To get the Gross form, we make the substitutions

$$\phi(x) = \psi(x)G_\nu(x); \qquad d\mu = G_\nu^{\,2}(x)d^\nu x \qquad (2.22)$$

in (2.21) to get

$$\int|\underline{\nabla}\psi|^2 d\mu \geq \int|\psi|^2\log(|\psi|)d\mu; \qquad \int|\psi|^2 d\mu = 1 \qquad (2.23)$$

If ψ is not normalised, the inequality (2.23) becomes

$$\int|\underline{\nabla}\psi|^2 d\mu \geq \int|\psi|^2\log(|\psi|)d\mu - \|\psi\|_2^2\log(\|\psi\|_2) \qquad (2.24)$$

where the right hand side is homogeneous of degree 2 in ψ.

The p=2 version of the logarithmic Sobolev inequality, equation 2.21, follows also from a strong version of the uncertainty principle inequality obtained by Białynicki-Birula and Mycielski (1975). This latter inequality takes the form

$$-\int\rho(x)\log(\rho(x))d^\nu x - \int\tilde{\rho}(k)\log(\tilde{\rho}(k))d^\nu x \geq \nu(1+\log(\pi)) \qquad (2.25)$$

where $\rho = |\phi|^2$ and $\tilde{\rho} = |\tilde{\phi}|^2$ are probability density functions in position and momentum space respectively. The momentum space wave function $\tilde{\phi}(k)$ is the unitary Fourier transform of $\phi(x)$. We can interpret the integrals on the left hand side of (2.25) as information theoretic entropies which measure the degree of localisation of ρ in position space and $\tilde{\rho}$ in momentum space. Białynicki-Birula and Mycielski (1975) use a standard variational argument to show that, for a fixed variance, the extremum of the entropy functional is attained by a Gaussian wave function and (2.21) follows as a straightforward consequence.

As shown by Gross (1976), the main importance of the logarithmic form of the Sobolev inequality in quantum field theory lies in the fact that, unlike the standard Sobolev inequality, it remains non-trivial even when $\nu \longrightarrow \infty$. He showed that the inequality was equivalent to certain estimates of E. Nelson (1966, 1973) arising in constructive quantum field theory, where the passage to infinite dimensions is essential. Nelson's inequalities assert that, for $1 < q \leq p < \infty$, the operator e^{-tN} is a contraction from L^q to L^p if and only if $t \geq (1/2)\log((p-1)/(q-1))$. The operator N is defined on a suitable domain by the quadratic form

$$(Nf,f) = \int |\underline{\nabla}f|^2 \, d\mu \qquad (2.26)$$

and -N is the infinitesimal generator of the semigroup $\{e^{-tN}: t\geq 0\}$. These inequalities, which would nowadays be called hypercontractive, have given rise to the extensive modern theory of hypercontractive semigroups, a brief review of which appears in Simon (1982). Further connections with Sobolev inequalities and Schrödinger spectra are given in Rothaus (1978, 1980, 1981, 1985), Bakry and Emery (1986). Davies and Simon (1984) and Davies (1986) use logarithmic Sobolev inequalities to obtain sharp pointwise upper bounds on heat kernals and hence on Green functions and eigenfunctions. Logarithmic Sobolev inequalities for other measure spaces have been treated by Weissler (1978, 1980), Korzeniowski and Stroock (1985) and Korzeniowski (1987).

3. Majorization inequalities.

As shown by Uhlmann (1970), Wehrl (1978) and Thirring (1980), we can impose a natural partial order on the set of states of a quantum mechanical system which makes precise the heuristic concepts of a state being less pure or more mixed than another state. In the von Neumann picture a state is a normalised positive linear functional on the algebra of observables, which we take to be $\mathcal{B}(\mathcal{H})$, the algebra of bounded operators on a separable Hilbert space $\mathcal{H}$. A state w is conveniently represented by a trace class operator ρ, or density matrix, such that

$$w(A) = tr(\rho A), \quad \rho \geq 0, \quad tr(\rho) = 1, \qquad A \in \mathcal{B}(\mathcal{H}) \tag{3.1}$$

For general properties of the ideal $\mathcal{J}_1$ of trace class operators see Simon (1979a). Any $T \in \mathcal{J}_1$ possesses a canonical norm-convergent expansion

$$T = \sum_i \mu_i(T)\ (\cdot,\phi_i)\psi_i \tag{3.2}$$

where $\{\phi_i\}$ and $\{\psi_i\}$ are orthonormal sets in $\mathcal{H}$. The coefficients $\mu_i(T)$ in (3.2) are the non-null eigenvalues of $|T| = (T^{\dagger}T)^{1/2}$, known as the singular values of T. The singular values are labelled in decreasing order, i.e. $\mu_1(T) \geq \mu_2(T) \geq \ldots$. For the case of a density matrix ρ, the singular values of ρ are just the (positive) eigenvalues of ρ, and satisfy $\sum_i \lambda_i(\rho) = 1$. The two orthonormal sets in (3.2) are then also essentially the same.

We can now define the indicated partial order. A state ρ is less pure (or more mixed) than another state ρ', if

$$\sum_{i=1}^{n} \mu_i(\rho) \leq \sum_{i=1}^{n} \mu_i(\rho') \qquad n = 1,2,\ldots \tag{3.3}$$

We denote this by $\rho \prec \rho'$, a notation which is consistent with that introduced into the theory of majorization by Hardy, Littlewood and

Pólya (1934), and differs from the notations used by Wehrl (1978) and by Thirring (1980). For states we also have

$$\sum_i \mu_i(\rho) = \sum_i \mu_i(\rho') \tag{3.4}$$

because of the normalisation condition. The partial order can be extended to the whole of $\mathcal{T}_1$ in two main ways. If both (3.3) and (3.4) hold then we still set $\rho \prec \rho'$, i.e. ρ is majorized by ρ'. If (3.4) is dropped then we say that ρ is weakly majorized by ρ', and use the notation of Marshall and Olkin (1979): $\rho \prec_w \rho'$. The latter order can be extended to the ideal of all compact operators. Thirring (1980) gives the following equivalent conditions in

Theorem 3.1. *If ρ, ρ' are density matrix states, then the following statements are equivalent*

(a) $\rho \prec \rho'$,

(b) *For every convex function* k *on* $[0,\infty)$, *we have* $\mathrm{tr}(k(\rho)) \le \mathrm{tr}(k(\rho'))$,

(c) *The state ρ is in the weakly closed convex hull of the set* $\{U\rho' U^{-1} \mid U \text{ unitary}\}$.

The most important choice of convex function in (b) occurs for $k(\rho) = \rho\log(\rho)$, which gives the von Neumann expression for entropy in quantum mechanics

$$S(\rho) = -\mathrm{tr}(\rho\log(\rho)) \tag{3.5}$$

and consequently the fundamental property that the purest states possess the lowest entropy.

There are multiplicative versions of these partial orders, introduced by Lenard (1971) and Thompson (1971). For A and B of the form (I + a compact operator), we define

$$A \prec_w^m B \quad \Leftrightarrow \quad \prod_{i=1}^n \sigma_i(A) \le \prod_{i=1}^n \sigma_i(B) \qquad n = 1,2,\ldots \tag{3.6}$$

and if, in addition, we have

$$\prod_i \sigma_i(A) = \prod_i \sigma_i(B) \tag{3.7}$$

then we set $A \prec^m B$.

For general trace class operators, the singular values and eigenvalues may be distinct. The basic comparison result is due to Weyl (1949) and Horn (1950), as given by Simon (1979a)

Theorem 3.2 *Let* A *and* B *be compact operators*

(a) $$\prod_{i=1}^n |\lambda_i(A)| \le \prod_{i=1}^n \sigma_i(A) \qquad n = 1,2\ldots \tag{3.8}$$

(b) *if* f *is a non-negative monotone increasing function on* $[0,\infty)$

such that $t \longmapsto f(e^t)$ *is convex, then*

$$\sum_{i=1}^{n} f(|\lambda_i(A)|) \le \sum_{i=1}^{n} f(\sigma_i(A)) \qquad n = 1,2\ldots \qquad (3.9)$$

and in particular, for $p > 1$,

$$\sum_{i=1}^{n} |\lambda_i(A)|^p \le \sum_{i=1}^{n} (\sigma_i(A))^p \qquad n = 1,2\ldots \qquad (3.10)$$

(c) $$\sum_{i=1}^{n} \sigma_i(AB) \le \sum_{i=1}^{n} \sigma_i(A)\sigma_i(B) \qquad n = 1,2\ldots \qquad (3.11)$$

This theorem is the basis of many majorization inequalities with applications to quantum statistical mechanics. For example, we have

Theorem 3.3. *For any compact operators* A *and* B *we have*

(a) $A^2 \prec_w |A|^2$,

(b) *if* AB *is self-adjoint then* $AB \prec_w BA$.

Proof.

(a) From Horn's inequality (3.11), we get

$$\sum_{i=1}^{n} \sigma_i(A) \le \sum_{i=1}^{n} (\sigma_i(A))^2 = \sum_{i=1}^{n} (\sigma_i(|A|))^2$$

$$= \sum_{i=1}^{n} \sigma_i(|A|^2) \qquad n = 1,2\ldots$$

whence the result.

(b) AB and BA possess the same non-zero eigenvalues $\{\lambda_i\}$. Since AB is self-adjoint, its singular values are given by $\sigma_i = |\lambda_i|$, $i = 1,2,\ldots$ Applying Weyl's inequality (3.9) to BA, we get

$$\sum_{i=1}^{n} \sigma_i(AB) \le \sum_{i=1}^{n} |\lambda_i(AB)| = \sum_{i=1}^{n} |\lambda_i(BA)|$$

$$= \sum_{i=1}^{n} \sigma_i(BA) \qquad n = 1,2\ldots$$

and the result follows.

Corollary 3.4. If $-A$ and $-B$ are lower semibounded self-adjoint operators and $e^{B/2}e^{A}e^{B/2}$ is compact, then e^{A+B} is compact and

$$e^{A+B} \prec_w e^{B/2}e^{A}e^{B/2} \prec_w e^{A}e^{B} \qquad (3.12)$$

Proof. If X and Y are compact and self-adjoint then theorem 3.3 gives

$$(XY)^2 \prec_w |XY|^2 = YX^2Y \prec_w X^2Y^2$$

and by iteration

$$(XY)^{2^k} \prec_w Y^{2^{k-1}}X^{2^k}Y^{2^{k-1}} \prec_w X^{2^k}Y^{2^k}$$

Setting $X = e^{A/2^k}$ and $Y = e^{B/2^k}$, we use the Trotter formula (see theorem 1.1 of Simon (1979b))

$$\text{s-lim}_{k \to \infty} \left(e^{A/2^k} e^{B/2^k}\right)^{2^k} = e^{A+B} \tag{3.13}$$

to complete the proof.

The ordering (3.12) is a generalisation of the celebrated inequality

$$\operatorname{tr}(e^{A+B}) \le \operatorname{tr}(e^A e^B) \tag{3.14}$$

proved originally by Golden (1965) and Thompson (1965) and used to obtain bounds on the free energy in quantum statistical mechanics. If f is convex then, as in theorem 3.1, we can show

$$\operatorname{tr}(f(e^{A+B})) \le \operatorname{tr}(f(e^A e^B)) \tag{3.15}$$

Starting from the multiplicative form of Weyl's inequality (3.8), Thompson (1965) showed that (3.15) holds also for functions f monotonic increasing on $[0,\infty)$ and for which $t \longmapsto f(e^t)$ is convex.

4. Convexity inequalities.

The natural occurrence of convex functions such as exp(x) and xlog(x) in equilibrium statistical mechanics ensures the central importance of convexity inequalities in the development of the subject. In addition, for quantum statistical mechanics, we often need operator versions of the classical convexity inequalities. Many examples appear in the reviews of Huber (1970) and Wehrl (1978), and the books of Ruelle (1969) and Thirring (1983).

Convexity of a function f on a convex subset D in $\mathbb{R}^\nu$ is usually defined in terms of the inequality

$$f(\alpha x + (1-\alpha)y) \le \alpha f(x) + (1-\alpha)f(y) \tag{4.1}$$

for $0 \le \alpha \le 1$ and $x,y \in D$. When generalised to arbitrary convex combinations, we get the inequality usually called Jensen's inequality (see section 16.C of Marshall and Olkin (1979)) as

Theorem 4.1 *Let $(\Omega,\mathcal{B},\mu)$ be a measure space where μ is a probability measure. Let x be a random vector on taking values in an open convex domain $D \in \mathbb{R}^\nu$ and $\|\int x \, d\mu\| < \infty$. If $f\colon D \longrightarrow \mathbb{R}$ is convex, then*

$$f\left(\int_\Omega x \, d\mu\right) \le \int_\Omega f(x) \, d\mu \tag{4.2}$$

This reduces to (4.1) for a two-point measure space. In

applications to quantum mechanics, the Jensen inequality forms the basis of the Berezin-Lieb inequalities for families of coherent states. These inequalities are very useful in the study of the classical limits and first appeared in papers by Berezin (1972) and Lieb (1973c). Later reformulations have been given by Baumgartner (1980), Yaffe (1982), Thirring (1983) and by Simon (1980), which we follow. Let $\mathcal{H}$ be a separable Hilbert space and $(\Omega,\mathcal{B},\mu)$ a measure space. Then we define a family of coherent projectors on $\mathcal{H}$ by

Definition 4.2. *A family of coherent projectors is a weakly measurable map* $x \longmapsto P(x)$ *from* Ω *to the one-dimensional orthogonal projectors on* $\mathcal{H}$, *such that*

$$\int_\Omega P(x)\, d\mu(x) = I \qquad \textit{weakly} \tag{4.3}$$

From (4.3), the measure μ must satisfy $\int d\mu = \mu(\Omega) = \dim(\mathcal{H})$. For any bounded operator A on $\mathcal{H}$, there are at least two ways of associating A with measurable functions on Ω, called upper and lower symbols by Simon (1980) or contravariant and covariant symbols by Berezin (1972). These are summarized in

Definition 4.3. *Let* $A \in B(H)$ *and* $\{P(x)\colon x \in \Omega\}$ *be a family of coherent projectors. Then*

(a) the lower symbol A^{ℓ} *of* A *is the function in* $L^{\infty}(\Omega,d\mu)$ *given by the expectation values*

$$A^{\ell}(x) = \operatorname{tr}(AP(x)) \tag{4.4}$$

(b) an upper symbol A^{u} *of* A *is any function in* $L^{\infty}(\Omega,d\mu)$ *such that*

$$A = \int_\Omega A^{u}(x)\, P(x)\, d\mu(x) \tag{4.5}$$

Neither existence nor uniqueness of $A^{u}(x)$ is guaranteed in general. The duality inherent in the concepts is expressed in the following readily proved

Theorem 4.4. *Let* $A,B \in \mathcal{B}(\mathcal{H})$ *be such that* A *is of trace class and* B *possesses an upper symbol* B^{u}. *Then*

$$\operatorname{tr}(AB) = \int_\Omega A^{\ell}(x)\, B^{u}(x)\, d\mu(x) \tag{4.6}$$

from which we conclude that the lower symbol of a trace class operator is μ-integrable.

A double application of Jensen's inequality then leads to the upper and lower bounds of the Berezin-Lieb inequalities in

Theorem 4.5. *Let* A *be a self-adjoint operator on* $\mathcal{H}$ *and* Φ *be convex on* $\mathbb{R}$. *Then*

$$\int_\Omega \Phi(A^{\ell}(x))\, d\mu(x) \le \mathrm{tr}(\Phi(A)) \le \int_\Omega \Phi(A^{u}(x))\, d\mu(x) \qquad (4.7)$$

where $A^u(x)$ *is any real-valued upper symbol for* A *(if one exists).*

If an upper symbol can be found then the inequalities (4.7) can provide matched upper and lower bounds on a quantum partition function, which is typically of the form tr(exp(-H)) where H is a suitable quantum Hamiltonian. Many earlier inequalities of this type are special cases of (4.7). For example, a simple family of coherent projectors can be formed from the one-dimensional orthogonal projectors of a complete orthonormal set $\{e_i\}$ in the Hilbert space $\mathcal{H}$. The lower inequality in (4.7) then gives the well-known Peierls inequality

$$\sum_i \exp(-(Ae_i,e_i)) \le \mathrm{tr}(\exp(-A)) \qquad (4.8)$$

on setting $\Phi(x) = \exp(-x)$. For $\Phi(x) = x\log(x)$, we obtain inequalities relating classical and quantum-mechanical entropies. Following Wehrl (1978), we can associate a classical phase space density function with a quantum-mechanical density matrix state ρ (see section 3) in two main ways. If, for example, we use p,q to denote the conjugate phase space variables of the classical limit, we choose a suitable family of coherent state projectors P(p,q), such as that determined by the (overcomplete) set of Gaussian wave packets or other states of minimal uncertainty. The classical density function $\rho^{(1)}_{cl}(p,q)$ is then defined by

$$\rho^{(1)}_{cl}(p,q) = \mathrm{tr}(\rho\, P(p,q)) \qquad (4.9)$$

Note that the range of this density function lies in the interval [0,1]. Alternatively, we can regard the quantum state ρ as a mixture of coherent state projectors weighted according to some classical probability density function $\rho^{(2)}_{cl}(p,q)$ in

$$\rho = \int \rho^{(2)}_{cl}(p,q)\, P(p,q)\, dpdq \qquad (4.10)$$

This density function can take values greater than 1 and can be supported on a set small enough to violate uncertainty principle inequalities. The quantum entropy $S_{qu}(\rho) = -\mathrm{tr}(\rho\log\rho)$ then satifies the inequalities

$$S_{cl}^{(2)} = -\int \rho_{cl}^{(2)} \log(\rho_{cl}^{(2)})\, dpdq \leq S_{qu}(\rho)$$
$$\leq -\int \rho_{cl}^{(1)} \log(\rho_{cl}^{(1)})\, dpdq = S_{cl}^{(1)} \qquad (4.11)$$

as a direct consequence of theorem 4.5.

Further convexity inequalities can be obtained by reexpressing (4.1) in the form

$$\frac{f(z) - f(x)}{z - x} \leq \frac{f(y) - f(z)}{y - z} \qquad x < z < y \qquad (4.12)$$

If f(x) is differentiable at z, then we get, by a limiting process

$$f(x) - f(z) - (x - z)f'(z) \geq 0 \qquad (4.13)$$

Following an argument of O. Klein, this was shown by Huber (1970) to give the corresponding operator inequality

$$\mathrm{tr}(f(A) - f(B) - (A - B)f'(B)) \geq 0 \qquad (4.14)$$

where A,B are self-adjoint operators with complete sets of eigenfunctions, f is convex on an interval containing the spectra of A and B and the operator argument is of trace class. Taking f to be the exponential function and setting $A = C + D$, $B = C + \langle D \rangle$, where C and D are self-adjoint and $\langle D \rangle = \mathrm{tr}(D\exp(C))/\mathrm{tr}(\exp(C))$, in (4.14) gives the Peierls-Bogoliubov inequality

$$\mathrm{tr}(e^{C+D}) \geq \mathrm{tr}(e^{C+\langle D \rangle}) \qquad (4.15)$$

Taking f to be the function $x \longmapsto x \log x$ and A,B to be positive semidefinite operators, we get

$$\mathrm{tr}(A \log A - A \log B - A + B) \geq 0 \qquad (4.16)$$

In the case of two density matrices ρ_1 and ρ_2 then, for $0 \leq \alpha \leq 1$, $\rho = \alpha\rho_1 + (1-\alpha)\rho_2$ is a density matrix and from (4.16) we get

$$\mathrm{tr}(\rho \log \rho) = \alpha\, \mathrm{tr}(\rho_1 \log \rho) + (1-\alpha)\, \mathrm{tr}(\rho_2 \log \rho)$$
$$\leq \alpha\, \mathrm{tr}(\rho_1 \log \rho_1) + (1-\alpha)\, \mathrm{tr}(\rho_2 \log \rho_2) \qquad (4.17)$$

which shows that $-S(\rho)$ is convex on the set of all density matrices. The same trace inequality (4.16) can also be used to prove a stronger result, that of <u>subadditivity</u> of entropy. This compares the entropy for a state of a system on a Hilbert state space $\mathcal{H}$ with the entropies of component subsystems possessing Hilbert state spaces $\mathcal{H}_1$ and $\mathcal{H}_2$, such that $\mathcal{H} = \mathcal{H}_1 \otimes \mathcal{H}_2$. Corresponding to this decomposition one can define the operation of <u>partial trace</u>, denoted tr_1, for any trace class operator A on $\mathcal{H}$, giving a trace class operator $\mathrm{tr}_1(A)$ on $\mathcal{H}_2$ defined by

$$(tr_1(A)f, g) = \sum_j (A\, e_j \otimes f,\, e_j \otimes g) \qquad f,g \in \mathcal{H} \tag{4.18}$$

where $\{e_j\}$ is any complete orthonormal set in $\mathcal{H}_1$. The expression $tr_2(A)$ is defined similarly with $1 \longleftrightarrow 2$. Given any density matrix ρ for the whole system, we then define reduced density matrices for the subsystems by $\rho_1 = tr_2(\rho)$ and $\rho_2 = tr_1(\rho)$. On setting $A = \rho$ and $B = \rho_1 \otimes \rho_2$ in (4.16), we obtain the subadditivity property

$$S(\rho) \le S(\rho_1) + S(\rho_2) \tag{4.19}$$

with equality holding only if $\rho = \rho_1 \otimes \rho_2$.

We can also show that the entropy difference $S(\rho_1) - S(\rho)$ is concave as a function of ρ. Taking any convex combination $\rho = \alpha\rho' + (1-\alpha)\rho''$ and writing, correspondingly, $\rho = \alpha\rho_1' + (1-\alpha)\rho_1''$ we require to show that

$$-S(\rho) + \alpha S(\rho') + (1-\alpha)S(\rho'') + S(\rho_1) - \alpha S(\rho_1') - (1-\alpha)S(\rho_1'') \ge 0 \tag{4.20}$$

From the definitions, the left hand side of (4.20) can be expressed as $\alpha\Delta + (1-\alpha)\Delta''$, where

$$\begin{aligned} \Delta' &= tr(\rho'(\log\rho - \log\rho' - \log\rho_1 + \log\rho_1')) \\ \Delta'' &= tr(\rho''(\log\rho - \log\rho'' - \log\rho_1 + \log\rho_1'')) \end{aligned} \tag{4.21}$$

Inequality (4.15) with $C = \log\rho'$ and $D = (\log\rho - \log\rho' - \log\rho_1 + \log\rho_1')$, yields

$$\exp(\Delta') \le tr(\exp(\log\rho - \log\rho_1 + \log\rho_1')) \tag{4.22}$$

and a similar inequality for $\exp(\Delta'')$. Finally (4.20) is demonstrated through the chain of inequalities

$$\begin{aligned} \exp(\alpha\Delta + (1-\alpha)\Delta'') &\le \alpha tr(\exp(\log\rho - \log\rho_1 + \log\rho_1')) \\ &\quad + (1-\alpha)tr(\exp(\log\rho - \log\rho_1 + \log\rho_1'')) \\ &\le tr(\exp(\log\rho - \log\rho_1 + \log(\alpha\rho_1' + (1-\alpha)\rho_1''))) \le tr(\exp(\log\rho)) = 1 \end{aligned} \tag{4.23}$$

The only non-obvious step in (4.23) is handled by a delicate convexity result first proved by Lieb (1973a) and extended by Epstein (1973). This states that the function $A \longmapsto h(A) = tr(\exp(B + \log(A)))$, for fixed self-adjoint B, is concave on the cone of positive definite operators, i.e.

$$h(\alpha A_1 + (1-\alpha)A_2) \ge \alpha h(A_1) + (1-\alpha)h(A_2) \qquad A_1, A_2 > 0, \quad 0 \le \alpha \le 1 \tag{4.24}$$

Because of the readily verified homogeneity property $h(\lambda A) = \lambda h(A)$ for $\lambda > 0$, it is sufficient to show, for any fixed A_1, $A_2 > 0$, that $h(A_1 + xA_2)$ is concave on the interval $x > 0$. Extending x to a complex variable z, defining

$$F(z) = h(A_1 + zA_2) \tag{4.25}$$

and setting Im z > 0, we use the observation that the spectrum of $\log(A_1 + zA_2)$ lies in $\{\lambda \in \mathbb{C}: 0 < \text{Im}\lambda < \pi\}$ to conclude that Im F(z) > 0. In addition F(z) is real for z real and positive and so is an R-function (Herglotz function, Pick function, Nevanlinna function) with a representation

$$F(z) = \alpha + \beta z + \int_{-\infty}^{0} \left(\frac{1}{t - z} - \frac{t}{t^2 + 1} \right) d\sigma(t) \qquad (4.26)$$

where $\beta \geq 0$ and $\sigma(t)$ is non-decreasing (see Kac and Krein (1974)). The Lieb concavity (4.24) then follows from the concavity of $1/(t - x)$ for x > t.

The stronger inequality, proved by Lieb and Ruskai (1973b),

$$S(\rho) - S(\rho_{12}) - S(\rho_{23}) + S(\rho_2) \leq 0 \qquad (4.27)$$

known as <u>strong subadditivity</u>, is a direct consequence of the concavity of the entropy difference which we have shown just above. Strong subadditivity compares the entropy of the state ρ for a system on a state space $\mathcal{H}_1 \otimes \mathcal{H}_2 \otimes \mathcal{H}_3$ with the reduced entropies for the subsystems on $\mathcal{H}_1$, $\mathcal{H}_3$ and $\mathcal{H}_1 \otimes \mathcal{H}_3$. It is useful in the study of the thermodynamic limit and in particular ensures the existence of a finite entropy density.

The Berezin-Lieb inequalities (4.6) can be generalised to the case of composite systems,as was was noted in passing by Lieb (1973c). Cegła, Lewis and Raggio (1987) have given a simple proof and application of this in their treatment of the Dicke maser model. Let $\mathcal{H} = \mathcal{H}_1 \otimes \mathcal{H}_2$ and $\{P(x)\}$ be a family of coherent projectors defined in $\mathcal{H}_2$ on a measure space $(\Omega_2, \mathcal{B}_2, \mu_2)$. In analogy with (4.4) and (4.5) we define, for a self-adjoint operator A on $\mathcal{H}$,

$$A^{\ell}(x) = \text{tr}_2(AP(x)) \qquad A = \int_{\Omega} A^{u}(x) \otimes P(x)\, d\mu(x) \qquad (4.28)$$

where the values taken by A^{ℓ} and A^{u} are self-adjoint operators on $\mathcal{H}_1$. An adaptation of the proof of Simon (1980) then gives

$$\int_{\Omega} \text{tr}_1(\Phi(A^{\ell}(x)))d\mu_2(x) \leq \text{tr}_{12}(\Phi(A)) \leq \int_{\Omega} \text{tr}_1(\Phi(A^{u}(x)))d\mu_2(x) \qquad (4.29)$$

for any convex function Φ.

5. Rearrangement inequalities.

In a typical rearrangement inequality, we compare numerical

values of some expression, depending on a function f, with values of the same expression of the rearranged function f^*, i.e. a function in which the values of f have been rearranged over the domain of f according to some suitable reordering prescription. Chapter X of Hardy, Littlewood and Pólya (1934) is devoted to such inequalities. Apart from their use in proving some of the Sobolev inequalities used in section 2, they have been used in a variety of related variational problems of mathematical physics, particularly in establishing symmetry properties, such as spherical symmetry, of the solutions to such problems. Typical examples occur in the work of Strauss (1977), Coleman, Glaser and Martin (1978) and chapter II of Bandle (1980).

For numerical valued functions on $\mathbb{R}^\nu$, we have the following

Definition 5.1 *(Schwarz symmetrization). Let* f *be a measurable function on* $\mathbb{R}^\nu$, *and* μ *denote the usual Lebesgue measure. If there is some finite* c *for which the set* $\{x: |f(x)| > c\}$ *has finite Lebesgue measure, then there is an essentially unique non-negative function* f^* *that is equimeasurable with* f

$$\mu(\{x: |f(x)| > c\}) = \mu(\{x; f^*(x) > c\}) \qquad c > 0 \tag{5.1}$$

and is symmetric decreasing in the sense that f^* *is a function depending only on the radial coordinate* $r = |x|$ *and is monotonic decreasing in* r.

We call f^* the (spherically) symmetric decreasing rearrangement or Schwarz symmetrization of f. If required, uniqueness can be imposed by requiring outward continuity, i.e. $f^*(x) = \lim_{\lambda\uparrow 1} f^*(\lambda x)$.

Note that we have not insisted that f itself is non-negative, so that f^* is more accurately regarded as a rearrangement of the absolute value of f. A general definition of rearrangement relative to a nested family of measurable sets in a measure space $\{\Omega, \mathcal{B}, \mu\}$ has been given by Crowe, Zweibel and Rosenbloom (1986), in the form

Definition 5.2 *Let* $\mathcal{F}$ *be a family of measurable sets for* $\{\Omega, \mathcal{B}, \mu\}$ *simply ordered with respect to inclusion and satisfying*

(a) *for every* $B \in \mathcal{B}$ *there is an* $A \in \mathcal{F}$ *with* $\mu(B) = \mu(A)$,

(b) *for every* A

$$A = \bigcap_{B \in \mathcal{F}} \{B: \mu(B) > \mu(A)\}$$

Let M *denote the map from* $\mathcal{B}$ *to* $\mathcal{F}$ *determined by condition* (a). *Let* f *be a measurable function on* Ω. *If there is some finite* c *for which the set* $\{x: |f(x)| > c\}$ *has finite* μ*-measure, then we define the* *rearranged function* f^* *by*

$$f^*(x) = \sup\{c: x \in M(\{x: |f(x)| > c\})\} \tag{5.2}$$

In Schwarz symmetrization we use the family of all closed balls in $\mathbb{R}^\nu$ centered on the origin. By using the family of sets which consist of all points within a distance d of a fixed hyperplane through the origin, we obtain the Steiner symmetrization map as defined by Brascamp et al (1974). For functions in $L^p(\Omega, d\mu)$, we have the fundamental rearrangement results

$$\|f\|_p = \|f^*\|_p; \qquad \|f^* - g^*\|_p \le \|f - g\|_p; \qquad 1 \le p \le \infty \tag{5.3}$$

In other words, the rearrangement map is a norm-preserving contraction on any L^p space. The first equality is easily proved. The contraction property is a consequence of the L-superadditivity, as defined by Marshall and Olkin (1979), of the function $g(x,y) = |x-y|^p$ for $p \ge 1$:

Definition 5.3 *A real function* $g: \mathbb{R}^m \longmapsto \mathbb{R}$ *is L-superadditive if*

$$g(x \vee y) + g(x \wedge y) \ge g(x) + g(y) \tag{5.4}$$

for $(x \wedge y) = \min(x_i,y_i)$; $(x \vee y) = \max(x_i,y_i)$, $i = 1,2,...,n$.

The prefix L indicates the appearance of $\mathbb{R}^m$ as a lattice under componentwise ordering. The related basic rearrangement inequality for sequences $x^{(1)}, x^{(2)}, \ldots, x^{(m)} \in \mathbb{R}^N$ is

$$\sum_{j=1}^{N} g(x_j^{(1)},\ldots,x_j^{(m)}) \le \sum_{j=1}^{N} g(x_j^{*(1)},\ldots,x_j^{*(m)}) \tag{5.5}$$

where $x^{*(i)}$ is the decreasing rearrangement of $x^{(i)}$, $i = 1,2,..,m$. Setting $m = 2$, this readily leads to the L^p-contraction property for rearrangements of sequences. In the case $m = 2$, L-superadditivity of a non-negative function is just the condition that distribution functions must satisfy in order to give a non-negative Lebesgue-Stieltjes measure on $\mathbb{R}^2$. This is used by Crowe at al (1986) to prove the general result (5.3).

A much deeper result, valid only for Schwarz symmetrisation, is the inequality, proved by Pólya and Szegö (1951) and Talenti (1976)

$$\|\underline{\nabla} f^*\|_p \le \|\underline{\nabla} f\|_p \qquad 1 \le p \le \infty \tag{5.6}$$

This is closely related to the classical isoperimetric inequality in ν dimensions. For this and a more recent discussion see chapter II of Bandle (1980). In the Hilbert space case ($p = 2$), an elegant proof based on the heat semigroup has been given by Lieb (1983b). Under certain technical assumptions, Friedman and McLeod (1985) and Kawohl (1985, 1986) have shown that equality in (5.6) is attained only for functions which are already rearranged, i.e. the inequality (5.6) is strict.

The most general integral inequality of the Riesz type is due to Brascamp, Lieb and Luttinger (1974), who proved

Theorem 5.4 *Let* $\{f_j : 1 \le j \le k\}$ *be measurable functions on* $\mathbb{R}^\nu$ *with Schwarz symmetrizations* $\{f_j^* : 1 \le j \le k\}$. *Let* $\{a_{jm} : 1 \le j \le k,\ 1 \le m \le N\}$ *be a real matrix. Then*

$$\left|\int \prod_{j=1}^k f_j\left(\sum_{m=1}^N a_{jm} x_m\right) \prod_{m=1}^N d^\nu x_m\right| \le \int \prod_{j=1}^k f_j^*\left(\sum_{m=1}^N a_{jm} x_m\right) \prod_{m=1}^N d^\nu x_m \qquad (5.7)$$

This was applied by Luttinger (1973) to obtain inequalities for Wiener integrals in quantum mechanics. It includes as a special case the Riesz (1930) rearrangement inequality for the multiple convolution of several functions on $\mathbb{R}^\nu$.

For applications to vector and matrix field theories, we need generalisations of the above inequalities to matrix-valued functions on $\mathbb{R}^\nu$. If $\mathcal{M}_n$ denotes the space of all n × n complex matrices, we can construct L^p spaces of functions $F: \mathbb{R}^\nu \longmapsto \mathcal{M}_n$, in terms of the norms

$$\|F\|_p = \begin{cases} \left(\int \operatorname{tr}(|F(x)|^p) d^\nu x\right)^{1/p} & 1 \le p < \infty \\ \sup_{x \in \mathbb{R}^\nu} \|F(x)\| & p = \infty \end{cases} \qquad (5.8)$$

where $A \longmapsto |A| = (A^\dagger A)^{1/2}$ denotes the absolute value map on $\mathcal{M}_n$. The required generalisations can now be expressed using the <u>singular value diagonalisation map</u> S. This is a map from $\mathcal{M}_n$ to the subspace of diagonal matrices for which the diagonal elements of S(A), $A \in \mathcal{M}_n$, are the singular values of A (see section 3) taken in decreasing order down the diagonal. In analogy to (5.3), we can show

$$\|F\|_p = \|S(F)\|_p; \quad \|S(F) - S(G)\|_p \le \|F - G\|_p; \quad 1 \le p \le \infty \qquad (5.9)$$

and in analogy to (5.6)

$$\|\underline{\nabla}\, S(F)\|_p \le \|\underline{\nabla}\, F\|_p \qquad (5.10)$$

where

$$\|\underline{\nabla}\, F\|_p = \left(\int_{\mathbb{R}^\nu} \operatorname{tr}\left(\left(\sum_{\mu=1}^{\nu} \frac{\partial F^\dagger}{\partial x_\mu}\frac{\partial F}{\partial x_\mu}\right)^{p/2}\right) d^\nu x\right)^{1/p} \tag{5.11}$$

If we now define the Schwarz symmetrisation map $F \longmapsto F^*$ for matrix-valued functions in terms of the map S and symmetrisation in the sense of definition (5.1) by

$$\left(F^*\right)_{ii} = \left(S(F)\right)_{ii}^{*} \tag{5.12}$$

then the natural generalisations of (5.3) and (5.6) hold. In addition, by using an inequality of Fan given in section 20.B.2 of Marshall and Olkin (1979)

$$|\operatorname{tr}(F_1 F_2 \ldots F_N)| \le \operatorname{tr}(S(F_1)S(F_2)\ldots S(F_N)) \tag{5.13}$$

we obtain the following matrix generalisation of the inequality (5.7)

$$\left|\int \operatorname{tr}\left(\prod_{j=1}^{k} F_j\left(\sum_{m=1}^{N} a_{jm} x_m\right)\right)\prod_{m=1}^{N} d^\nu x_m\right| \le \int \operatorname{tr}\left(\prod_{j=1}^{k} F_j^*\left(\sum_{m=1}^{N} a_{jm} x_m\right)\right)\prod_{m=1}^{N} d^\nu x_m$$

in which the matrices under the trace sign can be taken in any order.

Acknowledgements.

I would like to thank the editor, Professor W.N.Everitt, for the invitation to write this review. Many helpful comments on the manuscript were provided by E.B. Davies, J.T. Lewis and E.H. Lieb.

References.

Adams R.A. (1975) Sobolev spaces. Academic Press, N.Y.

Adams R.A., Clarke F.H. (1979) Gross's logarithmic Sobolev inequality: a simple proof. Amer. J. Math. 101 1265-1269.

Araki H., Lieb E.H. (1970) Entropy inequalities. Commun. Math. Phys. 18 160-170.

Aubin T. (1976) Problèmes isopérimétriques et espaces de Sobolev. J. Differential Geom. 11 573-598.

Bandle C. (1980) Isoperimetric inequalities and applications. Pitman, London.

Baumgartner B. (1980) Classical bounds on Quantum partition functions. Commun. Math. Phys. 75 25-71.

Berezin F.A. (1972) Covariant and contravariant symbols of operators. Math. USSR-Izv. 6 1117-1151.

Białynicki-Birula I., Mycielski J. (1975) Uncertainty relations for information entropy in wave mechanics. Commun Math. Phys. 44 129-132.

Brascamp H.J., Lieb E.H., Luttinger J.M. (1974) A general rearrangement inequality for multiple integrals. J. Funct. Anal. 17 227-237.

Brascamp H.J., Lieb E.H. (1976a) Best constant in Young's inequality, its converse and its generalisations to more than three functions. Adv. Math. 20 151-173.

Breen S. (1983) Leading order asymptotics for $(\phi^4)_2$ perturbation theory. Commun. Math. Phys. 92 179-194.

Brezis H., Lieb E.H. (1983a) A relation between pointwise convergence of functions and convergence of functionals. Proc. Amer. Math. Soc. 88 486-490.

Brezis H., Lieb E.H. (1984) Minimum action solutions of some vector field equations. Commun.Math.Phys. 94 439-458.

Cegła W., Lewis J.T., Raggio G.A. (1987) Bounds for the free energy of quantum spin systems. Dublin IAS preprint, to be published.

Coleman S., Glaser V., Martin A. (1978) Action minima among solutions to a class of Euclidean scalar field equations. Commun. Math. Phys. 58 211-221.

Conlon J.G. (1984) The ground state energy of a classical gas. Commun. Math. Phys. 94 439-458.

Conlon J.G. (1985) The ground state energy of a Bose gas with Coulomb interactions. Commun. Math. Phys. 100 356-397.

Conlon J.G. (1987a) The ground state energy of a Bose gas with Coulomb interactions II. Commun. Math. Phys. 108 363-374.

Conlon J.G., Lieb E.H., Yau H-T. (1987b) The $N^{7/5}$ law for bosons. (To be published).

Crowe J.A., Zweibel J.A., Rosenbloom P.C. (1986) Rearrangements of functions. J.Funct.Anal. 66 432-438.

Daubechies J. (1983a) An Uncertainty Principle for Fermions with generalised kinetic energy. Commun. Math. Phys. 90 511-520.

Daubechies J., Lieb E.H. (1983b) One-electron relativistic molecules with Coulomb interaction. Commun. Math. Phys. 90 497-510.

Davies E.B., Simon B. (1984) Ultracontractivity and the heat kernel for Schrödinger operators and Dirichlet Laplacians. J. Funct. Anal. 59 335-395.

Davies E.B. (1986) Perturbations of ultracontractive semigroups. Quart. J. Math. Oxford 37 167-176.

Dhesi G., Gunson J. (1987) Instantons in matrix valued ϕ^4 quantum field theory. J. Phys. A20 1185-1192.

Epstein H. (1973) Remarks on two theorems of E. Lieb. Commun. Math. Phys. 31 317-325.

Faris W.G. (1978) Inequalities and the Uncertainty principle. J. Math. Phys. 19 461-466.

Fefferman C. (1983) The uncertainty principle. Bull. Amer. Math. Soc. 9 129-206.

Fefferman C. (1985) The atomic and molecular nature of matter. Rev. Math. Iberoamericana 1 1-44.

Fefferman C. (1986) The N-body problem in Quantum Mechanics. Commun. on Pure and Applied Math. 34 S67-S109.

Friedman A., McLeod B. (1985) Strict inequalities of integrals of decreasing rearranged functions. Proc. Roy. Soc. Edin. 102A 271-289.

Golden S. (1965) Lower bounds for Helmholtz functions. Phys.Rev. B137 1127-1128.

Gross L. (1976) Logarithmic Sobolev inequalities. Amer.J.Math. 97 1061-1083.

Hardy G.H., Littlewood J.E., Pólya G. (1934) Inequalities. Cambridge University Press.

Horn A. (1950) On the singular values of a product of completely continuous operators. Proc. Nat. Acad. Sci. 36 374-375.

Huber A. (1970) An inequality for traces and its application to extremum principles and related theorems in quantum statistical mechanics. Methods and Problems of Theoretical Physics. North-Holland, Amsterdam.

Israel R.B. (1979) Convexity in the theory of lattice gases. Princeton Univ. Press, N.J.

Kac I.S., Krein M.G. (1974) R-functions - analytic functions mapping the upper halfplane into itself. Amer. Math. Soc. Transl. (2) 103 1-18.

Kawohl B. (1985) Rearrangements and convexity of level sets in PDE. Lect. Notes in Math. 1150 Springer-Verlag, Berlin.

Kawohl B. (1986) On the Isoperimetric nature of a Rearrangement Inequality and its consequences for some Variational Problems. Arch. Rat. Mech. Anal. 94 227-244.

Korzeniowski A., Stroock D.W. (1985) An example in the theory of hypercontractive semigroups. Proc. Amer. Math. Soc. 94 87-90.

Korzeniowski A. (1987) On logarithmic Sobolev constants for diffusion semigroups. J. Funct. Anal. 71 363-370.

Lebowitz J., Lieb E.H. (1972) The Constitution of Matter: Existence of thermodynamics for systems composed of electrons and nuclei. Adv. Math. 9 316-398.

Levine H.A. (1980) An estimate for the best constant in a Sobolev inequality involving three integral norms. Annali di Mat. pura ed appl. 124 181-197.

Lenard A. (1971) Generalisation of the Golden-Thompson inequality Ind.Math.J. 21 457-468.

Li P, Yau S-T. (1983) On the Schrödinger equations and the Eigenvalue problem. Commun. Math. Phys. 88 309-318.

Lieb E.H. (1973a) Convex trace functions and the Wigner-Yanase-Dyson conjecture. Adv.in Maths 11 267-288.

Lieb E.H., Ruskai M.B. (1973b) Proof of the strong subadditivity of quantum mechanical entropy. J.Math.Phys. 14 1938-1941.

Lieb E.H. (1973c) The classical limit of quantum spin systems. Commun. Math. Phys. 31 327-340.

Lieb E.H. (1975) Some convexity and subadditivity properties of entropy. Bull. Amer. Math. Soc. 81 1-13.

Lieb E.H., Thirring W. (1976a) Inequalities for the moments of the eigenvalues of the Schrödinger Hamiltonian and their relation to Sobolev inequalities. 268-304, Studies in Mathematical Physics in honor of Valentine Bargmann, Princeton Univ. Press, N.J.

Lieb E.H. (1976b) Inequalities for some operator and matrix functions Adv.in Maths. 20 174-178.

Lieb E.H. (1976c) The Stability of Matter. Rev. Mod. Phys. 48 553-569.

Lieb E.H., Simon B. (1977a) The Thomas-Fermi theory of Atoms, Molecules and Solids. Adv. Math. 23 22-116.

Lieb E.H. (1977b) Existence and uniqueness of the minimizing solution of Choquard's nonlinear equation. Studies in Applied Mathematics 57 93-105.

Lieb E.H. (1978) Proof of an entropy conjecture of Wehrl. Commun. Math. Phys. 62 35-61.

Lieb E.H. (1980) The number of bound states of one-body Schrödinger operators and the Weyl problem. Proc. Symp. Pure. Math. 36 241-252.

Lieb E.H. (1981) Thomas-Fermi and related theories of atoms and molecules. Rev. Mod. Phys. 63 603-642.

Lieb E.H. (1983a) An L^p bound for the Riesz and Bessel potentials of orthonormal functions. J. Funct. Anal. 51 159-165.

Lieb E.H. (1983b) Sharp constants in the Hardy-Littlewood-Sobolev and related inequalities. Ann. Math. 118 349-374.

Lieb E.H. (1984a) On characteristic exponents in turbulence. Commun. Math. Phys. 92 473-480.

Lieb E.H., Thirring W. (1984b) Gravitational collapse in quantum mechanics with relativistic kinetic energy. Ann. Phys. (NY) 155 494-512.

Lieb E.H. (1986a) Bounds on Schrödinger operators and generalised Sobolev type inequalities. Seminaire Equations aux Dérivées Partielles 1985-86.

Lieb E.H., Loss M. (1986) Stability of Coulomb systems with magnetic fields II. The many-electron atom and the one-electron molecule. Commun. Math. Phys. 104 271-282.

Lions P.L. (1982) Symétrie et compacité dans les espaces de Sobolev. J. Funct. Anal. 49 315-334.

Lions P.L. (1985) The concentration-compactness principle in the calculus of variations I, II. Rev. Math. Iberoamericana 1 145-201, 2 45-121.

Luttinger J.M. (1973) Generalised isoperimetric inequalities I, II, III. J. Math. Phys. 14 586-593, 1444-1447, 1448-1450.

Magnen J. Rivasseau V. (1985) The Lipatov argument for ϕ^4_3 perturbation theory. Commun. Math. Phys. 102 59-88.

Magnen J., Nicolò F., Rivasseau V., Sénéor R. (1987) A Lipatov bound for ϕ^4_4 Euclidean quantum field theory. Commun. Math. Phys. 108 257-290.

Marshall A.W., Olkin I. (1979) Inequalities: Theory of Majorization and its Applications. Acacdemic Press, N.Y.

Nelson E. (1966) A quartic theory in two dimensions. Mathematical theory of elementary particles. M.I.T. Press, Cambridge, Mass.

Nelson E. (1973) The free Markoff field. J. Funct. Anal. 12 211-227.

Pólya G., Szegö G. (1951) Isoperimetric inequalities in mathematical physics Ann.Math.Studies 27 Princeton Univ. Press, N.J.

Riesz F. (1930) Sur une inequalité integrale. J. Lond. Math. Soc. 5 162-168.

Rothaus O.S. (1978) Lower bounds for eigenvalues of regular Sturm-Liouville operators and the logarithmic Sobolev inequality. Duke Math.J. 45 351-362.

Rothaus O.S. (1980) Logarithmic Sobolev inequalities and the spectrum of Sturm-Liouville operators. J. Funct. Anal. 39 42-56.

Rothaus O.S. (1981) Logarithmic Sobolev inequalities and the spectrum of Schrödinger operators. J. Funct. Anal. 42 110-120.

Rothaus O.S. (1985) Analytic inequalities, Isoperimetric inequalities and logarithmic Sobolev inequalities. J. Funct. Anal. 64 296-313.

Ruelle D. (1969) Statistical Mechanics. Benjamin, N.Y.

Simon B. (1979a) Trace ideals and their applications. Cambridge Univ. Press Cambridge, England.

Simon B. (1979b) Functional integration and quantum physics. Academic Press, N.Y.

Simon B. (1980) The classical limit of quantum partition functions. Commun. Math. Phys. 71 247-276.

Simon B. (1982) Schrödinger semigroups. Bull. Amer. Math. Soc. 7 447-526.

Strauss W.A. (1977) Existence of solitary waves in higher dimensions. Commun. Math. Phys. 55 149-162.

Talenti G. (1976) Best constant in Sobolev inequality. Annali di Mat. pura ed appl. 110 353-372.

Thirring W. (1980) A course in Mathematical Physics 3: Quantum Mechanics of Atoms and Molecules. Springer-verlag, N.Y.

Thirring W. (1983) A course in Mathematical Physics 4: Quantum Mechanics of large systems. Springer-verlag, N.Y.

Thompson C. (1965) Inequality with applications in statistical mechanics. J.Math.Phys. 6 1812-1813.

Thompson C. (1971) Inequalities and partial orders on matrix spaces. Ind.Math.J. 21 469-480.

Uhlmann A. (1970) Relative Entropy and the Wigner-Yanase-Dyson-Lieb Concavity in an interpolation theory. Commun. Math. Phys. 18 160-170.

Veling E.J.M. (1984) A relation between the infimum of the spectrum of the Schrödinger equation in R^N and the Sobolev inequalities. University of Birmingham preprint.

Wehrl A. (1978) General properties of entropy. Rev. Mod. Phys. 50 221-260.

Weinstein M.I. (1983) Nonlinear Schrödinger equations and sharp interpolation estimates. Commun. Math. Phys. 87 567-576.

Weissler F.B. (1979) Two-point inequalities, the Hermite semigroup and the Gauss-Weierstrass semigroup. J. Funct. Anal. 32 102-121.

Weissler F.B. (1980) Logarithmic Sobolev inequalities and hypercontractive estimates on the circle. J. Funct. Anal. 37 218-234.

Weyl H. (1949) Inequalities between two kinds of eigenvalues of a linear transformation. Proc. Nat. Acad. Sci. 35 408-411.

Yaffe L.G. (1982) Large N limits as classical mechanics. Rev. Mod. Phys. 54 407-436.

5 Growth Lemmas in Function Theory

Walter K. Hayman University of York, Heslington, York, England

1. Introduction

Suppose that $G(x)$ is positive, continuous and increasing for $x \ge x_0$. It is often necessary to estimate other quantities $g(x)$ at a point x_1 in terms of the value of G at a larger value $x_1 + h$. Suppose for instance that $g(x)$ is convex and that

$$(1) \qquad 0 \le g(x) \le G(x), \quad x \ge x_0.$$

What bounds can we hope to achieve for $g'(x_1)$, where $x_0 \le x_1$?

We evidently have, since $g'(x)$ is increasing and $g(x_1) \ge 0$,

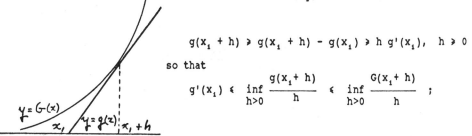

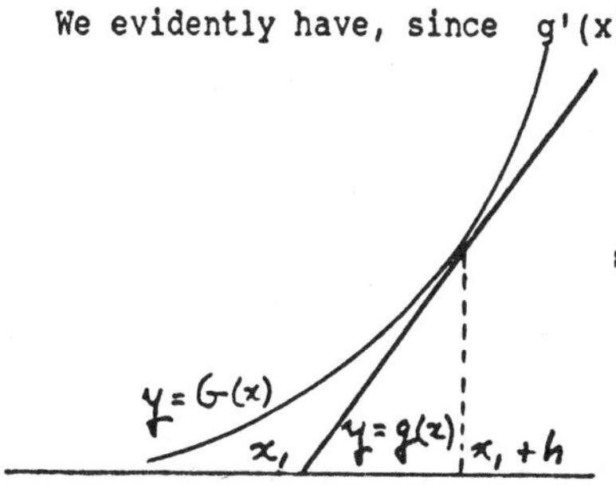

$$g(x_1 + h) \ge g(x_1 + h) - g(x_1) \ge h\, g'(x_1), \quad h \ge 0$$

so that

$$g'(x_1) \le \inf_{h>0} \frac{g(x_1 + h)}{h} \le \inf_{h>0} \frac{G(x_1 + h)}{h} \quad ;$$

and this bound is sharp for every fixed x_1, such that $x_1 \ge x_0$ and the class of all positive convex functions $g(x)$ satisfying (1).

If we assume that $g(x)$ has continuous positive and nondecreasing derivatives of orders up to n for $x \ge x_0$, we have

$$g(x_1+h) = g(x_1)+h\, g'(x_1) + \frac{h^{n-1}}{(n-1)!} g^{(n-1)}(x_1) + \frac{h^n}{n!}\, g^{(n)}(x_1+\theta h)$$

$$\ge \frac{h^n}{n!}\, g^{(n)}(x_1)$$

similarly. This yields, when we apply (1)

$$g^{(n)}(x_1) \le n! \inf_{h>0} \frac{g(x_1 + h)}{h^n} \le n! \inf_{h>0} \frac{G(x_1 + h)}{h^n} .$$

We now set

$$G_n(x) = \inf_{h>0} \frac{G(x + h)}{h^n} ,$$

and show that under favourable circumstances $G_n(x)$ is comparable with the nth derivative $G^{(n)}(x)$ for certain values of $x = x_1$. This enables us, somewhat surprisingly, to estimate $g^{(n)}(x_1)$ in terms of the derivative $G^{(n)}(x_1)$ of the dominating function $G(x)$. It is important in applications that the results are valid simultaneous for all the functions g at such points x.

2. Statement of results

Theorem 1. If $G^{(n)}(x_1)$ is continuous for $x \geq x_0$ and positive for some arbitrarily large x then, if $K > 1$,

$$(2) \quad G_n(x_1) < \left(\frac{eK}{n}\right)^n G^{(n)}(x_1)$$

holds for some arbitrarily large x_1.

It is often convenient to have an estimate of how frequently the inequality (2) occurs. Let the E be a measurable set on the positive real axis and let $\ell(X)$ be the measure of the part of E in the interval $[0,X]$. Then

$$\delta(E) = \lim_{X\to\infty} \frac{\ell(X)}{X}$$

is called the lower density of E. We have

Theorem 2. If $G(x)$ satisfies the hypothesis of Theorem 1 and in addition $G^{(n)}(x)$ increases with x, then the set E of x_n for which (2) holds, satisfies

$$(3) \quad \delta(E) \geq \frac{K-1}{K-1+n}.$$

Theorem 1 was proved by me for $n = 1,2$[2] and for general n by Stewart and myself [7]. We also had a weaker version of Theorem 2, with $\delta(E) > 0$ only. The present version of Theorem 2 occurs in paper with Rossi[6]. We see that as $K \to \infty$ in (2) the lower bound for $\delta(E)$ in (3) tends to one.

To test the sharpness of (2) we consider $G(x) = e^x$. Then

$$G_n(x) = \inf_{h>0} \frac{e^{x+h}}{h^n} = \frac{e^{x+n}}{n^n} = \left(\frac{e}{n}\right)^n G^{(n)}(x).$$

Thus (2) need not hold for any value of x if $K = 1$. We note the following

Corollary. If $g(x)$ satisfies (1) and $g, g', \dots g^{(n)}(x)$ and positive and increasing for $x \geq x_0$ then, if E is the set of all x such that $x \geq x_0$ and (2) holds, we have in E

$$g^{(n)}(x) \leq n! \left(\frac{eK}{n}\right) G^{(n)}(x).$$

The constant $C_n = n!(e/n)^n$ is sharp in the Corollary if we desire an inequality which holds, for a given x, simultaneously for all dominated functions g. Stirling's formula, shows that

$$C_n \sim \sqrt{(2\pi n)} \text{ as } n \to \infty, \qquad C_1 = e, \ C_2 = \frac{e^2}{2}, \ldots$$

In a recent joint paper with Miles [5] it turned out that instead of taking a fixed x_1 and a variable $x_2 = x_1 + h$, we needed to take a fixed x_2 and a variable x_1. I confine myself to the case $n = 1$, which is all we needed. Suppose that $G(x)$ is as above, write

$$\check{G}(x) = \sup_{0<h<x-x_0} h\, G'(x-h),$$

and define

$$\check{E}(K) = \{x \mid x > x_0 \text{ and } G(x) < e\, K\, \check{G}(x)\}.$$

Thus $\check{E}(K)$ is the set of all x for which we can find x_1, $x_0 < x_1 < x$ and

$$\frac{G(x)}{x-x_1} < e\, K\, G'(x_1).$$

Hence, if E is the set of $x = x_1$ in Theorems 1, 2 for $n = 1$ and $h = h(x_1)$ is a corresponding value of h, $\check{E}$ is the set of all $x_2 = x_1 + h(x_1)$. Thus it is immediate that if E is unbounded so is $\check{E}$.

Theorem 3. With the hypotheses of Theorem 1, and $n = 1$, $\check{E}(K)$ is unbounded.

However the bounds for the density of $\check{E}(K)$ turn out to be much better than the corresponding bounds for $\delta(E)$. We quote [5, Lemma 3]

Theorem 4. With the hypotheses of Theorem 2, and $n = 1$, we have

$$\delta(\check{E}) \geq 1 - \min\{(2\, e^{K-1}-1)^{-1}, (1 + e(K-1))\exp e(1-K)\}.$$

in other words $1-\delta(\check{E}) = O\{e^{-(e-\epsilon)K}\}$ if K is large so that $1-\delta(\check{E})$ is exponentially small if K is large. The order of magnitude $K = O\{\log \frac{1}{1-\delta(\check{E})}\}$ turns out to be sharp though the constant e in the exponent is probably not.

3. Proof of Theorem 1 for $n = 1$. The proofs of all the above results, while not terribly long, would take us too far. However in order to indicate the ideas, I would like to give a proof of the simplest case, $n = 1$ of Theorem 1, and so Theorem 3. The basic idea is that $\alpha(x) = G'(x)/G(x)$ cannot increase too quickly in

a short space.

Lemma. With the hypotheses of Theorem 1, for $n = 1$, we have for some arbitrarily large x, such that $\alpha(x) > 0$,

(4) $\alpha(x+h) < K\,\alpha(x)$, for $0 \leq h \leq \{K\,\alpha(x)\}^{-1}$.

Suppose that the Lemma is false, so that (4) is false for $x \geq X$. We start with x_1, such that $x_1 \geq X$ and $\alpha(x_1) > 0$. If x_n has been defined, we define $h = h_n$ to be the smallest number for which (4) fails.

Thus

$$0 < h_n \leq \{K\,\alpha(x_n)\}^{-1}, \text{ and } \alpha(x_n + h_n) = K\,\alpha(x_n).$$

We then define $x_{n+1} = x_n + h_n$. Thus

$$\alpha(x_{n+1}) = K\,\alpha(x_n) = \ldots = K^n\,\alpha(x_1)$$

and

$$x_{n+1} - x_n = h_n \leq \{K\,\alpha(x_n)\}^{-1} = K^{-n}/\alpha(x_1),\ n=1,2,\ldots$$

Thus x_n converges to a finite limit

$$(5)\quad x = x_1 + \sum_{n=1}^{\infty}(x_{n+1} - x_n) \leq x_1 + \frac{1}{\alpha(x_1)}\sum_{1}^{\infty} K^{-n} = x_1 + \frac{1}{(K-1)\,\alpha(x_1)},$$

and $\alpha(x_n) \to +\infty$ as $n \to \infty$. This contradicts the hypothesis that $\alpha(x)$ is continuous and so the Lemma is proved. Our argument also shows that if $\alpha(x_1) > 0$, we can find x satisfying (4) and (5).

To prove Theorem 1, suppose now that (4) holds, and choose $h = \{K\,\alpha(x)\}^{-1}$. Then

$$\frac{G(x+h)}{G(x)} = \exp\left\{\int_x^{x+h}\frac{G'(t)}{G(x)}\,dt\right\} = \exp\left\{\int_x^{x+h}\alpha(t)dt\right\} < \exp\left\{Kh\,\alpha(x)\right\} = e\,.$$

Thus

$$\frac{G(x+h)}{h} < \frac{e\,G(x)}{h} = eK\,\alpha(x)\,G(x) = eK\,G'(x).$$

This proves Theorem 1 for $n = 1$.

Let me say briefly what modifications are necessary to prove Theorems 1 and 2. To deal with the cases $n=2,3,\ldots$ we have to work with

$$\beta(x) = \sup_{0\leq\nu\leq n-1}\left\{\frac{G^{(\nu)}(x)}{G^{(n)}(x)}\right\}^{1/(n-\nu)} \quad \text{instead of } 1/\alpha(x),$$

and to replace the range for h by $0 \leq h \leq n\,\beta(x)/K$ in the Lemma. Under the hypotheses of Theorem 2 it turns out that $x - \beta(x)$ is nondecreasing. Thus $x - \beta(x)$ cannot increase by too much too often and so one can obtain a density result for the values x for which the conclusion of the Lemma holds. The conclusion of Theorem 3 holds for all those values y for which the equation $x + \beta(x) = y$ has a solution with x in E and the exceptional set of y turns out to be very small.

4. <u>An open problem</u> This concerns non integral n. It is convenient to take $G^{(n)}(x) = H(x)$ as our basic function, so that $G(x)$ is an n'th integral of $H(x)$. We write

$$H^{(-n)}(x) = \frac{1}{\Gamma(n)} \int_{x_0}^{x} (x-t)^{n-1} H(t)dt.$$

If n is an integer $H^{(-n)}(x) = G(x) + P(x)$, where $P(x)$ is a polynomial of degree less than n. Thus (2) may be written as

$$(2') \quad \inf_{h>0} \frac{H^{(-n)}(x+h)}{h^n} < \left(\frac{eK}{n}\right)^n H(x),$$

for some arbitrarily large x if $H(x)$ is continuous and $H(x) > 0$ for some large x and on a set E whose density satisfies (3), if $H(x)$ is in addition increasing. This inequality makes perfect sense for any positive n, but our proof breaks down. In an application to the minimum modulus of entire functions [3], which gave rise to the original theory it would be possible to improve a constant slightly if we had (2') for $2 < n < 3$ on a set of positive density. I would be very interested in a proof of (2') for general n and suitable values of x.

5. <u>Some applications</u>.

There are several applications in the original papers of Stewart and myself [7] as well as the inequality [3] for a set of r of positive lower logarithmic density,

$$(6) \quad m(r) > M(r)^{-A \log\log\log M(r)},$$

relating the minimum modulus $m(r)$ and the maximum modulus $M(r)$ of an entire function. Here $n=3$ gives the best lower bound $16e^3/(27\pi\sqrt{3})$ for A in (6). An example in [3] shows that (6) fails for $A = .09$.

For a much simpler application [7] suppose that f is a trascendental entiree function and that

$$M(r,f) = \sup_{|z|=r} |f(z)| \leqslant G(R).$$

The Cauchy integral formula yields for $|z|=r < R$

$$|f^{(n)}(z)| = \left| \frac{n!}{2\pi i} \int_{|\zeta|=R} \frac{f(\zeta)d\zeta}{(\zeta-z)^{n+1}} \right| \leqslant \frac{n!G(r)}{2\pi} \int_{-\pi}^{\pi} \frac{Rd\theta}{(R^2-2Rr\cos\theta+r^2)^{\frac{n+1}{2}}}$$

$$\leqslant \frac{n!\ G(R)}{(R-r)^{n-1}(R^2-r^2)} \frac{1}{2\pi} \int_{-\pi}^{\pi} \frac{(R^2-r^2)Rd\theta}{(R^2-2Rr\cos\theta+r^2)} = \frac{n!\ G(R)\ R}{(R-r)^n\ (R+r)}$$

$$\leqslant \frac{n!\ G(R)}{(R-r)^n} .$$

Thus

$$M(r,f^{(n)}) \leqslant n!\ G_n(r) \leqslant n! \left(\frac{eK}{n}\right)^n G^{(n)}(r) = K^n\ C_n\ G^{(n)}(r)$$

at least for a sequence of r tending to ∞. This is a complex analogue of the Corollary. Possibly the constant C_n can be improved here.

For a third application consider the Ahlfors-Shimizu characteristic $T_o(r)$ of a function $f(z)$ memomorphic in the plane. Then [4, p.12]

$$(7) \qquad T_o(r) = \int_o^r \frac{A(t)dt}{t} ,$$

where $\pi\ A(r)$ is the area of the image of $|z| < r$ by $f(z)$ considered as a map onto the Riemann sphere. We may write

$$(8) \qquad A(r) = \frac{1}{\pi} \int n(r,w)\ d\mu(w),$$

where $d\mu(w)$ density surface area on the Riemann sphere and $n(r,w)$ is the number of roots of the equation $f(z) = w$ in $|z| < r$. Thus $A(r)$ is a sort of average of $n(r,w)$ over all w.

It is reasonable to ask whether we can compare $A(r)$ with

$$n(r) = \sup_{w\in\mathbb{C}} n(r,w).$$

Clearly $n(r) \geq A(r)$. However we have [7; 4, p.14].

Theorem 4. $\overline{\lim_{r\to\infty}} \frac{n(r)}{A(r)} \leq e.$

In fact Nevanlinna's first fundamental Theorem yields for each fixed w

$$\int_1^r \frac{n(t,w)dt}{t} \leq T_0(r) + C = \int_1^r \frac{A(t)dt}{t} + C',$$

where C' is a constant independent of w. We write $r = e^x$, $N(r,w) = g(x)$, $T_0(r) + C' = G(x)$. Then $G(x)$ is a convex function of x and so the Corollary yields, for a set E of x depending only on G and so independent of w

$$g'(x) \leq K\,e\,G'(x),$$

where E has positive lower density. (For Theorem 4 it suffices that E is unbounded, and this follows from Theorem 1) If F is the corresponding set of r, we have on F, for all w,

$$n(r,w) \leq Ke\,A(r) \text{ and so } n(r) \leq K\,e\,A(r).$$

Since K is any number such that $K > 1$, Theorem 4 follows. It has been shown by Toppila [9] that $\underline{\lim}\, n(r)/A(r)$ can be greater than 80/79 so it is hard to guess what the right constant might be.

Finally let me quote an application [5], where we compare the characteristic of a meromorphic function $F(z)$ with that of its q'th derivative.

Theorem 5. Suppose that F is a transcendental meromorphic function. Then we have for $q = 1, 2, \ldots$

$$\overline{\lim_{r\to\infty}} \frac{T(r,F)}{T(r,F^{(q)})} \leq 3e.$$

If F is entire we can replace 3e by 2e.

There are also corresponding density theorems. Again an example of Toppila [10] shows that the left hand side can be greater than $1 + 7.\,10^{-7}$, even for an entire function F. It had been an open question ever since Nevanlinna developed his theory in the 1920's, whether $T(r,F)/T(r,F')$ can tend to $+\infty$ with r. If $\epsilon > 0$, it follows from the Lemma on the logarithmic derivative [4,p.41] that

$$T(r,F) > (\tfrac{1}{2} - \epsilon)\, T(r,F')$$

and, if F is entire,

$$T(r,F) > (1 - \epsilon)\, T(r,F'),$$

for all large r if F has finite order or all large r apart from an exceptional set of finite logarithmic measure, if F has infinite order.Nevanlinna [8,p.104] conjectured that, if F is entire, $T(r,F) \sim T(r,F')$ ourtside an exceptional set; but Toppila's example [10] shows this to be false. The density form of Theorem 5 shows however that, for "most" r, $T(r,F)/T(r,F')$ stays between positive constants.

Let me indicate the main ideas in the proof of Theorem 5. We apply Theorem 3 with $G(x) = T_0(e^x, F)$ and deduce that, if $K > 1$ and x is outside an exceptional set, we can find y such that $0 < y < x$ and

$$G(x) < e\, K(x-y)\, G'(y).$$

Setting $e^x = r$, $e^y = \rho$, we deduce from (7) for an unbounded set E of r

(9) $T_0(r,F) < eK(\log r/\rho)\, A(\rho,F)$, for some $\rho = \rho(r)$, such that $1 < \rho < r$.

Using (8) we can find a complex w, such that

$$A(\rho,F) \leqslant n(\rho,w) = n(\rho,w) - n(\rho,\infty) + n(\rho,\infty).$$

The argument principle yields

$$\left| n(\rho,w) - n(\rho,\infty) \right| = \left| \frac{1}{2\pi i} \int_{-\pi}^{\pi} \frac{F'(z)dz}{F(z)-w} \right| = \left| \frac{1}{2\pi i} \int_{-\pi}^{\pi} \left[\mathrm{Re}\, \frac{F'(\rho e^{i\theta})}{F(\rho e^{i\theta})-w} \right] d\theta \right|$$

We define

$$I(\rho,F) = \left| \frac{1}{2\pi} \int_{-\pi}^{\pi} \left| \mathrm{Re}\, \frac{F'(\rho e^{i\theta})}{F(\rho e^{i\theta})} \right| d\theta \right|,$$

and obtain

(10) $A(\rho,F) \leqslant n(\rho,\infty) + I(\rho,F-w)$.

We now use a Lemma of Hall and Ruscheweyh [1] to deduce that

(11) $I(\rho,F-w) \leqslant I(\rho,F') + 1 \leqslant \dots \leqslant I(\rho,F^{(q)}) + q$.

Finally an application of the Poisson-Jensen formula yields

$$(12) \qquad I(\rho, F^{(q)}) \leqslant \frac{2}{\log(r/\rho)} (T_0(r, F^{(q)}) + \text{constant}),$$

while Nevanlinna's first fundamental theorem gives

$$(13) \quad n(\rho,\infty,F) \leqslant n\left[\rho,\infty,F^{(q)}\right] \leqslant \frac{1}{\log(r/\rho)} \int_\rho^r \frac{n(t,\infty,F^{(q)})dt}{t} \leqslant \frac{T_0(r,F^{(q)})+O(1)}{\log(r/\rho)} .$$

Putting (9) to (13) together we obtain on E

$$T_0(r,F) < eK\{3\ T_0(r,F^{(q)}) + O(\log r)\},$$

and this yields Theorem 5.

References

1. R.R. Hall and S. Ruscheweyh, On transforms of functions with bounded boundary rotation, Indian J. Pure Appl. Math. 16 (1985), 1317-1325.

2. W.K. Hayman, An inequality for real positive functions, Proc. Cambridge Philos. Soc. 48 (1952), 93-105.

3. W.K. Hayman, The minimum modulus of large integral functions, Proc. London Math. Soc. (3) 2 (1952), 469-512.

4. W.K. Hayman, Meromorphic functions (Oxford 1964).

5. W.K. Hayman and Joseph Miles, On the growth of a meromorphic function and its derivatives, Complex Variables, to appear.

6. W.K. Hayman and J.F.Rossi, Characteristic, Maximum modulus and value distribution, Trans. Amer. Math. Soc., 284(1984), 651-664.

7. W.K. Hayman and F.M. Stewart, Real inequalities with applications for function theory, Proc. Cambridge Philos.Soc. 50 (1954), 250-260.

8. R. Nevanlinna, Le theoreme de Picard-Borel et la theorie des fonctions meromorphes (Paris, 1929).

9. S. Toppila, On the counting function for the a-values of a meromorphic function, Ann. Acad. Sci. Fenn.A.I. Math. 2(1976), 565-572.

10. S.Toppila, On Nevanlinna's characteristic function of entire functions and their derivatives, Ann. Acad.Sci. Fenn. Ser. A.I. Math. 3 (1977), nr.1, 131-134.

6 Norm Inequalities for Derivatives and Differences

Man Kam Kwong Argonne National Laboratory, Argonne, Illinois
Anton Zettl Northern Illinois University, DeKalb, Illinois

Introduction

Norm inequalities of product form relating (i) a function and two of its derivatives and (ii) a sequence and two of its differences are discussed and compared. In the continuous case (i) these inequalities include those often associated with the names of Landau, Hardy-Littlewood, Kolmogorov and Schoenberg-Cavaretta. Under (ii) are included the discrete versions of these.

Section 1 contains a discussion of the possible values of the p and q norms of a function y and its nth derivative $y^{(n)}$. In sections 2 and 3 are given detailed elementary proofs of the basic inequality for functions and a number of related inequalities. Some of these are of independent interest. Section 4 contains a discussion of the growth at infinity of derivatives and finally, in section 5, we give a summary of cases when the best constant is known explicitly for both the continuous and the discrete versions of the basic inequality. Also an extensive list of references is provided which we hope will be useful. We have tried to include the Soviet literature on these topics but have had considerable difficulty in trying to keep current with the vast Soviet and Eastern literature.

1. The Norms of y and $y^{(n)}$

The classical p norms are defined by

$$\|y\|_p = \left[\int_J |y(t)|^p dt\right]^{1/p}, \quad 1 \le p < \infty$$

$$\|y\|_\infty = \text{ess. sup} |y(t)| , \quad t \in J , p = \infty .$$

Here J is any nondegenerate interval of the real line bounded or unbounded.

This work was supported in part by the Applied Mathematical Sciences subprogram of the Office of Energy Research, U.S. Department of Energy, under contract W-31-109-Eng-38.

The set of equivalence classes (with respect to Lebesgue measure) of functions whose p norms are equal and finite is a Banach space denoted by $L^p(J)$, $1 \le p \le \infty$. Below $y^{(n)}$ denotes the nth derivative of y and $y^{(n)} \in L^p(J)$ means that $y^{(n-1)}$ is absolutely continuous on any compact subinterval of J, so that $y^{(n)}$ exists a.e. and is locally integrable, and $\|y^{(n)}\|_p$ is finite. The symbol $\|y\|_{p,J}$ will be used when we wish to emphasize the dependence of the norm on the interval J. If p or J are fixed in a result or argument we may merely use the symbol $\|y\|$. Throughout p is assumed to satisfy $1 \le p \le \infty$. The symbol $C^n(J)$ denotes the set of complex valued functions with a continuous n^{th} derivative on J, $C^\infty(J)$ is the set of infinitely differentiable functions on J, $C_0^\infty(J)$ denotes the infinitely differentiable functions with compact support in the interior of J.

THEOREM 1. Let $1 \le p, q \le \infty$, let n be a positive integer, and let J be any interval on the real line, bounded or unbounded. Given any positive numbers u and v there exists a function $y \in C^n(J)$ such that

$$\|y\|_p = u, \quad \|y^{(n)}\|_q = v\ . \tag{1.1}$$

Proof. Case 1. $1 \le p, q < \infty$, J an unbounded interval.

Let $J = R = (-\infty,\infty)$ or $J = R^+ = (0,\infty)$. Choose y in $C_0^\infty(R)$ such that $y \ge 0$ but not identically zero and y has compact support. Consider

$$y_{ab}(t) = ay(bt)\ , \quad a > 0\ , b > 0$$

and note that with $x = bt$

$$\|y_{ab}\|_p^p = \int_R a^p |y(bt)|^p dt = a^p b^{-1} \int_R |y(x)|^p dx = a^p b^{-1} \|y\|_p^p$$

$$\|y_{ab}^{(n)}\|_q^q = \int_R a^q b^{nq} |y^{(n)}(bt)|^q dt = a^q b^{nq-1} \int_R |y^{(n)}(x)|^q dx = a^q b^{nq-1} \|y^{(n)}\|_q^q\ .$$

Choose b such that

$$b^{nq-1+q/p} = u_n^q\,(\|y\|_p^p)^{q/p} u_0^{-q}\,\|y^{(n)}\|_q^{-q} \tag{1.2}$$

(observe that $y^{(n)}$ is not identically zero since y has compact support and is not the zero function) and then choose $a = (b\ u^p/\|y\|_p^p)^{1/p}$.

The proof for the case $J = R^+$ is similar.

Case 2. $1 \le p, q < \infty$, $J = (a,b)$, $-\infty < a < b < \infty$.

First consider the case $J = (0,1)$. Define

$$Q(y) = \|y^{(n)}\|_q / \|y\|_p\ , \quad y \in X = W_{p,q}^n(J) = \{\ y \in L^p(J) : y^{(n)} \in L^q(J)\ \}.$$

Since the norm is a continuous function from X into $(0,\infty)$ it follows that Q is a continuous function from $X-\{0\}$ into $[0,\infty)$. We show that Q is onto. Let $Q(y) = a$, $Q(z) = b$, $0 \le a < b$. Then y and z are linearly independent. Thus $S = \text{span}\{y,z\}-\{0\}$ is a two-dimensional connected subset of X. Since the continuous image of a connected set is connected it follows that $[a,b] \subset$ range Q. To show that b can be chosen arbitrarily large consider

$$y(t) = t^d , \quad 0 < t < 1 .$$

Then

$$\|y\|_p^p = \int_0^1 t^{pd} dt = 1/(pd+1)$$

$$\|y^{(n)}\|_q^q = h(d)\int_0^1 t^{(d-n)q} dt = h(d)/(q(d-n)+1) ,$$

$$h(d) = d(d-1) \cdots [d-(n-1)]$$

and

$$Q(y) = (h(d))^{1/q}(pd+1)^{1/p}(q(d-n)+1)^{-1/q} \to \infty \text{ as } d \to \infty .$$

We conclude that the range of Q is $[0,\infty)$. Let $r = v/u$. From the above argument we know there is a $y \in W^n_{p,q}(0,1)$ such that $Q(y) = r$. Choose the constant c so that $\|cy\|_p = u$, then $\|cy^{(n)}\|_q = v$. This completes the proof for the case $J = (0,1)$. The general case of a bounded interval J follows from this case and a transformation of the form $t \to ct + d = x$.

The proofs of the remaining cases are elementary and left to the reader as exercises.

2. The Norms of y, $y^{(k)}$ and $y^{(n)}$

The special case $p = q$, $n = 1$ of Theorem 1 says that the norm of a function and its derivative on any interval, bounded or unbounded, can assume arbitrary positive values. Can the norms of y, y' and y'' assume arbitrary positive values? Below we will see that the answer is no. But first we discuss some preliminary results.

LEMMA 1. Let $1 \le p \le \infty$. Assume $J = [a,b]$ is a compact interval of length $L = b-a$. If $y \in L^p(J)$ and $y^{(k)}$ exists on J and

$$\alpha_k(J) = \inf . |y^{(k)}(t)| , \quad t \in J ; \tag{2.1}$$

then

$$\alpha_k = \alpha_k(J) \le A \, \|y\|_{p,J} \quad k = 1,2,3,... \tag{2.2}$$

where A is a constant independent of y given by

$$A = A(k,p,L) = 2^k \cdot 3^{x(k)} L^{1/q-k-1} \tag{2.3}$$

where $x(k)$ is defined recursively by

$$x(1) = 1 - 1/q, \; x(k+1) = 1 + x_k + k - 1/q, \quad k = 1,2,3,... .$$

Proof. The proof uses a "triple interval" argument and induction on k.

Case 1. $p = \infty$. Let $J = J_1 \cup J_2 \cup J_3$ where J_i are nonoverlapping compact intervals, $l(J_i) = L/3$,

$i = 1,2,3$ and J_2 is between J_1 and J_3. By the mean value theorem for any t_1 in J_1 and t_3 in J_3 there exist t^* such that

$$\alpha_1 \le |y'(t^*)| = |(y(t_1)-y(t_3))/(t_1-t_3)| \le 3L^{-1}\,(|y(t_1)| + |y(t_3)|) \le 6\,L^{-1}\,\|y\|_{\infty,J}\,.$$

This establishes the case $k = 1$. To establish the inductive step it is convenient to use the notation $\alpha_k(J)$ to denote the dependence of α_k on the interval J. For the sake of clarity we consider the case $k = 2$. Using the above notation choose t_i in J_i so that $\alpha_1(J_i) = |y'(t_i)|$, i=1,3. By the mean value theorem we have

$$\alpha_2 \le |y''(t^*)| = |(y'(t_1) - y'(t_3))/(t_1-t_3)| \le 3L^{-1}(\alpha_1(J_1) + \alpha_1(J_3)) \tag{2.4}$$

$$\le 3L^{-1}(6(L/3)^{-1}\|y\|_{\infty,J_1} + 6(L/3)^{-1}\|y\|_{\infty,J_3} \le 3^2 6\,L^{-2}\|y\|_{\infty,J}\,.$$

Assume (2.2), (2.3) hold. Let $J = J_1 \cup J_2 \cup J_3$ as above in step $k = 1$. Then as above using the mean value theorem with $t_1 \in J_1$, $t_3 \in J_3$ chosen so that $\alpha_k(J_i) = |y^{(k)}(t_i)|$, i=1,3 we get

$$\alpha_{k+1}(J) \le |y^{(k+1)}(t^*)| = |(y^{(k)}(t_1) - y^{(k)}(t_3))/(t_1-t_3) \le L^{-1}3(\alpha_k(J_1) + \alpha_k(J_3))$$

$$\le 3L^{-1}2^k 3^{x(k)}(L/3)^{-k}(\|y\|_{\infty,J_1} + \|y\|_{\infty,J_3}) \le L^{-k-1}2^{k+1}3^{1+x(k)+k}\|y\|_{\infty,J}\,.$$

This completes the proof of (2.2), (2.3) for $p = \infty$.

Case 2. $1 \le p < \infty$. Let $p^{-1} + q^{-1} = 1$ with the usual convention that $q^{-1} = 0$ when $p = 1$. Let $J = J_1 \cup J_2 \cup J_3$ be as in case 1 above. Choose $t_i \in J_i$ such that $|y(t_i)| = \min |y(t)|$, $t \in J_i$, i=1,3. From the mean value theorem and Holder's inequality we have for some t^* between t_1 and t_3

$$\alpha_1 \le |y(t^*)| \le L^{-1}3(|y(t_1)| + |y(t_3)|)\,.$$

Also

$$L\cdot 3^{-1}|y(t_i)| = \int_{J_i} |y(t_i)|dt \le \int_{J_i} |y(t)|dt \le (L/3)^{1/q}\,\|y\|_{p,J_i}\,,\quad i=1,3\,.$$

From this we get

$$\alpha_1 \le 3^{1-1/q}L^{1/q-2}2\,\|y\|_{p,J}\,.$$

This is (2.2), (2.3) for $k = 1$. Assume (2.2), (2.3) holds for k. Decompose J as above and choose $t_i \in J_i$ such that $|y^{(k)}(t_1)| = \inf.|y^{(k)}(t)|$ for $t \in J_i$, i=1,3. Then as above,

$$\alpha_{k+1}(J) \le |y^{(k+1)}(t^*)| \le 3L^{-1}(|y^{(k)}(t_1)| + |y^{(k)}(t_3)|) = 3L^{-1}(\alpha_k(J_1) + \alpha_k(J_3)) \le 3L^{-1}2(2^k 3^{x(k)}L^{1/q-k})\|y\|_{p,J}$$

and the proof of Lemma 1 is complete.

LEMMA 2. Let $1 \le p, q, r \le \infty$, $l(J) = L < \infty$. If $y \in L^p(J)$ and $y'' \in L^r(J)$ then $y' \in L^q(J)$ and

$$\|y'\|_q \le A\,L^{1/r'+1/q}\,\|y''\|_r + B\,L^{-2+1/p'+1/q}\,\|y\|_p \tag{2.5}$$

where $1/r' + 1/r = 1$ with $r' = 1$ when $r = \infty$, $r' = \infty$ when $r = 1$; and

$$A = 2^{1-1/q},\ B = 2^{2-1/q}3^{1-1/p'}\,. \tag{2.6}$$

Proof. We may assume, without loss of generality, that J is compact. With the notation of Lemma 3 we have

$$|y'(t)| \le \alpha_1 + |\int_{t_1}^{t} y''| \le 2^1 \cdot 3^{1-1/p'} L^{1/p'-2} \|y\|_p + L^{1/r'} \|y''\|_r . \tag{2.7}$$

If $q = \infty$ (2.5), (2.6) follows from (2.7) since $1 < 2$, $2 \cdot 3^{1-1/p'} < B$. If $q < \infty$, (2.7) implies

$$\int_J |y'(t)|^q dt \le 2^{q-1}(2^q \cdot 3^{q-q/p'} L^{q/p'-2q+1} \|y\|_p^q + L^{q/r'+1} \|y''\|_r^q) . \tag{2.8}$$

Now (2.5) follows from (2.8) and the elementary inequality

$$a^r + b^r \le (a+b)^r \le 2^{r-1}(a^r+b^r),\ 1 \le r,\ a > 0,\ b > 0 . \tag{2.9}$$

For later reference we also state the elementary inequality

$$2^{r-1}(a^r+b^r) \le (a+b)^r \le a^r + b^r\ ,\ 0 \le r \le 1\ ,\ a > 0\ ,\ b > 0 . \tag{2.10}$$

LEMMA 3. Let $1 \le p \le \infty$, $l(J) = L \le \infty$. Given $\varepsilon > 0$ there exists a $K(\varepsilon) > 0$ such that if $y \in L^p(J)$, y' is locally absolutely continuous on J, $y'' \in L^p(J)$ then $y' \in L^p(J)$ and

$$\|y'\|_p \le \varepsilon \|y''\|_p + K(\varepsilon) \|y\|_p . \tag{2.11}$$

Furthermore, for fixed ε, $K(\varepsilon)$ can be chosen to be a nonincreasing function of the length of J.

Proof. Case 1. Assume $L < \infty$. Let $\varepsilon > 0$. If $L \le \varepsilon/A$ then (2.11) follows from (2.5) with

$$K(\varepsilon) = BL^{-1}\ ,\ B = 2^{2-1/p} . \tag{2.12}$$

If $\varepsilon_1 = \varepsilon/A < L < \infty$ let $J = \bigcup_{i=1}^{n} J_i$ where J_i are nonoverlapping, $l(J_i) = \varepsilon_1/2$, $i = 1,\dots,n-1$ and $\varepsilon_1/2 \le l(J_n) \le \varepsilon_1$. From (2.8) we get

$$\int_I |y'|^p \le \varepsilon^p \int_I |y''|^p + (2AB)^p \varepsilon^{-p} \int_I |y|^p\ ,\ 1 \le p < \infty \tag{2.13}$$

holding on each interval $I = J_i$, $i = 1,\dots,n-1$. On $I = J_n$ we get from (2.8)

$$\int_I |y'|^p \le \varepsilon^p \int_I |y''|^p + B^p L^{-p} \int_I |y|^p \le \varepsilon^p \int_I |y''|^p + (2AB)^p \varepsilon^{-p} \int_I |y|^p . \tag{2.14}$$

Summing inequalities (2.13) and (2.14) over all the intervals J_i, $i = 1, \dots, n$ and then taking the pth root we obtain (2.11) with

$$K(\varepsilon) = 2AB\varepsilon^{-1} \tag{2.18}$$

Case 2. $L = \infty$. Let $J = \bigcup_{i=1}^{\infty} J_i$ where J_i are nonoverlapping intervals each of length $\varepsilon_1/2$. Proceeding as above we get inequality (2.13) on each interval $I = J_i$, $i=1,2,3,\dots$. Summing over the intervals J_i and taking the pth root yields (2.11) with $K(\varepsilon)$ given by (2.18).

For fixed ε it is clear that $K(\varepsilon)$, chosen according to (2.12) and (2.18) is a nonincreasing function of the length of the interval J. This completes the proof for $p < \infty$. The modifications needed when $p = \infty$ are straightforward and hence omitted.

THEOREM 1. Let $1 \le p \le \infty$, let n, k be integers with $1 \le k < n$, and let J be any interval of the real line bounded or unbounded. Given any $\varepsilon > 0$ there exists a positive $K(\varepsilon)$ such that if $y \in L^p(J)$, $y^{(n-1)}$ is locally absolutely continuous and $y^{(n)} \in L^p(J)$ then $y^{(k)} \in L^p(J)$ and

$$\|y^{(k)}\|_p \le \varepsilon \, \|y^{(n)}\|_p + K(\varepsilon) \, \|y\|_p \, . \tag{2.19}$$

Furthermore for a given $\varepsilon > 0$ the constant $K(\varepsilon)$ can be chosen to be a nonincreasing function of the length of the interval J.

Proof. The proof is by induction on n. Since p is fixed throughout the proof we will suppress the subscript p on the norm symbol. The case $n = 2$ is Lemma 3. Assume Theorem 1 is true for $n = k$ (and $r = 1,2,\dots,k-1$) and suppose that $y \in L^p(J)$, y is locally absolutely continuous and $y^{(k+1)} \in L^p(J)$. We need to show that $y^{(r)} \in L^p(J)$ for $1 \le r < k+1$ and (2.19) holds with $n = k+1$, $k = r$. Note that it does not follow immediately from the induction hypothesis that $y^{(r)} \in L^p(J)$ for $r = 1,\dots,k$. But $y^{(r)} \in L^p(I)$ for any compact subinterval I of J since $y^{(r)}$ is absolutely continuous on I, $r = 1,\dots,k$. Hence by Lemma 3 given $\varepsilon_1 > 0$ there exists a $K(\varepsilon_1) > 0$ such that

$$\|y^{(k)}\|_I \le \varepsilon_1 \, \|y^{(k+1)}\|_I + K(\varepsilon_1) \, \|y^{(k-1)}\|_I \le \varepsilon_1 \, \|y^{(k+1)}\|_J + K(\varepsilon_1) \, (\varepsilon_2 \, \|y^{(k)}\|_I + K(\varepsilon_2) \, \|y\|_I) \, .$$

Here we used the inductive hypothesis in the last step. Rearranging terms we get

$$(1-K(\varepsilon_1) \, \varepsilon_2 \, \|y^{(k)}\|_I \le \varepsilon_1 \, \|y^{(k+1)}\|_J + K(\varepsilon_1) \, K(\varepsilon_2) \, \|y\|_I \, .$$

Choose $\varepsilon_1 < \varepsilon/2$ and ε_2 such that $1/2 \le 1-K(\varepsilon_1) \, \varepsilon_2 < 1$. Then dividing by $1-K(\varepsilon_1)\varepsilon_2$ and using $\|y\|_I \le \|y\|_J$ we get

$$\|y^{(k)}\|_I \le \varepsilon \, \|y^{(k+1)}\|_J + 2 \, K(\varepsilon_1) \, K(\varepsilon_2) \, \|y\|_J \, . \tag{2.20}$$

Since this inequality holds for each compact subinterval I of J we conclude that $y^{(k)}$ is in $L^p(J)$ and (2.19) holds with $n = k+1$ and $K(\varepsilon) = 2K(\varepsilon_1)K(\varepsilon_2)$. The rest of the argument follows from (2.20), the induction hypothesis and similar computations.

The furthermore part of Theorem 1 follows from the choice $K(\varepsilon) = 2 \, K(\varepsilon_1) \, K(\varepsilon_2)$ and the fact that $K(\varepsilon_1)$ and $K(\varepsilon_2)$ both can be chosen as nonincreasing functions of the length of J; $K(\varepsilon_1)$ by Lemma 3 and $K(\varepsilon_2)$ by the induction hypothesis. This completes the proof of Theorem 1.

In Theorem 1 let

$$\mu_1 = \inf\,(\varepsilon, K(\varepsilon)) \, , \quad 0 < \varepsilon < \infty \tag{2.21}$$

then

$$\|y^{(k)}\|_p \le \mu_1 \left[\|y\|_p + \|y^{(n)}\|_p \right] . \tag{2.22}$$

Let $\mu = \mu(k,n,p,J)$ denote the best i.e. smallest constant in (2.22). The exact value of μ is not known except in a few special cases. Some of these we will mention in section 3. Clearly $\mu \le \mu_1$. From the proof of Theorem 1 one does not expect equality to hold here. In fact it may happen that $\mu < \mu_1$ even if the best value for $K(\varepsilon)$ for all $\varepsilon > 0$ is known in (2.19). For the case $k = 1$, $n = 1$, $p = 2$ $J = [a,b]$ Redheffer [1963] has found the best constant $K(\varepsilon)$ for each $\varepsilon > 0$:

QMW Library

PRE-CATALOGUING WORKSHEET

LOCATION (if not MAIN LIB)

CLASSMARK (+3 letter code if "special")

QA331.5

ADDED CLASSMARKS

LOAN STATUS (if "special")

ESSENTIAL KEYWORDS NOT IN TITLE

OTHER INFO. (e.g. not for acc.list/display; processing/ triggering requirements)

Subject Librarian's initials

$$K(\varepsilon) = 1/\varepsilon + 12/(b-a)^2 \ . \tag{2.23}$$

This yields $\mu_1 = 12$ in this case. On the other hand Phong [1981] developed an algorithm for the computation of μ according to which $K(1,2,2,[0,1]) = 6.45 < 12$. (This algorithm of Phong's seems not to have been implemented in the case of a bounded interval except for the case mentioned.)

Inequality (2.22) restricts the values of the norm of $y^{(k)}$ in terms of the norms of y and $y^{(n)}$. Do all values of $\|y^{(k)}\|$ subject to the constraint (2.22) occur? Note that $\|y^{(k)}\| = 0$ implies that y is a polynomial of degree $k-1$. So $\|y^{(k)}\|$ assumes the value zero if and only if J is bounded or J is unbounded and $p = \infty$ and $k = 1$. The next result shows that all values strictly between 0 and μ actually occur.

Let

$$Q(y) = \|y^{(k)}\| \, (\|y\| + \|y^{(n)}\|)^{-1} \tag{2.24}$$

for all

$$y \in D(Q) = \{y \in L^p(J) \colon y^{(n)} \in L^p(J),\ y \not\equiv 0\} \ . \tag{2.25}$$

Clearly $\mu = \sup Q(y),\ y \in D(Q)$.

THEOREM 2. Let p, J, n, k be as in Theorem 1. The range of Q contains the open interval $(0,\mu)$.

REMARK 2.8. We will see in section 3 that μ may or may not be in the range of Q. It is clear from (2.25) that Q assumes the value 0 only if (i) J is bounded (so that polynomials are in $L^p(J)$) or (ii) J is unbounded, $k = 1$ and $p = \infty$ in which case Q takes constant functions into 0.

Proof. Case 1. J is unbounded. The proof consists of two parts. First we show that if $a, b \in R(Q)$, $a < b$ then $[a,b] \subset R(Q)$. Suppose $Q(y_1) = a$, $Q(y_2) = b$, $a < b$. Then y_1 and y_2 are linearly independent since $Q(cy) = c\,Q(y)$, c constant. Consider $S = \mathrm{sp}(y_1,y_2) - \{0\}$. Since S is connected and Q is continuous the image of S under Q is a connected set and so it must contain the interval $[a,b]$.

From the definition of μ it follows that b can be chosen arbitrarily close to μ. The proof is complete if we can show that there are arbitrarily small positive numbers in the range of Q. For the case when J is bounded this follows from the fact that we can replace y in (2.24) by $y+c$ for an arbitrary constant c. So suppose J is unbounded. Consider a function y in $C^{(n)}(J)$ which is nonnegative, not identically zero, has compact support I in J and exactly one relative maximum i.e. the graph of y looks like this:

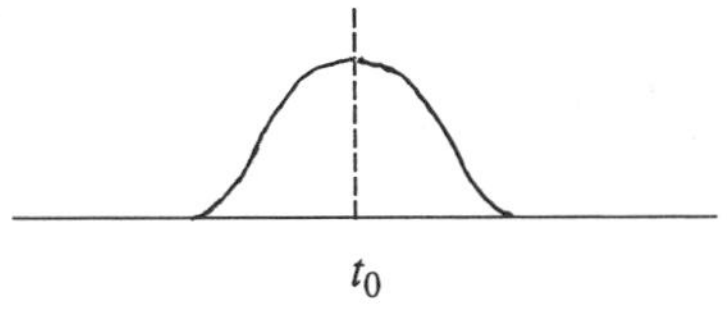

For any $h > 0$ let y_h be the function obtained from y by pulling its graph apart a distance h at the point $(t_0,\ y(t_0))$:

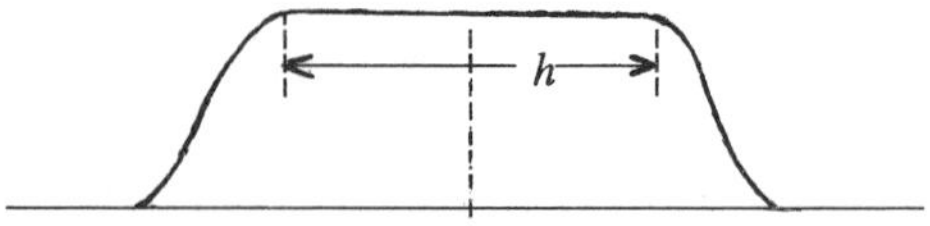

Then $\|y_h^{(i)}\| = \|y^{(i)}\|$ for $i = k, n, h > 0$. Hence $Q(y)$ can be made arbitrarily large by choosing h large enough and t_0 far enough away from the finite end point of J if J has a finite end point.

Case 2. The interval J is bounded. In this case to see that $Q(y)$ assumes arbitrarily small positive values we need only observe that y can be replaced by $y+c$ for any constant c. This leaves $y^{(k)}$, $y^{(n)}$ unchanged and makes $\|y+c\|$ arbitrarily large by choosing c large enough. The rest of the proof is the same as in case 1.

3. Inequalities of Product Form

Here we consider the inequality

$$\|y^{(k)}\|_q \le K\, \|y\|_p^\alpha\, \|y^{(n)}\|_r^\beta\,. \tag{3.1}$$

DEFINITION 1. For J a bounded or unbounded interval, n a positive integer and $1 \le p, r \le \infty$ let $W_{p,r}^n(J)$ denote the set of all functions y from J to R such that $y \in L^p(J)$, $y^{(n-1)}$ is absolutely continuous on compact subintervals of J and $y^{(n)} \in L^r(J)$. Let $\mathring{W}_{p,r}^n(J)$ denote all y in $W_{p,r}^n(J)$ whose support is contained in a compact subinterval (which may be different for different y) of the interior of J.

The simple example $n = 2$, $k = 1$, and $y' = 1$ shows that (3.1) does not hold on bounded intervals J. In case of an unbounded interval J we say that inequality (3.1) is valid whenever there is a positive number K such that (3.1) holds for all y in $W_{p,r}^n(J)$. Two basic questions about (3.1) are:

Question 1. Given an unbounded interval J, positive integers k, n with $1 \le k < n$ and numbers p, r satisfying $1 \le p, r \le \infty$, for what values of the parameters α, β and q is (3.1) valid?

Question 2. Given that (3.1) is valid there clearly is a smallest constant K. This best constant is denoted by $K = K(n, k, p, q, r; J)$ to highlight its dependence on these quantities. (Theorem 1 below will show that α and β are determined by n, k, p, q and r). What are the exact values of the constants $K(n, k, p, q, r; J)$ for $J = R$ or $J = R^+$?

The first question is answered completely by Theorem 1 below. The answer to question 2 is known only in a few special cases.

THEOREM 1. (Gabushin [1967].) Let $J = (a,\infty)$, $-\infty \le a < \infty$. Let k, n be integers with $0 \le k < n$. Let $1 \le p, r \le \infty$. There is a positive constant K such that

$$\|y^{(k)}\|_q \le K\, \|y\|_p^{\alpha}\, \|y^{(n)}\|_r^{\beta} \tag{3.1}$$

holds for all $y \in W^n_{p,r}(J)$ if and only if

$$nq^{-1} \le (n-k)p^{-1} + kr^{-1} \tag{3.2}$$

and

$$\alpha = (n-k-r^{-1}+q^{-1})/(n-r^{-1}+p^{-1}),\ \beta = 1-\alpha\ . \tag{3.3}$$

Before presenting a proof we make some observations. Given p, r, n, k satisfying the conditions of the theorem if q satisfies (3.2) then (3.1) holds for some constant K provided α,β are determined by (3.3). We follow the usual convention that $q^{-1} = 0$ when $q = \infty$. It follows from the change of variable $t \to t+a$ and $t \to -t$ that $K(n, k, p, q, r; J) = K(n, k, p, q, r; R^+)$ for any $J = (a,\infty)$ with $-\infty < a < \infty$. On the other hand, as we will see later, these constants are not the same in general for $J = R$ and $J = R^+$. Thus there are only two distinct cases to consider $J = R$ and $J = R^+$.

In the special case $p = r$ (3.2) reduces to $q \ge p$. If $p = r = \infty$ then only $q = \infty$ satisfies (3.2). For $p = \infty$ and r finite (3.2) holds when $nrk^{-1} \le q \le \infty$. Similarly when $r = \infty$ and p finite we must have $np/(n-k) \le q \le \infty$.

Proof of Theorem 1. The necessity part of the proof is taken from Gabushin [1967]. Our sufficiency proof, although somewhat related to Gabushin's, has some new features.

(a) Necessity. Suppose (3.1) holds for all $y \in W^n_{p,r}(J)$, $J = R$ or R^+ with some fixed constant K. Choose a non-trivial C^∞ function ϕ with compact support in (0,1). Consider the class of functions

$$\phi_{m,s,t}(x) = s\sum_{j=0}^{m-1} \phi(tx - j/t)$$

with s,t any positive numbers and m any positive integer. Clearly ϕ and each of $\phi_{m,s,t}$ is in $W^n_{p,r}(J)$. Hence (3.1) holds for each $\phi_{m,s,t}$. A computation yields that

$$\|\phi^{(i)}_{m,s,t}\|_p = m^{1/p}st^{i-1/p}\, \|\phi^{(i)}\|_p\ ,\quad i = 0,\dots,n\ . \tag{3.4}$$

By substituting (3.4) into (3.1) we get

$$m^{1/q}st^{k-1/q}\, \|\phi^{(k)}\|_q \le Km^{\alpha/p+\beta/r}s^{\alpha+\beta}t^{-\alpha/p+\beta(n-1/r)}\, \|\phi\|_p^{\alpha}\, \|\phi^{(n)}\|_r^{\beta}\ . \tag{3.5}$$

Since we can let s and t go to zero and to $+\infty$ and m can be an arbitrarily large positive integer it follows that

(i) $\alpha+\beta = 1$
(ii) $k-q^{-1} = -\alpha p^{-1}+\beta(n-r^{-1})$
(iii) $q^{-1} \le \alpha p^{-1}+\beta r^{-1}$

and (3.2), (3.3) follow.

(b) Sufficiency. This half of the proof of Theorem 1 is established with the help of several lemmas some of which may be of independent interest.

LEMMA 1. Suppose $J = [a, b]$ is a closed and bounded interval. If f is a differentiable function on J satisfying $f'(t) \geq c > 0$ or $-f'(t) \geq c > 0$ for all t in J, then there exists a subinterval $[a_1, b_1]$ of J of length $(b-a)/4$ such that $|f(t)| \geq c(b-a)/4$, $t \in [a_1, b_1]$.

Proof. Suppose $f'(t) \geq c > 0$ for t in J. Then f has at most one zero in J. If f has no zero in J then either f or $-f$ is positive on J. If f is positive then

$$f(t) \geq f(a) + c(t-a),\ t \in J$$

and hence

$$f(t) \geq c(b-a)/4\ ,\quad t \in [a+(b-a)/4,\ b]\ .$$

The case when $-f$ is positive is similar. The case when f has a zero at d can be reduced to the above by replacing the interval $[a,b]$ with $[a,d]$ or $[d,b]$ whichever has length $\geq (b-a)/2$.

In general knowing that y is in $L^p(a,\infty)$ does not provide any information about the pointwise asymptotic behavior of $y(t)$ as $t \to \infty$. However, knowing that y and some derivative of y are in $L^p(a,\infty)$ does provide such information.

LEMMA 2. If $y \in W^n_{p,r}(R^+)$, $1 \leq p < \infty$, $1 \leq r \leq \infty$, n a positive integer, then

$$y^{(k)}(t) \to 0 \text{ as } t \to \infty\ ,\quad k = 0,1,\ldots,n-1\ . \tag{3.6}$$

If $p = \infty$, $r < \infty$ and $n > 1$, then (3.6) holds for $k = 1,\ldots,n-1$.

Proof. Case 1. $p < \infty$, $r < \infty$. Suppose $y \in L^p(0,\infty)$, $y^{(n)} \in L^r(0,\infty)$ and (3.6) does not hold for $k = n-1$. Then there exists a $K > 0$ and a sequence $t_m \to \infty$ such that either $y^{(n-1)}(t_m) > 2K$ or $y^{(n-1)}(t_m) < -2K$. We may assume the former since in the latter case we can replace y by $-y$. For $t \in I_j = [t_j,\ t_j+1]$ we get from Holder's inequality

$$y^{(n-1)}(t) = y^{(n-1)}(t_j) - \int_t^{t_j} y^{(n)} \geq 2K - \|y^{(n)}\|_{r,j} \to 2K \text{ as } j \to \infty\ . \tag{3.7}$$

Here $\|\cdot\|_{r,j}$ denotes the usual L^r norm on the interval I_j. Hence for all sufficiently large j, $y^{(n-1)}(t) \geq K$ for $t \in I_j$. If $n=1$ this contradicts $y \in L^p(R^+)$ and the proof is complete. If $n > 1$ we may conclude from Lemma 1 that $y^{(n-2)}(t) \geq K/4$ in some subinterval J_j of I_j of length 1/4. Repeated applications of Lemma 1 yield that $y(t) \geq K/4^h$, $h = 1+2+\ldots+n-1$ for t in some subinterval of I_j of length 4^{1-n}. This contradicts $y \in L^p(R^+)$. Having proved that $y^{(n-1)}(t) \to 0$ as $t \to \infty$ we can use the same technique to show that $y^{(n-2)}(t) \to 0$ as $t \to \infty$. By repeating this argument we arrive at (3.6).

Case 2. $1 \leq p < \infty$, $r = \infty$. Here we proceed as in case 1 but use the intervals $I_j = [t_j,\ t_j + K/\|y^{(n)}\|_{\infty,j}]$. In place of (3.7) we have

$$y^{(n-1)}(t) = y^{(n-1)}(t_j) - \int_t^{t_j} y^{(n)} \geq 2K - K = K\ .$$

Case 3. $p = \infty$, $1 \leq r < \infty$. Again we use the same argument as in case 1 with the intervals

$$I_j = [t_j,\ t_j + a_j] \text{ where } a_j = \|y^{(n)}\|_{r,j}^{-1} .$$

In place of (3.7) we have

$$y^{(n-1)}(t) = y^{(n-1)}(t_j) - \int_t^{t_j} y^{(n)} \le 2K - \|y^{(n)}\|_{r,j}\, a_j^{1/r'}$$

$$= 2K - \|y^{(n)}\|_{r,j}^{1/r} \to 2K \text{ as } j \to \infty .$$

Hence $y^{(n-1)}(t) \ge K$ for all $t \in I_j$ and all j sufficiently large. From Lemma 1 we get $y'(t) \ge K4^{-h}$, $h = 1+2+\cdots+n-2$ for t in some subinterval of I_j of length $a_j 4^{1-n} \to \infty$ as $j \to \infty$. This contradicts the boundedness of y and completes the proof of Lemma 2.

Next we show that functions in $W^n_{p,r}(R^+)$ can be approximated by functions of compact support when not both of p and r are infinite.

LEMMA 3. Let $y \in W^n_{p,r}(R^+)$ with not both of p, r infinite and when $p = \infty$ either $r > 1$ or $n > 1$. Given any $\varepsilon > 0$ there is some $T > 0$ and a function $y_0 \in \mathring{W}^n_{p,r}(R^+)$ such that

$$y_0(t) = y(t) \text{ for } t \in (0,T]$$

and

$$\|y - y_0\|_p < \varepsilon \text{ if } p < \infty \tag{3.8}$$

$$\|y_0\|_\infty \le \|y\|_\infty \text{ if } p = \infty \tag{3.9}$$

$$\|y^{(k)} - y_0^{(k)}\|_r < \varepsilon \text{ if } r < \infty,\ k = 1,\dots,n \tag{3.10}$$

$$\|y_0^{(k)}\|_\infty \le \|y^{(k)}\|_\infty + \varepsilon \text{ if } r = \infty,\ k = 1,\dots,n . \tag{3.11}$$

Proof. First consider the case $1 \le p < \infty$. Let ϕ be a C^∞ function on [0,1] which is 1 in a right neighborhood of 0, 0 in a left neighborhood of 1, and satisfies $0 \le \phi(t) \le 1$. For each positive integer j define

$$\Psi_j(t) = \begin{cases} 1,\ 0 \le t \le j \\ \phi(t-j),\ j \le t \le j+1 \\ 0,\ j+1 \le t \end{cases} \tag{3.12}$$

and consider

$$y_j = y\, \Psi_j . \tag{3.13}$$

Then $y_j = y$ on $[0, j]$. Since $0 \le \Psi_j \le 1$ we have

$$\|y - y_j\|_p^p = \int_j^\infty |y - y_j|^p \le \int_j^\infty |y|^p \to 0 \text{ as } j \to \infty . \tag{3.14}$$

From Leibnitz's formula for the derivatives of a product we get

$$y_j^{(k)} = \sum_{i=0}^{k} \binom{k}{i} y^{(k-i)} \Psi_j^{(i)}, \quad k = 1,\dots,n .$$

Hence when $r < \infty$ we have

$$\|y^{(k)}-y_j^{(k)}\|_r \leq \|y^{(k)}-y^{(k)}\Psi_j\|_r + \sum_{i=1}^{k} \binom{k}{i} \|y^{(k-i)}\Psi_j^{(i)}\|_j . \tag{3.15}$$

The first term on the right goes to zero as $j \to \infty$ by an argument similar to that in (3.14). The convergence to zero as $j \to \infty$ of the second term on the right in (3.15) follows from the uniform boundedness of $\Psi_j^{(i)}$, $i = 1,\dots,n$; the fact that the support of $y^{(n-i)}\Psi_j^{(i)}$ is contained in $(j, j+1)$, $i=1,\dots,n$ and (3.6) of Lemma 2.

For $r = \infty$ and $\varepsilon > 0$ we have

$$|y_j^{(k)}(t)| \leq |y^{(k)}(t)\Psi_j(t)| + \|\sum_{i=1}^{k} \binom{k}{i} y^{(k-i)}\Psi_j^{(i)}\|_\infty \leq |y^{(k)}(t)| + \varepsilon$$

for j sufficiently large by (3.6) and (3.12). Hence

$$\|y_j^{(k)}\|_\infty \leq \|y^{(k)}\|_\infty + \varepsilon ,$$

for j sufficiently large, $k = 1,\dots,n$.

In case $p = \infty$ we define Ψ_j by

$$\Psi_j(t) = \begin{cases} 1, \ 0 \leq t \leq j \\ \phi((t-j)/j), \ j \leq t \leq 2j \\ 0, \ 2j \leq t . \end{cases} \tag{3.16}$$

Then

$$\|\Psi_j^{(k)}\|_r = j^{-k+1/r}\|\phi^{(k)}\|_r \quad \to 0 \text{ as } j \to \infty$$

for $k = 1,2,\dots,n$ except when $r = 1$. We define y_j by (3.13) using (3.16) and proceed as before to get (3.9) with $y_0 = y_j$ for j large enough.

Choosing $y_0 = y_j$ for j large enough we get (3.8), (3.9), (3.10), (3.11) for each case.

LEMMA 4. If $y \in W^1_{p,r}([a,b])$, $1 \leq p, r \leq \infty$, $b-a = 1$, then

$$\|y\|_\infty \leq \|y\|_p + \|y'\|_r . \tag{3.17}$$

Proof. Let $|y|$ attain its minimum at $t_1 \in [a,b]$. Then $|y(t_1)| \leq \|y\|_p$ and for any $t \in [a,b]$

$$|y(t)| \leq |y(t_1)| + |\int_{t_1}^{t} y'| \leq \|y\|_p + |t-t_1|^{1/s}\|y'\|_r \leq \|y\|_p + \|y'\|_r$$

and (3.7) follows. Here $s^{-1} + r^{-1} = 1$ with $s = 1$ if $r = \infty$.

LEMMA 5. Let $y \in W^1_{p,r}([a, b])$, a, b finite, $1 \leq p, r \leq \infty$. Suppose $y(c) = 0$ for some c in $[a,b]$. Then

$$\|y\|_\infty \le \|y\|_p + \|y'\|_r \, .$$

Here all norms are on $[a,b]$.

Proof. Let $[a,b] = \bigcup_{k=0}^{n} I_k$ where the intervals I_k have no interior point in common for $k = 1,...,n$; c is in I_0, the length of I_k is 1 for k=1,...,n and the length of $I_0 \le 1$. Then for $t \in I_k$ by Lemma 4

$$|y(t)| \le \|y\|_{p,I_k} + \|y'\|_{r,I_k} \le \|y\|_{p,[a,b]} + \|y'\|_{r,[a,b]}, \; k=1,...,n$$

and for $t \in I_0$

$$|y(t)| \le |y(c)| + |\int_c^t y'| \le |t-c|^{1/s} \, \|y'\|_{r,I_o} \le \|y'\|_{r,[a,b]}$$

where $s^{-1} + r^{-1} = 1$. The conclusion follows.

LEMMA 6. Let $1 \le p, r \le \infty$. Suppose $q \ge p$ and

$$\alpha = (1-r^{-1}+q^{-1})/(1-r^{-1}+p^{-1}), \; \beta = 1-\alpha \tag{3.18}$$

where $r^{-1} = 0$ if $r = \infty$. Then there exists a constant K such that for any $y \in W^1_{p,r}([a,b])$, $-\infty < a < b < \infty$, which has at least one zero in $[a,b]$ we have

$$\|y\|_q \le K \, \|y\|_p^\alpha \, \|y'\|_r^\beta \, . \tag{3.19}$$

Proof. Let $\lambda > 0$ and consider $f(t) = y(\lambda t)$ on $[a/\lambda, b/\lambda]$. Then

$$\|f\|_\infty = \|y\|_\infty, \quad \|f\|_p = \lambda^{-1/p} \, \|y\|_p, \quad \|f'\|_r = \lambda^{1-1/r} \, \|y'\|_r \tag{3.20}$$

where each norm is over the appropriate interval on which the function is defined. Applying Lemma 5 to f we get

$$\|y\|_\infty \le \lambda^{-1/p} \, \|y\|_p + \lambda^{1-1/r} \, \|y'\|_r \, , \; \lambda > 0 \, . \tag{3.21}$$

Choosing $\lambda = (\|y\|_p/\|y'\|_r)^{p(1-\alpha)}$ yields (3.19) for $q = \infty$. (Note that $\|y'\|_r \ne 0$ unless $y \equiv 0$). For $q < \infty$ we observe that

$$\int |y|^q = \int |y|^p |y|^{q-p} \le \|y\|_\infty^{q-p} \, \|y\|_p^p$$

and estimate $\|y\|_\infty$ using (3.21). This completes the proof.

We now return to the sufficiency proof of Theorem 1. First some simplifications. It suffices to prove (3.1) for $J = R^+$. The case $J = R$ will follow from Theorem 3 below.

Case 1. $p = \infty$, $n = 1$. In this exceptional case we must have $k = 0$, $q = \infty$, $\alpha = 1$, $\beta = 0$ and so (3.1) is satisfied trivially.

Case 2. $p = \infty$, $r = \infty$. By (3.2), (3.3) we have $q = \infty$, $\alpha = (n-k)/n$, $\beta = k/n$. It suffices to establish the case $n = 2$, $k = 1$. The case $k = 0$ is trivial and the case $n > 0$ follows by an induction argument.

Let $y \in W^2_{\infty,\infty}(R^+)$. By Taylor's formula

$$y(t+s) = y(t) + sy'(t) + \int_0^t (s-u)y''(t+u)du$$

or

$$y'(t) = s^{-1}[y(t+s)-y(t)] - s^{-1}\int_0^t (s-u)y''(t+u)du \tag{3.22}$$

for s, $t > 0$. The integral in (3.22) is dominated by $s^2\|y''\|_\infty/2$. Hence

$$\|y'\|_\infty \le 2s^{-1}\|y\|_\infty + (s/2)\|y''\|_\infty \; . \tag{3.23}$$

Now minimizing the right hand side over s in $(0,\infty)$ we find that the minimum value is attained when

$$s = 2\,\|y\|^{1/2}\,\|y''\|^{-1/2} \; .$$

This gives

$$\|y'\|_\infty^2 \le 4\,\|y\|_\infty\,\|y''\|_\infty \; . \tag{3.24}$$

Case 3. Either $1 \le p < \infty$ or $1 \le r < \infty$. First we show that we may restrict consideration to functions whose support lies in some interval $[0,T]$. Let $y \in W^n_{p,r}(R^+)$ and $\varepsilon > 0$. By Lemma 3 there is some $T > 0$ and $y_0 \in W^n_{p,r}(R^+)$ such that $y_0 = y$ on $(0,T]$, $y_0 = 0$ on $[T+1, \infty)$ and (3.8) to (3.11) hold. From

$$\|y_0^{(k)}\|_q \le K\,\|y_0\|_p^\alpha\,\|y_0^{(n)}\|_r^\beta \tag{3.25}$$

it follows that

$$\|y^{(k)}\|_{q,[0,T]} \le \|y_0^{(k)}\|_q \le K\,(\|y\|_p+\varepsilon)^\alpha\,(\|y^{(n)}\|_r+\varepsilon)^\beta \; .$$

Letting $\varepsilon \to 0$ and $T \to \infty$ yields (3.1). Thus if (3.1) holds for functions in $W^n_{p,r}(R^+)$ which are supported on intervals $(0,T)$, $T > 0$ then (3.1) holds for all $y \in W^n_{p,r}(R^+)$.

The proof is by induction on n. Let $y \in W^n_{p,r}(R^+)$. Suppose $n=1$, $k=0$. Define y_0 as in the proof of Lemma 3 i.e. $y_0 = y_j$ where y_j is given by (3.13) with Ψ_j defined by (3.12) for $p < \infty$ and by (3.16) with $p = \infty$. Then $y_0 = y$ on $(0, j)$ and $y_0(j+1) = 0$. Also $q > p$ by (3.2) and α, β are given by (3.18). Thus (3.25) holds by Lemma 6. Hence (3.1) holds by the above approximation argument. Now suppose $y \in W^n_{p,r}(R^+)$, $1 \le p$, $r \le \infty$ with at least one of p, r finite, $n = 2$, $k = 1$ and y is supported on some interval $(0,T)$, $T < \infty$. Then $y^{(k)} \in L^q(R^+)$ for all k, $0 \le k < n$ and all q, $1 \le q \le \infty$. Define

$$S = \{t > 0 \mid y'(t) \ne 0\} \; .$$

Then S is an open set in the relative topology of R^+. Hence $S = \cup_{n=1}^\infty I_n$ where the I_n's are disjoint and each I_n is an open interval in the relative topology of R^+. In each interval I_n, y' is of constant sign and vanishes at least one end point. We first show that inequality (3.1) holds if the norms are interpreted as being taken over I_n. We may assume that $y' > 0$ in I_n otherwise we replace y by $-y$. Then y is increasing. If y has a zero in $I_n = [a,b]$, then from (3.19) with $q = \infty$ we have

$$y(b)-y(a) \le 2\,\|y\|_\infty \le 2\,K\,\|y\|_p^{\alpha_1}\,\|y'\|_r^{\beta_1} \; .$$

If y has no zero in I_n, then (3.19) applied to $y-y(a)$ if $y(a) > 0$ or to $y(b)-y$ if $y(a) < 0$ yields

$$y(b)-y(a) = \|y-y(a)\|_\infty \text{ (or } \|y(b)-y\|_\infty) \le K \, \|y-y(a)\|_p^{\alpha_1} \, \|y'\|_r^{\beta_1} .$$

In either case

$$y(b)-y(a) \le 2\, K \, \|y\|_p^{\alpha_1} \, \|y'\|_r^{\beta_1} . \tag{3.26}$$

Applying (3.19) to y' yields

$$\|y'\|_q \le K \, \|y'\|_1^{\alpha_2} \, \|y''\|_r^{\beta_2} . \tag{3.27}$$

Noting that $\|y'\|_1 = y(b)-y(a)$, (3.26) and (3.27) yield (3.1). (Observe that α (and hence β) in (3.26) and (3.27) are not the same since $q = \infty$ in (3.26) and $q < \infty$ in (3.27); see Lemma 3.6 for definition of α.) To obtain inequality (3.1) on R^+ we first consider the two extreme cases: $q = \infty$ and $q = s$ where $2s^{-1} = p^{-1}+r^{-1}$. In the former case

$$\|y'\|_\infty = \sup_n \{\|y'\|_{\infty,I_n}\} \le \sup_n \{K \, \|y\|_{p,I_n}^\alpha\} \, \|y''\|_{r,I_n}^{1-\alpha} \le K \, \|y\|_p^\alpha \, \|y''\|_r^{1-\alpha} . \tag{3.28}$$

In the last step we used the fact that the constant K in (3.27) is independent of the interval I_n.

In the case $q = s$, $\alpha = 1/2$ and

$$\|y'\|_s^s = \sum_{n=1}^\infty \|y'\|_{s,I_n}^s \le K^s \sum_{n=1}^\infty (\|y\|_{p,I_n}^p)^{s/2p} (\|y''\|_{r,I_n}^r)^{s/2r} \le K^s \left[\sum_{n=1}^\infty \|y\|_{p,I_n}^p\right]^{s/2p} \left[\sum_{n=1}^\infty \|y''\|_{r,I_n}^r\right]^{s/2r} \tag{3.29}$$

$$= K^s \, \|y\|_p^{s/2} \, \|y''\|_r^{s/2} .$$

The last inequality follows from the discrete Holder inequality.

For $s < q < \infty$ inequality (3.1) follows from (3.28), (3.29) and

$$\int |y'|^q = \int |y'|^s \, |y'|^{q-s} \le \|y'\|_\infty^{q-s} \, \|y'\|_s^s .$$

Finally we establish (3.1) for functions y in $\mathring{W}_{p,r}^n(R^+)$ by induction on n. (It is here that we use the fact that y has compact support to ensure that y and its derivatives up to order $n-1$ are in $L^q(R^+)$ for all q, $1 \le q \le \infty$.) We have just proved the case $n = 2$. Suppose (3.1) holds for $n-1$ and all k, $0 \le k < n-1$. It suffices to prove only the case $k = n-1$ since the cases $k < n-1$ follow from the inductive hypothesis and this case. Let s be a real number such that $(n-1)s^{-1} = (n-2)p^{-1} + q^{-1}$. Then $(n-1)q^{-1} \le (n-2)r^{-1} + s^{-1}$. By (3.1) with $y \in \mathring{W}_{p,q}^{n-1}$ and p, s, q, $n-1$, 2 in place of p, q, r, n, k we get

$$\|y'\|_s \le K_1 \, \|y\|_p^{\alpha_1} \, \|y^{(n-1)}\|_q^{\beta_1} . \tag{3.30}$$

By (3.1) applied to $y' \in \mathring{W}_{s,r}^{n-1}$ with s, q, r, n, $n-1$ in place of p, q, r, n, k we get

$$\|y^{(n-1)}\|_q \le K_2 \, \|y'\|_s^{\alpha_2} \, \| y^{(n)}\|_r^{\beta_2} . \tag{3.31}$$

In (3.30) and (3.31) $\alpha_1,\alpha_2,\beta_1,\beta_2$ are chosen according to (3.2). Now (3.1) follows upon using (3.30) in (3.31). This completes the proof of Theorem 3.1.

Clearly inequality (3.1) cannot hold on a finite interval J since for $y^{(k)} = 1$ and $y^{(n)} = 0$ the right hand side is zero and the left hand side is not. However, note that the proof of Theorem 3.1 contains the following.

COROLLARY 1. Suppose J is a bounded interval, $1 \le p,\ r \le \infty$ and n, k are integers with $0 \le k < n$. Then there exists a constant K such that for all $y \in W^n_{p,r}(J)$ satisfying $y^{(k)}(c_k) = 0$ for some $c_k \in \overline{J}$, $k = 1,\ldots,n-1$ we have

$$\|y^{(k)}\|_q \le K\ \|y\|_p^\alpha\ \|y^{(n)}\|_r^\beta$$

where α, β, q satisfy (3.2) and (3.3) and the norms are taken over J.

THEOREM 2. Let $J = R$ or R^+, $1 \le p,\ r \le \infty$, n, k positive integers with $k < n$. Then the following statements are all equivalent:

1. $nq^{-1} \le (n-k)p^{-1} + kr^{-1}$ and $\alpha = (n-k-r^{-1}+q^{-1})/(n-r^{-1}+p^{-1})$, $\beta = 1-\alpha$.

2. There exists a constant K such that
$$\|y^{(k)}\|_q \le K\ \|y\|_p^\alpha\ \|y^{(n)}\|_r^\beta\ , \tag{3.1}$$
for all $y \in W^n_{p,r}(J)$.

3. There exists a constant K such that for all $\lambda > 0$ and all $y \in W^n_{p,r}(J)$ we have
$$\|y^{(k)}\|_q \le K\ (\lambda^a\ \|y\|_p + \lambda^b\ \|y^{(n)}\|_r) \tag{3.32}$$
where $a\alpha + b\beta = 0$.

4. There exists a constant K such that for all $y \in W^n_{p,r}(J)$
$$\|y^{(k)}\|_q \le K\ (\|y\|_p + \|y^{(n)}\|_r)\ . \tag{3.33}$$

5. If $y \in W^n_{p,r}(J)$ then $y^{(k)} \in L^q(J)$, $k=1,\ldots,n-1$.

The constant K need not be the same in these statements but in each case there is a smallest constant K which depends on p, q, r, n, k and J but not on y and in the case of (3.32) K does not depend on λ.

COROLLARY 2. Suppose the hypotheses and the conditions of statement (1) of Theorem 2 hold. In addition assume that $1 \le q < \infty$ if $r = 1$ and $k = n-1$. Then

(i) Given $\varepsilon > 0$ there exists a $K = K(\varepsilon) > 0$ such that for all $y \in W^n_{p,r}(J)$ we have
$$\|y^{(k)}\|_q \le \varepsilon\ \|y\|_p + K(\varepsilon)\ \|y^{(n)}\|_r\ .$$

(ii) Given $\varepsilon > 0$ there exists a $K = K(\varepsilon) > 0$ such that for all $y \in W^n_{p,r}(J)$
$$\|y^{(k)}\|_q \le K(\varepsilon)\ \|y\|_p + \varepsilon\ \|y^{(n)}\|_r\ .$$

Proof. This follows from part (3) of Theorem 2. In case (i) note that $\alpha > 0$ and choose $a = \alpha^{-1}$,

$b = -\beta^{-1}$. Then, for part (i), choose λ so small that $K\,\lambda^a < \varepsilon$. For part (ii) choose λ so large that $K\,\lambda^b < \varepsilon$.

Proof. The equivalence of (1) and (2) was established in Theorem 1. Inequality (3.32) follows from (3.1) and the general inequality between weighted arithmetic and geometric means: Hardy, Littlewood and Polya [1934]:

$$\|y^{(k)}\|_q \le K\,\|y\|_p^\alpha\,\|y^{(n)}\|_r^\beta = K\,(\lambda^a\,\|y\|_p)^\alpha\,(\lambda^b\,\|y^{(k)}\|_r)^\beta \le K\,(\alpha\lambda^a\,\|y\|_p + \beta\lambda^b\,\|y^{(n)}\|_r)\,.$$

Inequality (3.33) is a special case of (3.32) and statement (5) clearly follows from (4). Thus we have shown that (1) $\to$ (2) $\to$ (3) $\to$ (4) $\to$ (5).

To show that (5) $\to$ (4) we use the closed graph theorem. The linear space $W^n_{p,r}(J)$ can be made into a Banach space with norm

$$\|y\| = \|y\|_p + \|y^{(n)}\|_r\,. \tag{3.34}$$

The differentiation operator $A = d^k/dt^k$ is a closed linear operator from $W^n_{p,r}(J)$ into $L^q(J)$. By (5) A is defined on all of $W^n_{p,r}(J)$. Hence A is bounded by the closed graph theorem i.e. (3.33) holds.

To show that $W^n_{p,r}(J)$ is a Banach space with the norm (3.36) we consider the notion of a weak derivative. A locally integrable function y is n times weakly differentiable in $L^r(J)$ if there exists a function z in $L^r(J)$ such that

$$\int_J y\phi^{(n)} = (-1)^n \int_J z\phi$$

for all ϕ infinitely differentiable with compact support in the interior of J. Such a z is a weak n^{th} derivative of y. It is unique a.e., Soboljev's lemma, see Adams [1975], in the case of functions of one variable, states that weak differentiability implies strong differentiability i.e. there exists a function w such that $y = w$ a.e. and w has a classical n^{th} derivative. (Although Soboljev's lemma is usually stated only for functions y such that $y, y', \ldots, y^{(n)}$ all belong to L^p, it applies in our situation since the restrictions to compact subintervals will be in L^1.)

Suppose y_k is a Cauchy sequence in $W^n_{p,r}(J)$ in the norm (3.25). Then y_k and $y_k^{(n)}$ are Cauchy sequences in $L^p(J)$ and $L^r(J)$ respectively. Let $y_k \to y$ in $L^p(J)$ and $y_k^{(n)} \to z \in L^r(J)$. For any C_0^∞ function ϕ we have

$$\int_J y\phi^{(n)} = \lim_{n\to\infty} \int_J y_k\phi^{(n)} = (-1)^n \lim_{n\to\infty} \int_J y_k^{(n)}\phi = (-1)^n \int_J z\phi\,.$$

Thus z is the weak n^{th} derivative of y. Hence there exists a classically n times differentiable function $w = y$ a.e. such that $w^{(n)} = z$ a.e. Thus $w \in W^n_{p,r}(J)$ and is the limit of the Cauchy sequence y_k.

The proof that the linear operator $Ay = y^{(k)}$ from $W^n_{p,r}(J)$ into $L^q(J)$ is a closed operator is similar. This completes the proof of the implication (5) $\to$ (4).

To show that (4) implies (3) let

$$f(t) = y(\lambda t)\,, \quad \lambda > 0\,.$$

Then $f \in W^n_{p,r}(J)$ and

$$\|f^{(i)}\|_q^q = \lambda^{i-1} \int_J y^{(i)q}(\lambda t) d(\lambda t) = \lambda^{i-1} \|y^{(i)}\|_q^q, \quad 1 \le q < \infty .$$

Applying (3.33) to f we get

$$\lambda^{(k-1)/q} \|y^{(k)}\|_q \le K (\lambda^{-1/p} \|y\|_p + \lambda^{(n-1)/r} \|y^{(n)}\|_r)$$

which yields (3.32) with

$$a = -p^{-1} - kq^{-1} + q^{-1}, \quad b = nr^{-1} - r^{-1} - kq^{-1} + q^{-1} . \tag{3.35}$$

Using (3.2) a tedious computation shows that $a\alpha + b\beta = 0$. This establishes (3.32) for p, q, r all finite. The case when at least one of p, q, r is infinite is verified similarly. The implication (3) implies (2) is established by considering the right hand side of (3.32) as a function of λ, say $g(\lambda)$, and then minimizing g over all λ in $(0,\infty)$. That (2) implies (1) was shown in Theorem 1. This completes the proof of Theorem 2.

COROLLARY 3. Let $1 \le p, r \le \infty$, $J = R$ or R^+ and let n be a positive integer. Let k_j be an integer satisfying $1 \le k_j < n$ and suppose each q_j satisfies

$$nq_j^{-1} \le (n-k_j)\, p^{-1} + k_j\, r^{-1}, \quad j = 1,\dots,m .$$

For $Q = (p, q_1, \dots, q_m, r)$ let W_Q^n denote the set of all $y \in L^p(J)$ such that $y^{(n-1)}$ exists and is locally absolutely continuous and $y^{(k_j)} \in L^{q_j}(J)$, $j=1,\dots,m$, $y^{(n)} \in L^r(J)$. Then $W_{p,r}^n(J) = W_Q^n(J)$ and the two norms

$$\|y\|_1 = \|y\|_p + \|y^{(n)}\|_r$$

$$\|y\|_2 = \|y\|_p + \sum_{j=1}^{m} \|y^{(k_j)}\|_{q_j} + \|y^{(n)}\|_r$$

are equivalent.

Proof. This is an immediate consequence of (3.33).

COROLLARY 4. Let n be a positive integer, let $J = R$ or R^+, $1 \le p \le \infty$. Let $W_p^n(J)$ denote the set of all functions y such that $y \in L^p(J)$, $y^{(n-1)}$ is locally absolutely continuous and $y^{(k)} \in L^p(J)$, for $k=1,\dots,n$. Then $W_{p,p}^n(J) = W_p^n(J)$ and the two norms

$$\|y\|_1 = \|y\|_p + \|y^{(n)}\|_p$$

$$\|y\|_2 = \|y\|_p + \sum_{k=1}^{n} \|y^{(k)}\|_p$$

are equivalent.

Proof. This is a special case of Corollary 2 i.e. $p = r$, $m = n-1$, $k_j = j$, $q_j = p$, $j=1,\dots,n-1$, $Q = (p, p, \dots, p)$.

THEOREM 3. Suppose the hypotheses of Theorem 2 hold. Let K be the smallest constant in inequality (3.1). Assume that μ_0, μ_k, μ_n are positive numbers satisfying

$$\mu_k < K\, \mu_0^{\alpha}\, \mu_n^{\beta} .$$

Then there exists a function y in $W^n_{p,r}(J)$ such that

$$\|y\|_p = \mu_0, \quad \|y^{(k)}\|_q = \mu_k, \quad \|y^{(n)}\|_r = \mu_n .$$

Proof. Suppose $1 \le p < \infty$. Define

$$Q(y) = \|y^{(k)}\|_q / (\|y\|_p^{\alpha}\, \|y^{(n)}\|_r^{\beta}) , \quad y \in W^n_{p,r}(J) , \quad y \not\equiv 0 .$$

(Note that $\|y^{(n)}\|_r \neq 0$ since $\|y^{(n)}\|_r = 0$ implies $y^{(n)} \equiv 0$ implies y is a polynomial. Since J is an unbounded interval the only polynomial in $L^p(J)$ is the zero polynomial.) Then

$$K = \sup Q(y) \tag{3.36}$$

where the sup is taken over all $y \not\equiv 0$ in $W^n_{p,r}(J)$.

Claim. The range of Q contains the interval $(0,K)$. Let $Q(y_1) = a$, $Q(y_2) = b$, $a < b$. Then y_1 and y_2 are linearly independent. The interval $[a,b]$ is contained in the range of Q since the continuous image of a connected set is connected and Q restricted to the connected two-dimensional set $S = \text{span}\{y_1,y_2\}-\{0\}$ is continuous. Clearly b can be chosen arbitrarily close to K. The claim is established if we can show that a can be made arbitrarily small. To do this we consider two cases.

Case 1. $p = \infty$. In this case for any constant c, $y + c$ is in $W^n_{p,r}(J)$ if y is in this set. Also $\|y^{(i)}\| = \|(y+c)^{(i)}\|$ for $i = k,n$ and any c while $\|y+c\|_\infty \to \infty$ as $c \to \infty$. Thus the quotient $Q(y)$ can be made arbitrarily small.

Case 2. $1 \le p < \infty$. Choose a function $y \in C^n$ with compact support in (a,∞) whose graph is as follows:

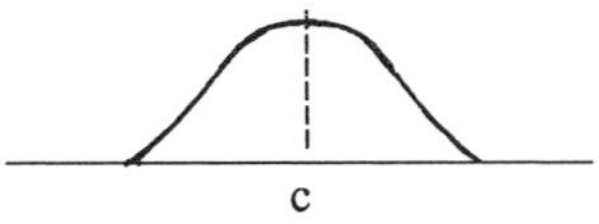

Figure 1

For any $h > 0$ let y_h be the function obtained from y by pulling its graph apart a distance h at the point $(c, y(c))$ as indicated in Figure 2:

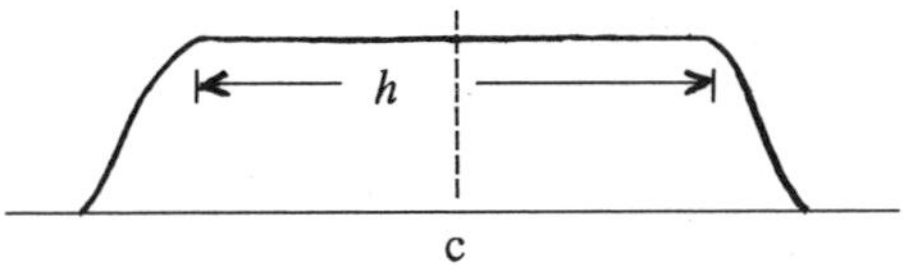

Figure 2

Then $\|y^{(k)}\|_q = \|y_h^{(k)}\|_q$ and $\|y^{(n)}\|_r = \|y_h^{(n)}\|_r$ but $\|y_h\|_p \to \infty$ as $h \to \infty$ (and $a \to \infty$ if $J = R^+$). This establishes our claim.

Now choose $z \in W^n_{p,r}(J)$ such that $Q(z) = \mu_k/(\mu_0^\alpha \mu_n^\beta)$. Consider

$$z_{ab}(t) = az(bt),\ t \in J,\ a > 0,\ b > 0\ .$$

Then $z_{ab} \in W^n_{p,r}(J)$. A direct calculation, as in the proof of Theorem 1 of section 1, shows that a and b can be so chosen that $\|z_{ab}\|_p = \mu_0$ and $\|z_{ab}^{(k)}\|_q = \mu_k$. Since $Q(a_{ab}) = Q(z) = \mu_k/(\mu_0^\alpha \mu_n^\beta)$, it follows that $\|z_{ab}^{(n)}\|_r = \mu_n$. This completes the proof of Theorem 3.

Remark. Kwong and Zettl [1980] showed that for $1 < p = q = r \le \infty$, $J = R^+$, $n = 2$, $k = 1$, so that $\alpha = \beta = 1/2$, there exists a function $y \in W^2_{p,p}(R^+)$ such that $y \not\equiv 0$ and $Q(y) = K$. Such a function is called an extremal. It is also shown in Kwong and Zettl [1980] that for $J = R$ and $1 < p < \infty$ such extremal functions do not exist. It is known also that for $p = 1$ and $J = R$ or $J = R^+$ extremals do not exist in $W^2_{1,1}$ but there are extremals in this case in an appropriate generalized sense.

4. Growth at Infinity

In this section we extend the asymptotic estimates of Lemma 3.2.

THEOREM 1. Let n denote a positive integer, let f, g be positive nondecreasing functions on R^+. If y is an n times (weakly) differentiable function on R^+ such that

$$fy \in L^p(R^+) \text{ and } gy^{(n)} \in L^r(R^+)$$

for some p, r, $1 \le p,\ r \le \infty$, then

$$y^{(k)}(t) = o(f^{-\alpha}(t)g^{-\beta}(t)),\ t \to \infty\ ,\ k = 0,1,\ldots,n-1 \tag{4.1}$$

where $\alpha = (n-k-r^{-1})/(n-r^{-1}+p^{-1})$, $\beta = 1-\alpha$, unless $p = r = \infty$ in which case we can only conclude that

$$y^{(k)}(t) = O(f^{-\alpha}(t)g^{-\beta}(t))\ ,\ t \to \infty\ ,\ k = 0,1,\ldots,n-1\ . \tag{4.2}$$

Proof. Assume $1 \le p,\ r < \infty$. Since f and g are nondecreasing we have

$$f^p(t)\int_t^\infty |y|^p \le \int_t^\infty |fy|^p\ ,\quad g^r(t)\int_t^\infty |y^{(n)}|^r \le \int_t^\infty |gy^{(n)}|^r\ .$$

Now apply inequality (3.1) to y restricted to the interval (t,∞) with $q = \infty$ to obtain

$$|y^{(k)}(t)| \le \|y^{(k)}\|_{\infty,(t,\infty)} \le K\left[\int_t^\infty |y|^p\right]^{\alpha/p}\left[\int_t^\infty |y^{(n)}|^r\right]^{\beta/r} \le Kf^{-\alpha}(t)g^{-\beta}(t)\left[\int_t^\infty |fy|^p\right]^{\alpha/p}\left[\int_t^\infty |gy^{(n)}|^r\right]^{\beta/r}\ . \tag{4.3}$$

The last two integrals $\to 0$ as $t \to \infty$. Here we have used the fact that the constant K in inequality (3.1) is the same for all half lines.

If $p = \infty$ and $1 \le r < \infty$ note that

$$f(t)\ \|y\|_{\infty,(t,\infty)} \le \|fy\|_{\infty,(t,\infty)}$$

and proceed as above. Similarly for $1 \le p < \infty$ and $r = \infty$. If $p = \infty = r$ then in place of (4.3) we get

$$|y^{(k)}(t)| \le K\, f^{-\alpha}(t)\ g^{-\beta}(t)\ \|fy\|_{\infty,(t,\infty)}\ \|gy^{(n)}\|_{\infty,(t,\infty)}\ . \tag{4.4}$$

Clearly the two norms on the right hand side of (4.4) are nonincreasing functions of t and hence bounded as $t \to \infty$. This completes the proof of Theorem 1.

COROLLARY 1. Let f be a positive nondecreasing function on R^+, n a positive integer and $1 \le p$, $r \le \infty$. If y is an n times (weakly) differentiable function such that $fy \in L^p(R^+)$ and $fy^{(n)} \in L^r(R^+)$ then

$$y^{(k)}(t) = o(1/f(t))\ ,\quad t \to \infty\ ,\quad K = 0,1,\ldots,n-1\ ,$$

unless $p = r = \infty$ in which case we can only conclude that

$$y^{(k)}(t) = O(1/f(t))\ ,\quad t \to \infty\ ,\quad k = 0,1,\ldots,n-1\ .$$

Proof. This is the case $f = g$ in Theorem 1.

The special case $p = q = r = \infty$, $n = 2$, $k = 1$– so that $\alpha = 1/2 = \beta$– of inequality (3.1):

$$\|y'\|_\infty^2 \le K\ \|y\|_\infty\ \|y''\|_\infty \tag{4.5}$$

is called Landau's inequality when $J = R^+$ and $K = 4$ and Hadamard's inequality when $J = R$ and $K = 2$. Our next result extends (4.5). Also our proof is elementary and may be new even when specialized to (4.5).

THEOREM 2. Let $0 \le a < 1$, $J = R$ or $J = R^+$. Suppose $y \in L(J)$, y' is locally absolutely continuous and $y''|y|^a \in L^\infty(J)$. Then $y' \in L^\infty(J)$ and

$$\|y'\|_\infty^2 \le K\ \|y\|_\infty^{1-a}\ \|y''|y|^a\|_\infty \tag{4.6}$$

with $K = 2/(1-a)$ when $J = R$ and $K = 4/(1-a)$ when $J = R^+$.

REMARK 1. When $a = 0$ (4.6), (4.7) reduces to (4.5) for both cases $J = R$ and $J = R^+$ with the same constant K. These constants are known to be best possible when $a = 0$, see the survey paper by Kwong and Zettl [1980b]. The proof given below involves improper integrals when $0 < a < 1$ but not for $a = 0$. Note also that Theorem 1 is an extension of the Landau and Hadamard inequalities since $y''|y|^a$ might be bounded for some a, $0 < a < 1$ without y'' being bounded.

Proof. First consider the case $J = R^+$. It suffices to show that for any $x_0 \in R^+$,

$$|y'(x_0)| \le K\ \|y\|_\infty^{1-a}\ \|y''|y|^a\|_\infty\ . \tag{4.7}$$

Suppose first that y' has a zero in R^+. If $y'(x_0) = 0$ then (4.2) clearly holds. If $y'(x_0) \ne 0$ there must be a zero of y', say x_1, nearest to x_0. Assume $x_0 < x_1$; the case $x_1 < x_0$ is treated similarly. Note that y' is of constant sign, say positive on $[x_0, x_1]$. (If y' is negative replace y by $-y$.) Now

$$|y'(x_0)|^2 = |y'(x_1)|^2 - \int_{x_0}^{x_1} 2y'(x)y''(x)dx \le 2\int_{x_0}^{x_1} y'(x)|y|^{-a}(x)|y''(x)||y|^a(x)|dx \tag{4.8}$$

$$\le 2\|y''|y|^a\|_\infty \int_{x_0}^{x_1} y'(x)|y|^{-a}(x)dx \,.$$

If y has no zero in $[x_0, x_1]$, then

$$\int_{x_0}^{x_1} y'(x)|y|^{-a}(x)dx = (1-a)^{-1}||y|^{1-a}(x_1) - |y|^{1-a}(x_0)| \,. \tag{4.9}$$

If y has a zero x_2 in $[x_0, x_1]$, then the integral in (4.8) is improper. By dividing the interval of integration into $[x_0, x_2]$ and $[x_2, x_1]$ we see that

$$\int_{x_0}^{x_1} y'(x)|y|^{-a}(x)dx = (1-a)^{-1}[|y|^{1-a}(x_1) + |y|^{1-a}(x_0)] \,. \tag{4.10}$$

In both cases we have

$$\int_{x_0}^{x_1} y'(x)|y|^{-a}(x)dx \le 2(1-a)^{-1} \|y\|_\infty^{1-a} \,. \tag{4.11}$$

Thus (4.2) is established in case y' has a zero in R^+.

Suppose y' has no zero in R^+. Then y' is of constant sign and so we may assume y' is positive. By the mean value theorem there is a point x_n in $[0, n]$ such that

$$y'(x_n) = n^{-1}(y(n) - y(0)) \le 2n^{-1} \|y\|_\infty \to 0 \text{ as } n \to \infty \,.$$

Now by repeating the arguments above using x_n in place of x_1 inequality (4.3) becomes

$$|y'(x_0)|^2 \le |y'(x_n)|^2 + 2\,\|y''|y|^a\| \int_{x_0}^{x_n} y'(x)|y|^{-a}(x)dx \le |y'(x_n)|^2 + K\,\|y''|y|^a\|_\infty \|y\|_\infty^{1-a} \,.$$

Letting $n \to \infty$ completes the proof for $J = R^+$.

In the case $J = R$ the arguments are similar with the additional observation that x_1 or x_n can always be chosen so that y has no zeros in (x_0, x_1) and (x_0, x_n) (or $x_1, x_0)$, (x_n, x_0)). Let us establish this claim for one case; the other cases are similar. Suppose y' has a zero both to the right and to the left of x_0. If $y'(x_0) = 0$, (4.2) holds. If $y'(x_0) \ne 0$, choose the two nearest zeros of y', $x_1 < x_0 < \overline{x_1}$, one on each side of x_0. We claim that either (x_1, x_0) or $(x_0, \overline{x_1})$ contains no zero of y. Suppose not, and $t_1 \in (x_1, x_0)$ $t_2 \in (x_0, x_1)$ are zeros of y. Rolle's theorem then yields a zero of y' in (t_1, t_2) contradicting the choice of x_1 and $\overline{x_1}$. With this observation we see that (4.5) does not occur in this case when x_1 and x_n are properly chosen and (4.4) becomes

$$\int_{x_0}^{x_1} y'(x)|y|^{-a}(x)dx \le (1-a)^{-1} \|y\|_\infty^{1-a} \,.$$

Theorem 2 follows from this and (4.3).

REMARK. Theorem 2 is not valid when $a = 1$. To see this consider the initial value problem

$$y''y = -2, \ y(0) = 1, \ y'(0) = 0 . \tag{4.12}$$

The solution of (4.12) is given implicitly by

$$t = \int_y^1 (-\log y)^{-1/2} dy$$

on $[0, t_0]$ with $t_0 = \int_0^1 (-\log y)^{-1/2} dy$. Now extend y to $[-t_0, t_0]$ as an even function and then to the whole real line as a periodic function of period $2t_0$. In $[0, t_0]$, we have $y' < 0$, $y'^2 = (-\log y)$. Hence y is decreasing on $[0, t_0]$ and so $\|y\|_\infty = y(0) = 1$. On the other hand y' is not bounded since $\lim |y'(t)| = \infty$ as $t \to t_0$.

5. The Difference Operator

In this section the difference operator is considered on the classical l^p spaces. Let $Z = \{...-2,-1,0,1,2,...\}$, let $Z^+ = \{0,1,2,...\}$. With $M = Z$ or $M = Z^+$ the difference operator Δ is defined by

$$x = (x_m)_{m\in M}, \ \Delta x = (x_{m+1} - x_m)_{m\in M} ,$$

$$\Delta^2 = \Delta(\Delta), \ \Delta^j = \Delta(\Delta^{j-1}), \ j = 2,3,\dots .$$

In contrast with the derivative operator, Δ is a bounded linear operator defined on all of $l^p(M)$ and onto $l^p(M)$ for any p with $1 \le p \le \infty$ and $M = Z$ or $M = Z^+$.

The discrete analogue of inequality (3.1) has apparently only quite recently been established.

THEOREM 1. (Kwong and Zettl [1988]) Let $M = Z$ or $M = Z^+$. Let k, n be integers with $1 \le k < n$. Let $1 \le p, r \le \infty$. There is a positive constant C such that

$$\|\Delta^k x\|_q \le C \ \|x\|_p^\alpha \ \|\Delta^n x\|_r^\beta \tag{5.1}$$

holds for all $x \in l^p(M)$ satisfying $\Delta^n x \in l^r(M)$ if and only if

$$nq^{-1} \le (n-k)p^{-1} + kr^{-1} \tag{5.2}$$

and

$$\alpha = (n-k-r^{-1}+q^{-1})/(n-r^{-1}+p^{-1}) , \ \beta = 1-\alpha . \tag{5.3}$$

Proof. See Kwong and Zettl [1988]. The proof for the discrete case is different from the one in the continuous case since the latter does not seem to be adaptable to the former.

Given p, q, r, $1 \le p, q, r \le \infty$, n, k, α, β satisfying the conditions of the theorem there clearly is a smallest positive constant C such that inequality (5.1) holds for all $x \in l^p(M)$ with $\Delta^n x \in l^r(M)$. This smallest i.e. best constant is denoted by $C = C(n, k, p, q, r; M)$? Surprisingly there seems to be less known about the discrete constants C than there is for the continuous ones K. Below we survey results about C. Many of these are quite recent. Let $C(n, k, p, p, p; M) = C(n, k, p; M)$.

THEOREM 2. Assume the hypotheses and notation of Theorem 1. Then

$$C(n, k, p, q, r; \genfrac{}{}{0pt}{}{Z}{Z^+}) \geq K(n, k, p, q, r; \genfrac{}{}{0pt}{}{R}{R^+}) . \tag{5.4}$$

Proof. Ditzian [1983] proved some special cases of (5.4). The general case can be proved similarly. The basic idea here is to approximate the derivatives with differences.

An obvious question is

Question. Does equality hold in (5.4)? In other words is it true that, under the conditions on p, q, r, n, k given in Theorem 1, we have

$$C(n, k, p, q, r; Z) = K(n, k, p, q, r; R)?$$

and

$$C(n, k, p, q, r; Z^+) = K(n, k, p, q, r; R^+)?$$

Ditzian [1983] showed that the answer, in general, is no. He found that

$$C(3, 1, \infty; Z) = 2^{1/3} > (9/8)^{1/3} = K(3, 1, \infty; R) . \tag{5.5}$$

Below we summarize the known cases of equality and inequality in (5.4) as far as we are aware of them. Proofs can be found in the references given for each case.

$$I. \;\; C(n, k, \infty; Z^+) = K(n, k, \infty; R^+), \; 1 \leq k < n . \tag{5.6}$$

Equality (5.6) follows from an abstract result. If A is an m-dissipative linear but not necessarily bounded operator on a Banach space X then

$$\|A^k x\| \leq K(n, k, \infty; R^+) \, \|x\|^{(n-k)/n} \, \|A^n x\|^{k/n} , \quad x \in D(A^n) , \; 1 \leq k < n . \tag{5.7}$$

A linear (bounded or unbounded) operator A on a Banach space X is m-dissipative if and only if it is the infinitesimal generator of a strongly continuous C_0 contraction semigroup. If A is the infinitesimal generator of a strongly continuous group of isometries then (5.7) can be strengthened to

$$\|A^k x\| \leq K(n, k, \infty; R) \, \|x\|^{(n-k)/n} \, \|A^n x\|^{k/n} , \quad x \in D(A^n) , \;\; 1 \leq k < n . \tag{5.8}$$

Inequalities (5.7) and (5.8) are established in Certain and Kurtz [1975]. The case $n = 2$ was previously established by Kallman and Rota [1971].

If X is a Hilbert space then (5.7) can be improved to

$$\|A^k x\| \leq K(n, k, 2, R^+) \, \|x\|^{(n-k)/n} \, \|A^n x\|^{k/n} , \quad x \in D(A^n) , \; 1 \leq k < n . \tag{5.9}$$

The case $n = 2$ of (5.9) was proven by Kato [1971] and the general case was proven independently by Chernoff [1979], Phong [1981] and Kwong and Zettl [1979].

Since Δ is an m-dissipative operator on any of the spaces $l^p(M)$, $1 \leq p \leq \infty$, $M = Z$ or $M = Z^+$, the inequalities (5.4), (5.7), (5.8), (5.9) yield some information about the constants $C(n, k, p; M)$. In particular equality (5.6) follows from (5.4) and (5.8).

$$II. \;\; C(n, k, 2; Z^+) = K(n, k, 2; R^+) , \; 1 \leq k < n . \tag{5.10}$$

Proof. This follows from (5.4) and (5.9).

$$III.\ \ C(n, k, 2; Z) = K(n, k, 2; R) = 1\ ,\ 1 \le k < n\ . \tag{5.11}$$

Proof. The case $n = 2$ of the equality on the right is established in the classic book by Hardy, Littlewood and Polya [1934]. The general case follows by induction. The case $n = 2$ of the left equality was proven by Copson [1977] and the general case follows by induction.

$$IV.\ \ C(n, n-1, \infty; Z) = K(n, n-1, \infty; R)\ ,\ \ n = 2,3,4,\ldots\ . \tag{5.12}$$

Proof. This was established by Ditzian [1983]. The proof depends on properties of B-splines.

$$V.\ \ C(n, n-1, 1; Z) = K(n, n-1, 1; R)\ ,\ \ n = 2,3,4,\ldots\ . \tag{5.13}$$

Proof. This follows from (5.4), IV, and VI.

$$VI.\ \ K(n, k, 1; R) = K(n, k, \infty; R)\ ,\ \ 1 \le k < n\ . \tag{5.14}$$

Proof. See Ditzian [1975].

$$VII.\ \ K(n, k, p; R^+) \le K(n, k, \infty; R^+)\ ,\ \ 1 \le p \le \infty\ . \tag{5.15}$$

Proof. This is a consequence of the abstract inequality (5.7).

$$VIII.\ \ K(n, k, p; R) \le K(n, k, \infty; R)\ ,\ \ 1 \le p \le \infty\ . \tag{5.16}$$

Proof. This is a consequence of (5.8).

$$IX.\ \ C(n, k, p; Z^+) \le C(n, k, \infty; Z^+)\ ,\ \ 1 \le p \le \infty\ . \tag{5.17}$$

Proof. This is a consequence of (5.7) and (5.6). See also Ditzian [1975] and Kwong and Zettl [1987].

$$X.\ \ C(n, k, p; Z) \le C(n, k, \infty; Z)\ ,\ \ 1 \le p \le \infty\ . \tag{5.18}$$

In contrast with (5.16) this result does not seem to follow from the theory of semigroups of operators. A proof is given in Ditzian [1975] — see also Kwong and Zettl [1987].

The discrete analogue of (5.14) does not seem to be known. We raise:

Question. Is $C(n, k, 1; Z) = C(n, k, \infty; Z)$ for all n, k satisfying $1 \le k < n$? The answer is yes when $k = n-1$. This follows from (5.13), (5.14) and (5.12). In addition the answer is also yes in all cases when both of these constants are known explicitly. See below.

Below we summarize cases for which the exact values of K or C are known explicitly. We do this only for the special case when $p = q = r$. See the survey paper by Kwong and Zettl [1980] for explicit values of K in some cases when not all three norms are the same. Explicit values of C do not seem to be known when not all three norms are equal.

				Known Cases	
n	k	p	J or M	K or C	Author reference
2	1	∞	R^+	2	Landau [1913]
2	1	∞	Z^+	2	Gindler and Goldstein [1981]
2	1	∞	R	$2^{1/2}$	Hadamard [1914]
2	1	∞	Z	$2^{1/2}$	Ditzian and Neumann [1986]. See also Kaper and Spellman [1987].
2	1	2	R^+	$2^{1/2}$	Hardy and Littlewood [1932]
2	1	2	Z^+	$2^{1/2}$	Copson [1979]
2	1	2	R	1	Hardy, Littlewood and Pólya [1934]
2	1	2	Z	1	Copson [1979]
2	1	1	R^+	$(5/2)^{1/2}$	Berdyshev [1971]
2	1	1	Z^+	$(5/2)^{1/2}$	Kwong and Zettl [1986]
2	1	1	R	$2^{1/2}$	Ditzian [1975], Berdyshev [1971]
2	1	1	Z	$2^{1/2}$	Kwong and Zettl [1986]
3	1	∞	R	$(9/8)^{1/3}$	Shilov [1937]
*3	1	∞	Z	$2^{1/3}$	Ditzian and Neumann [1986]
3	2	∞	R	$3^{1/3}$	Shilov [1937]
3	2	∞	Z	$3^{1/3}$	Ditzian [1983]
4	1	∞	R	$(512/375)^{1/4}$	Shilov [1937]
*4	1	∞	Z	$2^{1/2}$	Ditzian and Neumann [1986]. See also Kaper and Spellman [1987].
4	2	∞	R	$(36/25)^{1/4}$	Shilov [1937]
*4	2	∞	Z	$(4/3)^{1/2}$	Ditzian and Neumann [1986]. See also Kaper and Spellman [1987].
4	3	∞	R	$(24/5)^{1/4}$	Shilov [1937]
4	3	∞	Z	$(24/5)^{1/4}$	Ditzian [1983]
5	1	∞	R	$(1953125/1572864)^{1/5}$	Kolmogorov [1938]
*5	1	∞	Z	$4^{1/5}$	Ditzian and Neumann [1986]. See also Kaper and Spellman [1987].
5	2	∞	R	$(125/72)^{1/5}$	Shilov [1937]
*5	2	∞	Z	$(4/3)^{4/5}$	Ditzian and Neumann [1986]. See also Kaper and Spellman [1987].
5	3	∞	R	$(225/128)^{1/5}$	Kolmogorov
*5	3	∞	Z	$2^{1/5}$	Kwong and Zettl [1988]
5	4	∞	R	$(15/2)^{1/5}$	Kolmogorov [1938]
5	4	∞	Z	$(15/2)^{1/5}$	Ditzian [1983]
n	k	∞	R	K_1	Kolmogorov [1938]
n	n-1	∞	Z	K_1	Ditzian [1983]
3	1	∞	R^+	$(243/8)^{1/3}$	Schoenberg and Cavaretta [1970]
3	2	∞	R^+	$(24)^{1/3}$	Schoenberg and Cavaretta [1970]
n	k	∞	R^+	K_2	Schoenberg and Cavaretta [1970]
n	k	∞	Z^+	K_2	Certain and Kurtz [1977] See also Gindler and Goldstein [1981]
n	k	2	R	1	Hardy, Littlewood and Pólya [1934]
n	k	2	Z	1	Follows from special case $n = 2$ established by Copson [1979]
3	1	2	R^+	$3^{1/2}[2(2^{1/2}-1)]^{1/3}$	Ljubic [1960]
3	2	2	R	$3^{1/2}[2(2^{1/2}-l)]^{1/3}$	Ljubic [1960]
n	k	2	R^+	K_3	Ljubic [1960]
n	k	2	Z^+	$K(n, k, 2, R^+)$	Chernoff [1979], Phong [1981], Kwong and Zettl [1979]
n	k	1	R	K_1	Ditzian [1975]
3	1	1	Z	$2^{1/3}$	Kwong and Zettl [1988]
3	2	1	Z	$3^{1/3}$	Kwong and Zettl [1988]
4	1	1	Z	$2^{1/2}$	Kwong and Zettl [1988]
4	2	1	Z	$(4/3)^{1/2}$	Kwong and Zettl [1988]
4	3	1	Z	$(24/5)^{1/4}$	Kwong and Zettl [1988]
5	1	1	Z	$4^{1/5}$	Kwong and Zettl [1988]
5	2	1	Z	$(4/3)^{4/5}$	Kwong and Zettl [1988]
5	3	1	Z	$2^{1/5}$	Kwong and Zettl [1988]
5	4	1	Z	$(15/2)^{1/5}$	Kwong and Zettl [1988]

Here

$$K_1 = K(n, k, \infty; R) = k_{n-k}\, k_n^{-(n-k)/n} ,$$

$$k_i = 4\pi^{-1}\sum_{j=0}^{\infty}(-1)^j(2j+1)^{-i-1} \text{ for even } i$$

$$k_i = 4\pi^{-1}\sum_{j=0}^{\infty}(2j+1)^{-i-1} \text{ for odd } i .$$

$K_2 = K(n, k, \infty; R^+)$: These constants have been characterized by Schoenberg and Cavaretta in terms of norms of Euler splines but an explicit formula for all n and k is not available.

$K_3 = K(n, k, 2; R^+)$: An algorithm to compute these constants for all $k=1,\dots,n-1$, $n=2,3,\dots$ was developed by Ljubic [1960]. This was refined elegantly by Kupcov [1977]. See Franco, Kaper, Kwong and Zettl [1985] for a numerical computation of K_3 and for some asymptotic estimates.

Comments on the above list of best constants. See Kwong and Zettl [1980] for some information on best constants when not all three norms are equal. There are many gaps in our knowledge of K and C. We mention a few below.

1. The values of $K(n, k, 1; R^+)$ seem to be known explicitly only for $n = 2$, $k = 1$. Note that although $K(n, k, 1; R) = K(n, k, \infty; R)$ the corresponding result for R^+ is false: $(5/2)^{1/2} = K(2,1,1; R^+) \neq K(2,1,\infty;R^+) = 2^{1/2}$. Kwong and Zettl [1980] proved that $K(2, 1, p; R^+)$ is a continuous function of p. Hence $K(2, 1, p; R^+) \neq K(2, 1, q; R^+)$ where $p^{-1}+q^{-1} = 1$ for some values of p other than $p = 1$ and $p = \infty$.

2. The stars * indicate cases when $C > K$. It is interesting to observe that, although $C(n, n-1, \infty; Z) = K(n, n-1, \infty; R)$ (and $C(n, k, \infty; Z^+) = K(n, k, \infty; R^+)$ for all n, k, $1 \le k < n$), we have $C(n, k, \infty; Z) > K(n, k, \infty; R)$ in all cases when $1 \le k < n-1$ for which these two constants are explicitly known. So the question arises.

   Question. Is $C(n, k, \infty; Z) > K(n, k, \infty; R)$ for all n, k satisfying $1 \le k < n-1$?

   A closely related question is:

   Question. Is $C(n, k, 1; Z) > K(n, k, 1; r)$ for all n, k satisfying $1 \le k < n-1$?

3. For general n, k satisfying $1 \le k < n$ the constants $K(n, k, \infty; R)$, $K(n, k, 2; R)$, $K(n, k, 1; R)$, $C(n, k, 2; Z)$ are known explicitly; $K(n, k, \infty; R^+)$, $C(n, k, \infty; Z^+)$, $K(n, k, 2; R^+)$ and $C(n, k, 2; Z^+)$ can be computed numerically with known algorithms but $C(n, k, \infty; Z)$, $C(n, k, 1; Z)$, $K(n, k, 1; R^+)$ and $C(n, k, 1; Z^+)$ do not seem to be known.

References

R. A. Adams [1975], "Sobolev Spaces," Academic Press, New York (1975).

V. V. Arestov [1972], "Exact inequalities between norms of functions and their derivatives," Acta Scientiarum Mathematicarum, 33, 243-267 (1972).

__________ [1967], "The best approximation to differentiation operators," Matematicheskie Zametkie, 1, part 2, 149-154 (1967).

__________, [1972], "On the best approximation of the operators of differentiation and related questions," Approximation Theory, Proceedings of Conference in Poznan, Poland. Reidel Publishing Co., Boston (1972).

V. I. Berdyshev [1971], "The best approximation in L(0,∞) to the differentiation operator," Matematicheskie Zametki, 5, 477-481 (1971).

B. Bollabas and J. R. Partington [1984], "Inequalities for quadratic polynomials in hermitian and dissipative operators," Advances in Mathematics (1984), 51, 271-280 (1984).

Yu. G. Bosse (G. E. Shilov) [1937], "On neravenstvakh mezhdu proizvodnymi," Mosk. Univ. Sbornik raport nauchnykh studencheskikh kruzhkov, 17-27 (1937).

J. Bradley and W. N. Everitt [1974], "On the inequality $\|f''\|^2 \le K\,\|f\|\,\|f^{(4)}\|$," Quart. J. Math. 25, 241-252 (1974).

________________________ [1973], "Inequalities associated with regular and singular problems in the calculus of variations," Trans. Amer. Math. Soc. 182, 303-321 (1973).

K. W. Brodlie and W. N. Everitt [1975], "On an inequality of Hardy and Littlewood," Proc. Roy. Soc. Edinburgh A72, 179- 186 (1975).

H. Cartan [1939], "On inequalities between the maxima of the successive derivatives of a function," Comptes Rendus Sci. Acad. 208, 414-426 (1939).

A. S. Cavaretta [1974], "An elementary proof of Kolmogorov's theorem," Amer. Math. Monthly 81, 480-486 (1974).

____________ [1976], "One-sided inequalities for the successive derivatives of a function," Bull. Amer. Math. Soc. 82, 303-305 (1976).

____________, "A refinement of Kolmogorov's inequality," MRC Technical Summary Report #1788.

Ceisielski and J. Musielak [1972], "Approximation Theory," Proceedings of the Conference jointly organized by the Mathematical Institute of the Polish Academy of Sciences and the Institute of Mathematics of the Adam Mickiewicz University held in Proznan 22-26 August, 1972. D. Reidel Publishing Co., Boston (1972).

M. W. Certain and T. G. Kurtz [1977], "Landau-Kolmogorov inequalities for semigroups and groups," Proc. Amer. Math. Soc. 63, 226-230 (1977).

E. T. Copson [1977], "On two integral inequalities," Proc. Roy. Soc. Edinburgh, 77A, 325-328 (1977).

__________ [1977], "On two inequalities of Brodlie and Everitt," Proc. Roy. Soc. Edinburgh, 77A, 329-333 (1977).

__________ [1979], "Two series inequalities," Proc. Roy. Soc. Edinburgh 83A, 109-114 (1979).

Z. Ditzian [1975], "Some remarks on inequalities of Landau and Kolmogorov," Equ. Math. 12, 145-151 (1975).

______________ [1977], "Note on Hille's question," Aequationes Mathematicae 15, 143-144 (1977).

______________ [1983], "Discrete and shift Kolmogorov type inequalities," Proc. Roy. Soc. Edinburgh 93A, 307-317 (1983).

Z. Ditzian and D. J. Newman [1986], "Discrete Kolmogorov-type inequalities," preprint.

W. D. Evans and A. Zettl [1978], "Norm inequalities involving derivatives," Proc. Roy. Soc. Edinburgh, 82A, 51-70 (1978).

W. N. Everitt [1972], "On an extension to an integro- differential inequality of Hardy, Littlewood and Polya," Proc. Roy. Soc. Edinburgh (A) 69, 295-333 (1972).

W. N. Everitt and M. Giertz [1974], "On the integro-differential inequality $\|f'\|_2^2 \leq K\,\|f\|_p\,\|f''\|_q$," J. Math. Anal. and Appl. 45, 639-653 (1974).

______________________________ [1972], "Some inequalities associated with certain ordinary differential operators," Math. Z. 126, 308-326 (1972).

______________________________ [1974], "Inequalities and separation for certain ordinary differential operators," P. London Math. Soc. 3, 28, 352-372 (1974).

W. N. Everitt and D. S. Jones, "On an integral inequality," Proc. Roy. Soc. London (A), (to appear).

W. N. Everitt and A. Zettl [1978], "On a class of integral inequalities," J. London Math. Soc. (2), 17, 291-303 (1978).

A. M. Fink [1977], "Best possible approximation constants," Trans. Amer. Math. Soc. 226, 243-255 (1977).

Z. M. Franco, H. G. Kaper, M. K. Kwong and A. Zettl [1983], "Bounds for the best constant in Landau's inequality on the line," Proc. Roy. Soc. Edinburgh 95A, 257-262 (1983).

____ [1985], "Best constants in norm inequalities for derivatives on a half-line," Proc. Roy. Soc. Edinburgh 100A, 67-84 (1985).

A. Friedman [1969], "Partial Differential Equations," New York (1969).

V. N. Gabushin [1967], "Inequalities for norms of a function and its derivatives in L_p metrics," Mat. Zam. 1, 291-298 (1967).

____________ [1968], "Exact constants in inequalities between norms of derivatives of functions," Mat. Zam. 4, 221-232 (1968).

____________ [1969], "The best approximation for differentiation operators on the half-line," Mat. Zam. 6, 573-582 (1969).

A. Gindler and J. A. Goldstein [1975], "Dissipative operator versions of some classical inequalities," J. D'Analyse Math. 28, 213-238 (1975).

________________________ [1981], "Dissipative operators and series inequalities," Bull. Austr. Math. Soc. 23, 429-442 (1981).

J. Goldstein, "On improving the constants in the Kolmogorov inequalities," preprint.

A. Gorny [1939], "Contributions to the study of differentiable functions of a real variable," Acta Math. 71, 317-358 (1939).

J. Hadamard [1914], "Sur le module maximum d'une fonction et de ses derivees," C. R. des Seances de l'année 1914, Soc. Math. de France, 66-72 (1914).

G. H. Hardy and J. E. Littlewood [1932], "Some integral inequalities connected with the calculus of variations," Quart. J. Math. Oxford Ser. 2, 3, 241-252 (1932).

G. H. Hardy, J. E. Littlewood and G. Polya [1934], "Inequalities," Cambridge (1934).

E. Hille [1972], "Generalizations of Landau's inequality to linear operators," Linear operators and approximation (edited by P. L. Butzer, et al.), Birkhauser Verlag, Basel and Stuttgart (1972).

______ [1970], "Remark on the Landau-Kallman-Rota inequality," Aequationes Mat. 4, 239-240 (1970).

______ [1972], "On the Landau-Kallman-Rota inequality," J. of Approx. Theory 6, 117-122 (1972).

E. Hille and R. S. Phillips [1957], "Functional Analysis and Semigroups," Amer. Math. Soc. Coll. Publ. 31, Rev. ed. Providence (1957).

J. A. R. Holbrook, "A Kallman-Rota-Kato inequality for nearly euclidean spaces," preprint.

R. R. Kallman and G. C. Rota [1970], "On the inequality $\|f'\|^2 \leq 4\,\|f\|\,\|f''\|$,"" Inequalities II (O. Shisha, ed.), Academic Press, New York, 187-192 (1970).

H. G. Kaper and B. E. Spellman [1987], "Best constants in norm inequalities for the difference operator," Trans. of the Amer. Math. Soc. 299, No. 1 (1987).

T. Kato [1971], "On an inequality of Hardy, Littlewood and Polya," Advances in Math. 7, 217-218 (1971).

A. N. Kolmogorov [1962], "On inequalities between the upper bounds of the successive derivatives of an arbitrary function on an infinite interval," Amer. Math. Soc. transl. (1) 2, 233-243 (1962).

N. P. Kupcov [1977], "Kolmogorov estimates for derivatives in $L^2(0,\infty)$," Proceedings of the Steklov Institute of Mathematics 138, AMS transl., 101-125 (1977).

S. Kurepa [1970], "Remarks on the Landau inequality," Aequations Math. 4, 240-241 (1970).

M. K. Kwong and A. Zettl [1979], "Remarks on best constants for norm inequalities among power of an operator," J. Approx. Theory, 26, 249-258 (1979).

______________________ [1979], "Norm inequalities for dissipative operators on inner product spaces," Houston J. Math 5, 543-557 (1979).

______________________ [1979], "An extension of the Hardy-Littlewood inequality," Proc. Amer. Math. Soc. 77, 117-118 (1979).

______________________ [1980], "Ramifications of Landau's inequality," Proc. Roy. Soc. Edinburgh 86A, 175-212 (1980).

______________________ [1980], "Norm inequalities for derivatives," Lecture Notes in Mathematics, Springer Verlag 846, 227-243 (1980).

______________________, "Landau's inequality," Rocky Mountain J. Math. (to appear).

______________________, "Norm inequalities of product form is weighted L^p spaces."

______________________ [1986], "Landau's inequality for the differential and difference operators," General inequalities V., Birkhauser Verlag, Basel. (to appear). Proceedings of the fifth International Conference on General Inequalities.

______________________ [1987], "Best constants for discrete Kolmogorov inequalities," Houston J. Math. (to appear). ______________________ [1988], "Landau's inequality for the difference operator," Proc. Amer. Math. Soc. (to appear).

E. Landau [1913], "Einige Ungleichungen fur zweimal differenzierbare Funktionen," Proc. London Math. 13, 43-49 (1913).

Ju I. Ljubic (or Yu Lyubich) [1960 and 1964], "On inequalities between the powers of a linear operator," Transl. Amer. Math. Soc. Ser. (2) 40(1964), 39-84; translated from Irv. Akad. Nauk. SSSR Ser. Mat. 24(1960), 825-864.

G. G. Magaril-Il'jaev and V. M. Tihomirov [1981], "On the Kolmorogov inequality for fractional derivatives on the half-line," Analysis Mathematica 7, 37-47 (1981).

A. P. Matorin [1955], "Inequalities between the maximum absolute values of a function and its derivatives on the half-line," Ukrain. Mat. Zh. 7, 262-266 (1955).

D. S. Mitrinovic [1970], "Analytic Inequalities," Springer-Verlag, Berlin (1970).

L. Nirenberg [1955], "Remarks on strongly elliptic partial differential equations," Comm. Pure Appl. Math. 8, 648-674 (1955).

B. Sz. Nagy [1941], "Uber Integralungleichungen zwischen einer Funktion and ihrer Ableitung," Acta. Sci. Math. 10, 64-74 (1941).

B. Neta, "On determination of best possible constants in integral inequalities involving derivatives," preprint.

J. R. Partington [1981], "Hadamard-Landau inequalities in uniformly convex spaces," Math. Proc. Camb. Phil. Soc., 90, 259-264 (1981).

____________ [1979], "Constants relating a Hermitian operator and its square," Math. Proc. Camb. Phil. Soc. 85, 325-333 (1979).

R. M. Redheffer [1963], "Uber eine beste Ungleichung zwischen den Normen von f, f', f''," Math. Zeitschr. 80, 390-397 (1963).

I. J. Schonberg [1973], "The elementary cases of Landau's problem of inequalities between derivatives," Amer. Math. Monthly, 80, 121-158 (1973).

I. J. Schonberg and A. Cavaretta [1970], "Solution of Landau's problem concerning higher derivatives on the half line," MRCT. S. R. 1050, Madison, Wisconsin (1970).

G. E. Shilov [1937], "O neravenstvah merzdu proizvodnymi," Sbornik Rabot Studenceskih Naucnyh Kruzhov Moskovskogo Gosudarstvennogo Universiteta, 17-27 (1937).

V. G. Solyer [1976], "On an inequality between the norms of a function and its derivative," Izvestia Vysshikh Ucbebaykh Zavedemy Matematika 2 (165) (1976).

S. B. Stechkin [1965], "Inequalities between norms of derivatives of an arbitrary function," Acta. Sci. Math. 26, 225-230 (1965).

____________ [1967], "The inequalities between upper bounds for the derivatives of an arbitrary function on the half-line," Mat. Zametki I V. 6, 665-674 (1967).

E. M. Stein [1957], "Functions of exponential type," Annals of Math. (2), 65, 582-592 (1957).

L. V. Taikov [1968], "Inequalities of Kolmogorov type and the best formulae for numerical differentiation," R. Mat. Zam. 4, 233-238 (1968).

W. Trebels and V. I. Westphal [1972], "A note on the Landau-Kallman-Rota-Hille inequality," Linear Operators and Approximation (edited by P. L. Butzer et al), Birkhauser Verlag, Basel and Stuttgart (1972).

7 Bounds on Schrödinger Operators and Generalized Sobolev-Type Inequalities with Applications in Mathematics and Physics

Elliott H. Lieb Princeton University, Princeton, New Jersey

Start with the usual Sobolev inequality on $\mathbf{R}^n, n \geq 3$:

$$\int_{\mathbf{R}^n} |\nabla f|^2 \geq S_n \left\{ \int_{\mathbf{R}^n} |f|^{2n/(n-2)} \right\}^{(n-2)/n} = S_n \|f\|^2_{2n/(n-2)}. \tag{1}$$

Apply Hölder's inequality to the right side to obtain

$$\int_{\mathbf{R}^n} |\nabla f|^2 \geq K_n^1 \left\{ \int_{\mathbf{R}^n} \rho^{(n+2)/n} \right\} \Big/ \left\{ \int_{\mathbf{R}^n} \rho \right\}^{2/n} = K_n^1 \|\rho\|^{(n+2)/n}_{(n+2)/n} \|\rho\|_1^{-2/n} \tag{2}$$

with $\rho(x) \equiv |f(x)|^2$. The superscript 1 on K_n^1 indicates that in (2) we are considering only one function, f. Hölder's inequality implies that $K_n^1 \geq S_n$ but, in fact, the *sharp* value of K_n^1 (which can be obtained by solving a nonlinear PDE) is larger than S_n. In particular, $K_n^1 > 0$ for *all* $n \geq 1$, even though $S_n = 0$ for $n < 3$.

Inequality (2), unlike (1) has the following important property: The non-linear term $\int \rho^{(n+2)/n}$ enters with the power 1 (and not $(n-2)/n$) and is therefore "extensive." The price we have to pay for this is the factor $\|f\|_2^{4/n} = \|\rho\|_1^{2/n}$ in the denominator, but since we shall apply (2) to cases in which $\|f\|_2 = 1$ (L^2 normalization condition) this is not serious.

Inequality (2) is equivalent to the following: Consider the Schrödinger operator on $\mathbf{R}^n$

$$H = -\Delta - V(x) \tag{3}$$

and let $e_1 = \inf \operatorname{spec}(H)$. (We assume H is self-adjoint.) Let $V_+(x) \equiv \max\{V(x), 0\}$. Then

$$e_1 \geq -L_{1,n}^1 \int V_+(x)^{(n+2)/n} dx = -L_{1,n}^1 \|V_+\|^{(n+2)/2}_{(n+2)/2} \tag{4}$$

with

$$L_{1,n}^1 = \left(\frac{n}{2K_n^1} \right)^{n/2} \left(1 + \frac{n}{2} \right)^{-(n+2)/n} \tag{5}$$

The reason for the subscript 1 in $L^1_{1,n}$ will be clarified in eq. (8).

Here is the proof of the equivalence. We have

$$e_1 \geq \inf_f \left\{ \int |\nabla f|^2 - \int \rho V_+ \;\middle|\; \|f\|_2 = 1 \text{ and } \rho = |f|^2 \right\}.$$

Use (2) and Hölder to obtain (with $X = \|\rho\|_{(n+2)/n}$)

$$e_1 \geq \inf_X \left\{ K^1_n X^{(n+2)/n} - \|V_+\|_{(n+2)/2} X \right\} \tag{6}$$

Minimizing (6) with respect to X yields (4). To go from (4) to (2), take $V = V_+ = \alpha|f|^{4/n} = \alpha\rho^{2/n}$ in (3). Then

$$-L^1_{1,n}\alpha^{(n+2)/2} \int \rho^{(n+2)/n} \leq e_1 \leq (f, Hf) = \int |\nabla f|^2 - \alpha \int \rho^{(n+2)/n}.$$

Optimizing this with respect to α yields (2).

So far this is trivial, but now we turn to a more interesting question. Let $e_1 \leq e_2 \leq \ldots \leq 0$ be the negative spectrum of H (which may be empty). Is there a bound of the form

$$\sum e_i \geq -L_{1,n} \int V_+(x)^{(n+2)/2} dx \tag{7}$$

for some universal, V independent, constant $L_{1,n} > 0$ (which, of course, is $\geq L^1_{1,n}$)? The point is that the right side of (7) has the same form as the right side of (4). More generally, given $\gamma \geq 0$, does

$$\sum |e_i|^\gamma \leq L_{\gamma,n} \int V_+(x)^{\gamma+\frac{n}{2}} \tag{8}$$

hold for suitable $L_{\gamma,n}$? When $\gamma = 0, \sum |e_i|^0$ is interpreted as the number of $e_i \leq 0$.

The answer to these questions is yes in the following cases:

$\underline{n = 1}$: All $\gamma > \frac{1}{2}$. The case $\gamma = 1/2$ is unsettled. For $\gamma < \frac{1}{2}$, examples show there can be no bound of the form (8).

$\underline{n = 2}$: All $\gamma > 0$. There can be no bound when $\gamma = 0$.

$\underline{n \geq 3}$: All $\gamma \geq 0$.

The cases $\gamma > 0$ were first done in [10], [11]. The $\gamma = 0$ case for $n \geq 3$ was done in [3], [6], [14], with [6] giving the best estimate for $L_{0,n}$. For a review of what is currently known about these constants and conjectures about the sharp values of $L_{\gamma,n}$, see [8].

The proof of (8) is involved (especially when $\gamma = 0$) and will not be given here. It uses the Birman-Schwinger kernel, $V_+^{1/2}(-\Delta + \lambda)^{-1}V_+^{1/2}$.

There is a natural "guess" for $L_{\gamma,n}$ in terms of a semiclassical approximation (and which is not unrelated to the theory of pseudodifferential operators):

$$\sum |e_i|^\gamma \approx (2\pi)^{-n} \int_{\mathbf{R}^n \times \mathbf{R}^n, p^2 \leq V(x)} [V(x) - p^2]^\gamma dp dx \tag{9}$$

$$= L_{\gamma,n}^c \int V_+(x)^{\gamma+n/2} dx. \tag{10}$$

From (9),

$$L_{\gamma,n}^c = (4\pi)^{-n/2}\Gamma(\gamma+1)/\Gamma(1+\gamma+n/2). \tag{11}$$

It is easy to prove that

$$L_{\gamma,n} \geq L_{\gamma,n}^c. \tag{12}$$

The evaluation of the sharp $L_{\gamma,n}$ is an interesting open problem – especially $L_{1,n}$. In particular, for which γ, n is $L_{\gamma,n} = L_{\gamma,n}^c$? It is known [1] that for each fixed n, $L_{\gamma,n}/L_{\gamma,n}^c$ is nonincreasing in γ. Thus, if $L_{\gamma_0,n} = L_{\gamma_0,n}^c$ for some γ_0, then $L_{\gamma,n} = L_{\gamma,n}^c$ for all $\gamma > \gamma_0$. In particular, $L_{3/2,1} = L_{3/2,1}^c$ [11], so $L_{\gamma,1} = L_{\gamma,1}^c$ for $\gamma \geq 3/2$. No other sharp values of $L_{\gamma,n}$ are known. It is also known [11] that $L_{\gamma,1} > L_{\gamma,1}^c$ for $\gamma < 3/2$ and $L_{\gamma,n} > L_{\gamma,n}^c$ for $n = 2, 3$ and small γ.

Just as (4) is related to (2), inequality (7) is related to a generalization of (2). (The proof is basically the same.) Let $\phi_1, \ldots, \phi_N$ be any set of L^2 orthonormal functions on $\mathbf{R}^n (n \geq 1)$ and define

$$\rho(x) \equiv \sum_{i=1}^N |\phi_i(x)|^2. \tag{13}$$

$$T \equiv \sum_{i=1}^N \int |\nabla \phi_i|^2 \tag{14}$$

Then we have **The Main Inequality**

$$T \geq K_n \int \rho(x)^{1+2/n} dx \tag{15}$$

with K_n related to $L_{1,n}$ as in (5), i.e.

$$L_{1,n} = \left(\frac{n}{2K_n}\right)^{n/2} \left(1 + \frac{n}{2}\right)^{-(n+2)/2}. \tag{16}$$

The best current value of K_n, for $n = 1, 2, 3$ is in [8]; in particular $K_3 \geq 2.7709$. We might call (15) *a Sobolev type inequality for orthonormal functions.* The point is that if the ϕ_i are merely normalized, but not orthogonal, then the best one could say is

$$T \geq N^{-2/n} K_n^1 \int \rho(x)^{1+2/n} dx. \tag{17}$$

The orthogonality eliminates the factor $N^{-2/n}$, but replaces K_n^1 by the slightly smaller value K_n.

One should notice, especially, the N dependence in (15). The right side, loosely speaking, is proportional to $N^{(n+2)/n}$, whereas the right side of (17) appears, falsely, to be proportional to N^1, which is the best one could hope for without orthogonality. The difference is crucial for applications. In fact, if one is willing to settle for N^1 one can proceed directly from (1) (for $n \geq 3$). One then has (with $p = n/(n-2)$)

$$T \geq S_n \left\{ \int \rho(x)^p dx \right\}^{1/p}, \quad (n \geq 3). \tag{17a}$$

This follows from $\sum \|\phi_i^2\|_p \geq \left\| \sum |\phi_i|^2 \right\|_p$.

Eq. (11) gives a "classical guess" for $L_{1,n}$. Using that, together with (16), we have a "classical guess" for K_n, namely

$$\begin{aligned} K_n^c &= 4\pi n \Gamma \left(\frac{n+2}{2} \right)^{2/n} /(2+n) \\ &= \frac{3}{5}(6\pi^2)^{2/3} \text{ for } n = 3. \end{aligned} \tag{18}$$

Since $L_{1,n} \geq L_{1,n}^c$, we have $K_n \leq K_n^c$. A conjecture in [11] is that $K_3 = K_3^c$, and it would be important to settle this.

Inequality (15) can be easily extended to the following: Let $\psi(x_1, \ldots, x_N) \in L^2\left((\mathbf{R}^n)^N\right), x_i \in \mathbf{R}^n$. Suppose $\|\psi\|_2 = 1$ and ψ is *antisymmetric* in the N variables, i.e., $\psi(x_1, \ldots, x_i, \ldots, x_j, \ldots, x_N) = -\psi(x_1, \ldots x_j, \ldots, x_i, \ldots, x_N)$. Define

$$\rho_i(x) = \int \left| \psi(x_1, \ldots, x_{i-1}, x, x_{i+1}, \ldots, x_N) \right|^2 dx_1 \ldots \widehat{dx_i} \ldots dx_N \tag{19}$$

$$T_i(x) = \int |\nabla_i \psi|^2 dx_1 \ldots dx_N \tag{20}$$

$$\rho(x) = \sum_{i=1}^N \rho_i(x) \qquad T = \sum_{i=1}^N T_i. \tag{21}$$

(Note that $\rho(x) = N\rho_1(x)$ and $T = NT_1$ since ψ is antisymmetric, but the general form (19)-(21) will be used in the next paragraph.) **Then (15) holds with ρ and T given**

by (19)-(21) (with the same K_n as in (15)). This is a generalization of (13)-(15) since we can take

$$\psi(x_1, \ldots, x_N) = (N!)^{-1/2} \det \left\{\phi_i(x_j)\right\}_{i,j=1}^{N},$$

which leads to (13) and (14).

A variant of (15) is given in (52) below. It is a consequence of the fact that (17) and (17a) *also hold* with the definitions (19)-(21). *Antisymmetry of ψ is not required.* The proof of (17a) just uses (1) as before plus convexity, namely for $p \geq 1$

$$\int \left\{ \int |F(x,y)|^p dy \right\}^{1/p} dx \geq \left\{ \int \left\{ \int |F(x,y)| dx \right\}^p dy \right\}^{1/p}.$$

We turn now to some applications of these inequalities.

Application 1. Inequality (15) can be used to bound L^p **norms of Riesz and Bessel potentials of orthonormal functions** [7]. Again, $\phi_1, \ldots, \phi_N$ are L^2 orthonormal and let

$$u_i = (-\Delta + m^2)^{-1/2} \phi_i \tag{22}$$

$$\rho(x) \equiv \sum_{i=1}^{N} |u_i(x)|^2. \tag{23}$$

Then there are constants L, B_p, A_n (independent of m) such that

$$n = 1: \ \|\rho\|_\infty \leq L/m, \qquad m > 0 \tag{24}$$

$$n = 2: \ \|\rho\|_p \leq B_p m^{-2/p} N^{1/p}, \qquad 1 \leq p < \infty, m > 0 \tag{25}$$

$$n \geq 3: \ \|\rho\|_p \leq A_n N^{1/p}, \qquad p = n/(n-2), m \geq 0. \tag{26}$$

If the orthogonality condition is dropped then the right sides of (24)-(26) have to be multiplied by $N, N^{1-1/p}, N^{1-1/p}$ respectively. Possibly the absence of N in (24) is the most striking. Similar results can be derived [7] for $(-\Delta + m^2)^{-\alpha/2}$ in place of $(-\Delta + m^2)^{-1/2}$, with $\alpha < n$ when $m = 0$.

Inequality (15) also has applications in mathematical physics.

Application 2. (Navier-Stokes equation.) Suppose $\Omega \subset \mathbf{R}^n$ is an open set with finite volume $|\Omega|$ and consider

$$H = -\Delta - V(x)$$

on Ω with Dirichlet boundary conditions. Let $\lambda_1 \leq \lambda_2 \leq \ldots$ be the eigenvalues of H. Let $\bar{N}$ be the smallest integer, N, such that

$$E_N \equiv \sum_{i=1}^{N} \lambda_i \geq 0. \tag{27}$$

We want to find an upper bound for $\bar{N}$.

If $\phi_1, \phi_2, \ldots$ are the normalized eigenfunctions then, from (13)-(15) with $\phi_1, \ldots, \phi_N$,

$$E_N = T - \int \rho V \geq K_n \int \rho^{1+n/2} - \int V_+ \rho \geq G(\rho), \tag{28}$$

where (with $p = 1 + n/2$ and $q = 1 + 2/n$)

$$G(\rho) \equiv K_n \|\rho\|_p^p - \|V_+\|_q \|\rho\|_p. \tag{29}$$

Thus, for all N,

$$E_N \geq \inf\{G(\rho) | \, \|\rho\|_1 = N, \ \rho(x) \geq 0\}. \tag{30}$$

But $\|\rho\|_p |\Omega|^{1/q} \geq \|\rho\|_1 = N$ so, with $X \equiv \|\rho\|_p$,

$$E_N \geq \inf\{J(X) \mid X \geq N|\Omega|^{-1/q}\} \tag{31}$$

where

$$J(X) \equiv K_n X^p - \|V_+\|_q X. \tag{32}$$

Now $J(X) \geq 0$ for $X \geq X_0 = \{\|V_+\|_q / K_n\}^{1/(p-1)}$, whence we have the following implication:

$$N \geq |\Omega|^{1/q} \{\|V_+\|_q / K_n\}^{1/(p-1)} \Longrightarrow E_N \geq 0. \tag{33}$$

Therefore

$$\bar{N} \leq |\Omega|^{1/q} \{\|V_+\|_q / K_n\}^{1/(p-1)}. \tag{34}$$

The bound (34) can be applied [8] (following an idea of Ruelle) to the Navier-stokes equation. There, $\bar{N}$ is interpreted as the Hausdorff dimension of an attracting set for the N-S equation, while $V(x) \equiv \nu^{-3/2} \varepsilon(x)$, where $\varepsilon(x) = \nu |\nabla v(x)|^2$ is the average energy dissipation per unit mass in a flow v. ν is the viscosity.

Application 3. (Stability of matter.) This is the original application [10,11]. In the quantum mechanics of Coulomb systems (electrons and nuclei) one wants a lower bound for the Hamiltonian operator:

$$H = -\sum_{i=1}^{N} \Delta_i - \sum_{i=1}^{N} \sum_{j=1}^{K} z_j |x_i - R_j|^{-1} + \sum_{1 \leq i < j \leq N} |x_i - x_j|^{-1} + \sum_{1 \leq i < j \leq K} z_i z_j |R_i - R_j|^{-1} \tag{35}$$

on the L^2 space of *antisymmetric* functions $\psi(x_1, \ldots, x_N), x_i \in \mathbf{R}^3$. Here, N is the number of electrons (with coordinates x_i) and $R_1, \ldots, R_K \in \mathbf{R}^3$ are fixed vectors representing the locations of fixed nuclei of charges $z_1, \ldots, z_K > 0$. The desired bound is linear:

$$H \geq -A(N+K) \tag{36}$$

for some A independent of $N, K, R_1, \ldots, R_K$ (assuming all $z_i <$ some $\bar{z}$).

The main point is that antisymmetry of ψ is crucial for (36) and this is reflected in the fact that (15) holds with antisymmetry, but only (17) holds without it. Without the antisymmetry condition, H would grow as $-(N+K)^{5/3}$. This is discussed in Application 6 below. By using (15) one can eliminate the differential operators Δ_i. The functional $\psi \to (\psi, H\psi)$, with $(\psi, \psi) = 1$ can be bounded below using (15) by a functional (called the Thomas-Fermi functional) involving only $\rho(x)$ defined in (21). The minimization of this latter functional with respect to ρ is tractable and leads to (36).

Application 4. (Stellar structure.) Going from atoms to stars, we now consider N neutrons which attract each other gravitationally with a coupling constant $\kappa = Gm^2$, where G is the gravitational constant and m is the neutron mass. There are no Coulomb forces. Moreover, a "relativistic" form is assumed for the kinetic energy, which means that $-\Delta$ is replaced by $(-\Delta)^{1/2}$. Thus (35) is replaced by

$$H_N = \sum_{i=1}^{N} (-\Delta_i)^{1/2} - \kappa \sum_{1 \leq i < j \leq N} |x_i - x_j|^{-1} \tag{37}$$

(again on antisymmetric functions). One finds asymptotically for large N, that

$$\begin{aligned} \inf \operatorname{spec}(H_N) &= 0 \quad \text{if} \quad \kappa \leq CN^{-2/3} \\ &= -\infty \quad \text{if} \quad \kappa > CN^{-2/3} \end{aligned} \tag{38}$$

for some constant, C. Without antisymmetry, $N^{-2/3}$ must be replaced by N^{-1}. Equation (38) is proved in [12]. An important role is played by Daubechies's generalization [4] of (15) to the operator $(-\Delta)^{1/2}$ on $L^2(\mathbf{R}^N)$, namely (for antisymmetric ψ with $\|\psi\|_2 = 1$)

$$\left(\psi, \sum_{i=1}^{N} (-\Delta_i)^{1/2} \psi\right) \geq B_n \int \rho(x)^{1+1/n} \tag{39}$$

with ρ given by (19), (21). In general, one has

$$\left(\psi, \sum_{i=1}^{N} (-\Delta)^p \psi\right) \geq C_{p,n} \int \rho(x)^{1+2p/n} dx. \tag{40}$$

Recently [13] there has been progress in this problem beyond that in [12]. Among other results there is an evaluation of the sharp asymptotic C in (38), i.e. if we first define $\kappa^c(N)$ to be the precise value of κ at which $\inf \operatorname{spec}(H_N) = -\infty$, we then define

$$C = \lim_{N\to\infty} N^{2/3}\kappa^c(N). \tag{41}$$

Let B_n^c be the "classical guess' in (39). This can be calculated from the analogue of (9) (using $|p|$ instead of p^2, and which leads to $\sum |e_i| \approx \tilde{L}\int V_+^{1+n}$), and then from the analogue of (16), namely $\tilde{L} = C_n B_n^{-n}$. One finds $B_3^c = (3/4)(6\pi^2)^{1/3}$ (cf. (18)). Using B_3^c, we introduce the functional

$$\mathcal{E}(\rho) = B_3^c \int \rho^{4/3} - \frac{1}{2}\kappa \int\int \rho(x)\rho(y)|x-y|^{-1}dxdy \tag{42}$$

for $\rho \in L^1(\mathbf{R}^3) \cap L^{4/3}(\mathbf{R}^3)$ and define the energy

$$E^c(N) = \inf\{\mathcal{E}(\rho) | \int \rho = N\}. \tag{43}$$

One finds there is a finite $\alpha^c > 0$ such that $E^c(N) = 0$ if $\kappa N^{2/3} \leq \alpha^c$ and $E^c(N) > -\infty$ if $\kappa N^{2/3} > \alpha^c$. (This α^c is found by solving a Lane-Emden equation.)

Now (42) and (43) constitute the semiclassical approximation to H_N in the following sense. We expect that if we set $\kappa = \alpha N^{-2/3}$ in (37), with α fixed, then if $\alpha < \alpha^c$

$$\lim_{N\to\infty} \inf \operatorname{spec}(H_N) = 0 \tag{44}$$

while if $\alpha > \alpha^c$ there is an N_0 such that

$$\inf \operatorname{spec}(H_N) = -\infty \quad \text{if} \quad N > N_0. \tag{45}$$

Indeed, (44) and (45) are true [13], and thus α^c is the sharp asymptotic value of C in (38).

An interesting point to note is that Daubechies's B_3 in (39) is about half of B_3^c. The sharp value of B_3 is unknown. Nevertheless, with some additional tricks one can get from (37) to (42) with B_3^c and not B_3. Inequality (39) plays a role in [13], but it is not sufficient.

Application 5. (Stability of atoms in magnetic fields.) This is given in [9]. Here $\psi(x_1, \ldots, x_N)$ becomes a spinor-valued function, i.e. ψ is an antisymmetric function in $\bigwedge_1^N L^2(\mathbf{R}^3; \mathbf{C}^2)$. The operator H of interest is as in (35) but with the replacement

$$-\Delta \to \{\sigma \cdot (i\nabla - A(x))\}^2 \tag{46}$$

where $\sigma_1, \sigma_2, \sigma_3$ are the 2×2 Pauli matrices (i.e. generators of $SU(2)$) and $A(x)$ is a given vector field (called the magnetic vector potential). Let

$$E_0(A) = \inf \operatorname{spec}(H) \tag{47}$$

after the replacement of (46) in (35). As $A \to \infty$ (in a suitable sense), $E_0(A)$ can go to $-\infty$. The problem is this: Is

$$\tilde{E}(A) \equiv E_0(A) + \frac{1}{8\pi} \int (\text{curl } A)^2 \tag{48}$$

bounded below for all A? In [9] the problem is resolved for $K = 1$, all N and $N = 1$, all K. It turns out that $\tilde{E}(A)$ is bounded below in these cases if and only if all the z_i satisfy $z_i < z^c$ where z^c is some fixed constant independent of N and K. The problem is still open for all N and all K.

One of the main problems in bounding $\tilde{E}(A)$ is to find a lower bound for the kinetic energy (the first term in (35) after the replacement given in (46)) for an antisymmetric ψ. First, there is the identity

$$\left(\psi, \sum_{i=1}^{N} \{\sigma \cdot (i\nabla - A(x_i)\}^2 \psi\right) = T(\psi, A) - \left(\psi, \sum_{i=1}^{N} \sigma \cdot B(x_i)\psi\right) \tag{49}$$

with $B = \text{curl } A$ being the magnetic field and

$$T(\psi, A) = \left(\psi, \sum_{i=1}^{N} |i\nabla - A(x)|^2 \psi\right). \tag{50}$$

The last term on the right side of (49) can be controlled, so it will be ignored here. The important term is $T(\psi, A)$. Since Pauli matrices do not appear in (50) we can now let ψ be an ordinary complex valued (instead of spinor valued) function.

It turns out that (8), and hence (15), hold with some $\tilde{L}_{\gamma,n}$ which is *independent* of A. The T in (15) is replaced, of course, by the $T(\psi, A)$ of (50). To be more precise, the *sharp* constants $L_{\gamma,n}$ and $\tilde{L}_{\gamma,n}$ are unknown (except for $\gamma \geq 3/2, n = 1$ in the case of $L_{\gamma,n}$) and conceivably $\tilde{L}_{\gamma,n} > L_{\gamma,n}$. However, all the current bounds for $L_{\gamma,n}$ (see [8]) also hold for $\tilde{L}_{\gamma,n}$. Thus, for $n = 3$ we have

$$T(\psi, A) \geq K_3 \int \rho^{5/3} \tag{51}$$

with K_3 being the value given in [8], namely 2.7709.

However, in [9] another inequality is needed

$$T(\psi, A) \geq C \left\{ \int \rho^2 \right\}^{2/3} . \tag{52}$$

It seems surprising that we can go from an $L^{5/3}$ estimate to an L^2 estimate, but the surprise is diminished if (17a) with its L^3 estimate is recalled. First note that (1) holds (with the same S_n) if $|\nabla f|^2$ is replaced by $|[i\nabla - A(x)]f|^2$. (By writing $f = |f|e^{i\theta}$ one finds that $|\nabla|f||^2 \leq |[i\nabla - A(x)]f|^2$.) Then (17a) holds since only convexity was used. Thus, using the mean of (15) and (17a),

$$T(\psi, A) \geq (S_n K_n)^{1/2} \|\rho\|_3^{1/2} \|\rho\|_{5/3}^{5/6}. \tag{53}$$

An application of Hölders inequality yields (52) with $C^2 = S_n K_n$.

Application 6. (Instability of bosonic matter.) As remarked in Application 3, dropping the antisymmetry requirement on ψ (the particles are now bosons) makes $\inf \operatorname{spec}(H)$ diverge as $-(N+K)^{5/3}$. The extra power 2/3, relative to (36) can be traced directly to the factor $N^{-2/3}$ in (17).

An interesting problem is to allow the positive particles also to be movable and to have charge $z_i = 1$. This should raise $\inf \operatorname{spec} H$, but by how much? For $2N$ particles the new H is

$$\tilde{H} = -\sum_{i-1}^{2N} \Delta_i + \sum_{1 \leq i < j \leq 2N} e_i e_j |x_i - x_j|^{-1} \tag{54}$$

with $e_i = +1$ for $1 \leq i \leq N$ and $e_i = -1$ for $N+1 \leq i \leq 2N$. $\tilde{H}$ acts on $L^2(\mathbf{R}^{3N})$ without any symmetry requirements.

Twenty years ago Dyson [5] proved, by a variational calculation, that $\inf \operatorname{spec}(\tilde{H}) \leq -AN^{7/5}$ for some $A > 0$. Thus, stability (i.e. a linear law (36)) is not restored, but the question of whether the correct exponent is 7/5 or 5/3, or something in between, remained open. It has now been proved [2] that $N^{7/5}$ is correct, $\inf \operatorname{spec}(\tilde{H}) \geq -BN^{7/5}$. The proof is much harder than for (36) because no simple semiclassical theory (like Thomas-Fermi theory) is a good approximation to $\tilde{H}$. Correlations are crucial.

References

[1] M. Aizenman and E.H. Lieb, On semiclassical bounds for eigenvalues of Schrödinger operators, Phys. Lett. *66*A, 427-429 (1978).

[2] J. Conlon, E.H. Lieb and H.-T. Yau, The $N^{7/5}$ law for bosons, Commun. Math. Phys. (submitted).

[3] M. Cwikel, Weak type estimates for singular values and the number of bound states of Schrödinger operators, Ann. Math. *106*, 93-100 (1977).

[4] I. Daubechies, Commun. Math. Phys. *90*, 511-520 (1983).

[5] F.J. Dyson, Ground state energy of a finite systems of charged particles, J. Math. Phys. *8*, 1538-1545 (1967).

[6] E.H. Lieb, The number of bound states of one-body Schrödinger operators and the Weyl problem, A.M.S. Proc. Symp. in Pure Math. *36*, 241-251 (1980). The results were announced in Bull. Ann. Math. Soc. *82*, 751-753 (1976).

[7] E.H. Lieb, An L^p bound for the Riesz and Bessel potentials of orthonormal functions, J. Funct. Anal. *51*, 159-165 (1983).

[8] E.H. Lieb, On characteristic exponents in turbulence, Commun. Math. Phys. *92*, 473-480 (1984).

[9] E.H. Lieb and M. Loss, Stability of Coulomb systems with magnetic fields: II. The many-electron atom and the one-electron molecule, Commun. Math. Phys. *104*, 271-282 (1986).

[10] E.H. Lieb and W.E. Thirring, Bounds for the kinetic energy of fermions which proves the stability of matter, Phys. Rev. Lett. *35*, 687-689 (1975). Errata *35*, 1116 (1975).

[11] E.H. Lieb and W.E. Thirring, "Inequalities for the moments of the eigenvalues of the Schrödinger Hamiltonian and their relation to Sobolev inequalities" in ***Studies in Mathematical Physics*** (E. Lieb, B. Simon, A. Wightman eds.) Princeton University Press, 1976, pp. 269-304.

[12] E.H. Lieb and W.E. Thirring, Gravitational collapse in quantum mechanics with relativistic kinetic energy, Ann. of Phys. (NY) *155*, 494-512 (1984).

[13] E.H. Lieb and H.-T. Yau, The Chandrasekhar theory of stellar collapse as the limit of quantum mechanics, Commun. Math. Phys. *112*, 147-174 (1987).

[14] G.V. Rosenbljum, Distribution of the discrete spectrum of singular differential operators. Dokl. Akad. Nauk SSSR *202*, 1012-1015 (1972). (MR 45 #4216). The details are given in Izv. Vyss. Ucebn. Zaved. Matem. *164*, 75-86 (1976). (English trans. Sov. Math. (Iz VUZ) *20*, 63-71 (1976).)

8 Inequalities Related to Carleman's Inequality

E. Russell Love University of Melbourne, Parkville, Victoria, Australia

1. Introduction

The common ancestor of the inequalities discussed here is:

Carleman's Inequality (1923) [HLP Theorem 334] *If $x_n \geqslant 0$ then*

$$\sum_{m=1}^{\infty} \left(\prod_{n=1}^{m} x_n \right)^{1/m} \leqslant e \sum_{m=1}^{\infty} x_m, \tag{A}$$

with equality only if all x_n vanish. The constant e is best possible.

References to proofs of this and contemporary allied results of several writers are given in [HLP p. 249, footnote[a]]. The generalizations given below omit discussion of cases of equality.

Heinig's Inequality (1975) [2: Theorem 8] *If $p \geqslant 1$, $s > 0$, $p + s > \lambda \geqslant 0$ and $x_n \geqslant 0$, then*

$$\sum_{m=1}^{\infty} m^{\lambda - sp} \left(\prod_{n=1}^{m} x_n^{pn^{p-1}} \right)^{1/m^p} \leqslant e^{1/p} p \left(1 + \frac{1}{p + s - \lambda} \right) \sum_{m=1}^{\infty} m^{\lambda - s} x_m. \tag{B}$$

The case $p = s = \lambda = 1$ is (A), except that e is replaced by $2e$. The next inequality is to some extent built on (B). It looks simpler, but requires a fairly sophisticated proof.

Cochran and Lee's Inequality (1984) [1: Theorem 2] *If $p \geqslant 1$, $k \geqslant 1$ and $0 \leqslant x_n \leqslant 1$, then*

$$\sum_{m=1}^{\infty} m^{k-1} \left(\prod_{n=1}^{m} x_n^{pn^{p-1}} \right)^{1/m^p} \leqslant e^{k/p} \sum_{m=1}^{\infty} m^{k-1} x_m. \tag{C}$$

The constant $e^{k/p}$ is best possible.

The case $p = k = 1$ is (A).

These inequalities, and those which follow, all have integral analogues; these have been discussed elsewhere [5]. Knopp's (or Pólya's) Inequality [HLP Theorem 335] is analogous to (A), while analogues of (B) and (C) receive attention in [2] and [1] respectively. Analogues of Theorems 1 and 2 below are in [3] and [4].

2. Further Inequalities

The next two theorems continue in order of increasing generality (rather than in logical order).

Theorem 1. *If $p > 0$, $k \geqslant 1$, $x_n \geqslant 0$ and τ_n is decreasing and positive, then*

$$\sum_{m=1}^{\infty} m^{k-1}\tau_m \left[\prod_{n=1}^{m} x_n^{n^p-(n-1)^p}\right]^{1/m^p} \leqslant e^{k/p} \sum_{m=1}^{\infty} m^{k-1}\tau_m x_m. \tag{1}$$

This is clearly a close relative of (C), which inspired it. It evidently does not include (C); but it implies it, as we now show.

Corollary 1a. *If $p \geqslant 1$, $k \geqslant 1$, $0 \leqslant x_n \leqslant 1$ and τ_n is decreasing and positive, then*

$$\sum_{m=1}^{\infty} m^{k-1}\tau_m \left[\prod_{n=1}^{m} x_n^{pn^{p-1}}\right]^{1/m^p} \leqslant \sum_{m=1}^{\infty} m^{k-1}\tau_m \left[\prod_{n=1}^{m} x_n^{n^p-(n-1)^p}\right]^{1/m^p}$$

$$\leqslant e^{k/p} \sum_{m=1}^{\infty} m^{k-1}\tau_m x_m. \tag{2}$$

Proof. The second inequality is a special case of Theorem 1, while the first comes from

$$n^p - (n-1)^p \leqslant pn^{p-1}, \qquad x_n^{pn^{p-1}} \leqslant x_n^{n^p-(n-1)^p}.$$

Notice that the inequality of the extremes in (2) reduces to (C) in the case when $\tau_n = 1$, and generalizes (C) when τ_n is not so restricted. In the former case (2) also interpolates an expression between the two members of (C).

Corollary 1b. *If $p > 0$, $k < 1 \leqslant \lambda$, $x_n \geqslant 0$ and $n^{k-\lambda}\tau_n$ is decreasing and positive, then (1) holds with $e^{k/p}$ replaced by $e^{\lambda/p}$.*

Proof. Let $\tau_n' = n^{k-\lambda}\tau_n$; this is decreasing and positive. Then $n^{k-1}\tau_n = n^{\lambda-1}\tau_n'$, and Theorem 1 with k and τ_n replaced by λ and τ_n' gives the result.

This corollary complements Theorem 1 by relaxing the restrictions on k and τ_n.

Theorem 1 itself will eventually be shown to be a consequence of Theorem 2 below.

3. The Main Theorem

This is a theorem of a rather more general kind. For it we need to introduce some weighted geometric means, which are defined as follows [HLP §2.2 and §6.7]:

$$G_w x_m = (G_w x)_m = \left[\prod_{n=1}^{m} x_n^{w_{mn}}\right]^{1/W_m} \quad \text{where } W_m = \sum_{n=1}^{m} w_{mn},$$

$$= \exp\left[\sum_{n=1}^{m} w_{mn} \log x_n \Big/ \sum_{n=1}^{m} w_{mn}\right];$$

$$G_w\rho(s) = (G_w\rho)(s) = \exp\left[\int_0^s w(s,t)\log\rho(t)\,dt \,/\, \int_0^s w(s,t)\,dt\right].$$

The latter is exactly the integral analogue of the former, and in both the weights w are supposed positive.

Theorem 2. *Let $\alpha(t)$ and $\rho(t)$ be positive on $0 < t < 1$, and $\alpha(t)$ and $\alpha(t)\log\rho(t)$ be integrable thereon. Let $\lambda_n > 0$, Λ_n be strictly increasing, and $\Lambda_0 = 0$. If*

$$w_{mn} = \int_{\Lambda_{n-1}/\Lambda_m}^{\Lambda_n/\Lambda_m} \alpha(t)\,dt,$$

$$\sum_{m=n}^{\infty} \frac{\lambda_m}{\lambda_n} \int_{\Lambda_{n-1}/\Lambda_m}^{\Lambda_n/\Lambda_m} \frac{\alpha(t)}{\rho(t)}\,dt \leqslant H\int_0^1 \alpha(t)\,dt \tag{3}$$

and $x_n \geqslant 0$, then

$$\sum_{m=1}^{\infty} \lambda_m G_w x_m \leqslant HG_\alpha\rho(1)\sum_{m=1}^{\infty}\lambda_m x_m.$$

4. A Proof of Carleman's Inequality

Our proof of Theorem 2 may be more fully appreciated if its main features are first seen in the simpler situation of Carleman's Inequality (A). Accordingly we now give this simplified proof, although on any other grounds it is quite unnecessary.

Supposing all $x_n > 0$, let $f(t) = x_n$ for $n - 1 \leqslant t < n$.

$$\begin{aligned}
\sum_{m=1}^{\infty}\left(\prod_{n=1}^{m} x_n\right)^{1/m} &= \sum_{m=1}^{\infty}\exp\left(\frac{1}{m}\sum_{n=1}^{m}\log x_n\right)\\
&= \sum_{m=1}^{\infty}\exp\left(\sum_{n=1}^{m}\int_{(n-1)/m}^{n/m}\log f(mt)\,dt\right)\\
&= \sum_{m=1}^{\infty}\exp\left(\int_0^1\{\log(tf(mt)) - \log t\}dt\right)\\
&= e\sum_{m=1}^{\infty}\exp\left(\int_0^1\log(tf(mt))\,dt\right)\\
&\leqslant e\sum_{m=1}^{\infty}\int_0^1 tf(mt)\,dt \qquad (4)\\
&= e\sum_{m=1}^{\infty}\sum_{n=1}^{m}\int_{(n-1)/m}^{n/m} tx_n dt\\
&= e\sum_{m=1}^{\infty}\sum_{n=1}^{m} x_n\frac{2n-1}{2m^2}\\
&= e\sum_{n=1}^{\infty} x_n(n - ½)\sum_{m=n}^{\infty}\frac{1}{m^2}, < e\sum_{n=1}^{\infty} x_n
\end{aligned}$$

because

$$\sum_{m=n}^{\infty} \frac{1}{m^2} < \sum_{m=n}^{\infty} \frac{1}{m^2 - \frac{1}{4}} = \sum_{m=n}^{\infty} \left(\frac{1}{m - \frac{1}{2}} - \frac{1}{m + \frac{1}{2}} \right) = \frac{1}{n - \frac{1}{2}}.$$

At (4) the inequality of integral geometric and arithmetic means [HLP Theorem 184] has been used.

Only minor modifications are needed when some of the x_n vanish.

It is perhaps surprising that this proof of (A) is not particularly simple; for instance, it is probably not as simple as Pólya's proof [HLP p. 249].

5. Proof of Theorem 2

The integrability hypotheses imply that $\rho(t)$ is measurable, since it is a continuous function of $\log \rho(t)$ which is a quotient of the measurable functions $\alpha(t)$ and $\alpha(t) \log \rho(t)$. Thus $\alpha(t)/\rho(t)$ is positive and measurable, and so condition (3) is meaningful.

Let

$$W = \sum_{n=1}^{m} w_{mn} = \int_0^1 \alpha(t)\, dt, > 0; \text{ and let}$$

$$f(t) = x_n \text{ for } \Lambda_{n-1} \leqslant t < \Lambda_n.$$

Let x_{k+1} be the first x_n to vanish; then $G_w x_m = 0$ for all $m > k$. This integer k has no connection with the real number k occurring in other sections of this chapter. If no x_n vanishes, let $k = \infty$.

$$\sum_{m=1}^{\infty} \lambda_m G_w x_m = \sum_{m=1}^{k} \lambda_m G_w x_m$$

$$= \sum_{m=1}^{k} \lambda_m \exp \left(\frac{1}{W} \sum_{n=1}^{m} \int_{\Lambda_{n-1}/\Lambda_m}^{\Lambda_n/\Lambda_m} \alpha(t) \log x_n\, dt \right)$$

$$= \sum_{m=1}^{k} \lambda_m \exp \left(\frac{1}{W} \sum_{n=1}^{m} \int_{\Lambda_{n-1}/\Lambda_m}^{\Lambda_n/\Lambda_m} \alpha(t) \log f(\Lambda_m t)\, dt \right)$$

$$= \sum_{m=1}^{k} \lambda_m \exp \left(\frac{1}{W} \int_0^1 \alpha(t) \left(\log \frac{f(\Lambda_m t)}{\rho(t)} + \log \rho(t) \right) dt \right)$$

$$= \sum_{m=1}^{k} \lambda_m G_\alpha \rho(1) \exp \left(\frac{1}{W} \int_0^1 \alpha(t) \log \frac{f(\Lambda_m t)}{\rho(t)}\, dt \right)$$

$$\leqslant G_\alpha \rho(1) \sum_{m=1}^{k} \lambda_m \frac{1}{W} \int_0^1 \alpha(t) \frac{f(\Lambda_m t)}{\rho(t)}\, dt \tag{5}$$

$$= \frac{G_\alpha \rho(1)}{W} \sum_{m=1}^{k} \lambda_m \sum_{n=1}^{m} \int_{\Lambda_{n-1}/\Lambda_m}^{\Lambda_n/\Lambda_m} \frac{\alpha(t)}{\rho(t)} x_n\, dt$$

$$= \frac{G_\alpha \rho(1)}{W} \sum_{n=1}^{k} x_n \sum_{m=n}^{k} \lambda_m \int_{\Lambda_{n-1}/\Lambda_m}^{\Lambda_n/\Lambda_m} \frac{\alpha(t)}{\rho(t)}\, dt$$

$$\leqslant \frac{G_\alpha \rho(1)}{W} \sum_{n=1}^{\infty} \lambda_n x_n \sum_{m=n}^{\infty} \frac{\lambda_m}{\lambda_n} \int_{\Lambda_{n-1}/\Lambda_m}^{\Lambda_n/\Lambda_m} \frac{\alpha(t)}{\rho(t)}\, dt$$

$$\leqslant \frac{G_\alpha \rho(1)}{W} \sum_{n=1}^{\infty} \lambda_n x_n HW = HG_\alpha \rho(1) \sum_{n=1}^{\infty} \lambda_n x_n.$$

At (5) the inequality of geometric and arithmetic means has been used. This completes the proof of Theorem 2.

6. Deduction of Theorem 1 from Theorem 2

Lemma (Cochran and Lee). *If $p > 0$, $k \geqslant 1$ and n is a positive integer, then*

$$\sum_{m=n}^{\infty} \frac{1}{m^{p+1}} \leqslant \frac{p+k}{p} \, \frac{n^{k-1}}{n^{p+k} - (n-1)^{p+k}}.$$

This is [1: Lemma 3] with s, γ and i replaced by $p + 1$, $k - 1$ and n. That it is an improvement on the simple classical integral estimation of the left side is evident in the case when $k = 1$; for then the right side is

$$\frac{p+1}{p} \Big/ \int_{n-1}^{n} (p+1)x^p dx < \frac{1}{p(n-1)^p} = \int_{n-1}^{\infty} \frac{1}{x^{p+1}}\, dx.$$

Theorem 1 (restated for convenience). *If $p > 0$, $k \geqslant 1$, τ_n is decreasing and positive, and $x_n \geqslant 0$, then*

$$\sum_{m=1}^{\infty} m^{k-1} \tau_m \left[\prod_{n=1}^{m} x_n^{n^p - (n-1)^p} \right]^{1/m^p} \leqslant e^{k/p} \sum_{m=1}^{\infty} m^{k-1} \tau_m x_m. \tag{1}$$

Proof. Let $q = p/k$. In Theorem 2 let $\alpha(t) = qt^{q-1}$, $\rho(t) = qt^{-1}$, $\lambda_n = n^{k-1}\tau_n$ and $\Lambda_n = n^k$. Then

$$w_{mn} = \int_{(n-1)^k/m^k}^{n^k/m^k} qt^{q-1}\, dt = \frac{n^p - (n-1)^p}{m^p},$$

$$\sum_{m=n}^{\infty} \frac{\lambda_m}{\lambda_n} \int_{\Lambda_{n-1}/\Lambda_m}^{\Lambda_n/\Lambda_m} \frac{\alpha(t)}{\rho(t)}\, dt = \sum_{m=n}^{\infty} \frac{m^{k-1}}{n^{k-1}} \frac{\tau_m}{\tau_n} \int_{(n-1)^k/m^k}^{n^k/m^k} t^q\, dt$$

$$\leqslant \sum_{m=n}^{\infty} \frac{m^{k-1}}{n^{k-1}} \frac{n^{p+k} - (n-1)^{p+k}}{(q+1)m^{p+k}}$$

$$= \frac{n^{p+k} - (n-1)^{p+k}}{(q+1)n^{k-1}} \sum_{m=n}^{\infty} \frac{1}{m^{p+1}}$$

$$\leqslant \frac{p+k}{p(q+1)} = \frac{k}{p} = \frac{k}{p}\int_0^1 \alpha(t)\,dt \tag{6}$$

Comparing (6) with (3) shows that $H = k/p$ is suitable. Cochran and Lee's lemma (stated above) is used in (6).

Theorem 2 also involves the geometric means

$$G_w x_m = \prod_{n=1}^{m} x_n^{w_{mn}} = \left(\prod_{n=1}^{m} x_n^{n^p-(n-1)^p}\right)^{1/m^p},$$

$$G_\alpha \rho(1) = \exp\left(\int_0^1 \alpha(t)\log\rho(t)\,dt\right) = \exp\left(\log q + \frac{1}{q}\right) = \frac{p}{k}e^{k/p}.$$

These expressions substituted into Theorem 2 give Theorem 1.

7. Best Possible Constant in Theorem 1

Theorem 3. *If* $p > 0$, $k \geqslant 1$, τ_n *is decreasing and positive, and also* $\Sigma\tau_n/n$ *is divergent, then* $e^{k/p}$ *is the best possible constant in Theorem 1.*

Proof. The simplest of the Euler-Maclaurin sum formulae can be written

$$\sum_{n=1}^{m} f(n) = \int_1^m f(x)\,dx + f(1) + \int_1^m \{x-[x]\}f'(x)\,dx$$

for f having continuous derivative f'. It gives

$$\sum_{n=1}^{m} n^p \log\frac{n+1}{n} = \frac{m^p}{p} + o(m^p) \quad \text{as} \quad m\to\infty.$$

Thus

$$\sum_{n=1}^{m}\{n^p-(n-1)^p\}\log n = m^p\log m - \sum_{n=1}^{m-1} n^p\log\frac{n+1}{n}$$

$$= m^p\{\log m - (1/p) + o(1)\},$$

$$\log\left\{\left(\prod_{n=1}^{m} n^{-k\{n^p-(n-1)^p\}}\right)^{1/m^p}\right\} = -\frac{k}{m^p}\sum_{n=1}^{m}\{n^p-(n-1)^p\}\log n$$

$$= -k\{\log m - (1/p) + o(1)\},$$

$$\left(\prod_{n=1}^{m} n^{-k\{n^p-(n-1)^p\}}\right)^{1/m^p} = m^{-k}e^{k/p}u_m \tag{7}$$

where $u_m \to 1$ as $m\to\infty$.

Given $\epsilon \in (0, 1)$, there is an integer $M > 2/\epsilon$ such that

$$u_m > 1 - \tfrac{1}{2}\epsilon \quad \text{for all } m \geqslant M. \tag{8}$$

Further there is an integer $N > M$ such that

$$\sum_{n=M}^{N} \frac{\tau_n}{n} > (M-1) \sum_{n=1}^{m-1} \frac{\tau_n}{n}.$$

Consequently

$$\sum_{n=1}^{N} \frac{\tau_n}{n} = \sum_{n=1}^{M-1} \frac{\tau_n}{n} + \sum_{n=M}^{N} \frac{\tau_n}{n} < \left\{ \frac{1}{M-1} + 1 \right\} \sum_{n=M}^{N} \frac{\tau_n}{n} = \frac{M}{M-1} \sum_{n=M}^{N} \frac{\tau_n}{n}.$$

With (7) and (8) this gives

$$\begin{aligned}
\sum_{m=1}^{N} m^{k-1} \tau_m \left\{ \prod_{n=1}^{m} n^{-k\{n^p - (n-1)^p\}} \right\}^{1/m^p} &= \sum_{m=1}^{N} m^{-1} \tau_m e^{k/p} u_m \\
&> e^{k/p} \sum_{m=M}^{N} \frac{\tau_m u_m}{m} > e^{k/p} \sum_{m=M}^{N} \frac{\tau_m}{m} (1 - \tfrac{1}{2}\epsilon) \\
&> e^{k/p} (1 - \tfrac{1}{2}\epsilon) \frac{M-1}{M} \sum_{m=1}^{N} \frac{\tau_m}{m} \\
&> e^{k/p} (1 - \tfrac{1}{2}\epsilon)^2 \sum_{m=1}^{N} \frac{\tau_m}{m} \\
&> e^{k/p} (1 - \epsilon) \sum_{m=1}^{N} m^{k-1} \tau_m m^{-k}.
\end{aligned}$$

Let $x_n = n^{-k}$ for $0 < n \leqslant N$, and $x_n = 0$ for $n > N$.

$$\sum_{m=1}^{\infty} m^{k-1} \tau_m \left\{ \prod_{n=1}^{m} x_n^{n^p - (n-1)^p} \right\}^{1/m^p} > (1 - \epsilon) e^{k/p} \sum_{m=1}^{\infty} m^{k-1} \tau_m x_m.$$

A closely similar proof can be used to show that $e^{k/p}$ is also the best possible constant in (C). If this were done it would follow from Corollary 1a that $e^{k/p}$ is also the best possible constant in (1); but only provided that $p \geqslant 1$.

8. References

HLP. G. H. Hardy, J. E. Littlewood, G. Pólya, Inequalities. (Cambridge, 1934).

1. J. A. Cochran and C.-S. Lee, Inequalities related to Hardy's and Heinig's. Math. Proc. Camb. Phil. Soc. 96 (1984) 1-7.
2. H. P. Heinig, Some extensions of Hardy's inequality. SIAM J. Math. Anal. 6 (1975) 698-713.
3. E. R. Love, Inequalities related to those of Hardy and of Cochran and Lee. Math. Proc. Camb. Phil. Soc. 99 (1986) 395-408.
4. E. R. Love, Inequalities related to Knopp's Inequality. J. Math. Anal. Appl. 137 (1989) 173-180.

9 Some Comments on the Past Fifty Years of Isoperimetric Inequalities

Lawrence E. Payne Cornell University, Ithaca, New York

Introduction

At the time of publication of the book of Hardy, Littlewood, and Polya in 1934 the field of isoperimetric inequalities as we understand the term today was essentially nonexistent. It is true that a number of purely geometric results had appeared (see e.g. [2], [5], [11], [47]), but isoperimetric inequalities in physical contexts were very few. However, since the late 1940's there has been considerable research effort directed toward the development of isoperimetric inequalities for a number of physically important quantities. This study was initiated primarily by Pólya himself and his colleagues at Stanford, i.e., Szegö, Schiffer, Garabedian and others. A number of important conjectures were made, many of which have been either proved or disproved by now. However, there are still several outstanding, unproved conjectures some of which I shall mention later on.

Today we use the term isoperimetric inequality to mean any inequality relating two or more domain functionals (defined on the same domain) which is such that the equality sign actually holds for some domain of the class under consideration (or in some appropriate limiting sense). A typical isoperimetric inequality is the following: Of all clamped membranes of given area the circular membrane has the lowest fundamental mode. Mathematically this inequality states that if λ is the eigenvalue corresponding to the solution u of

$$\begin{aligned} &\Delta u + \lambda u = 0 \quad \text{in} \quad \Omega \subset \mathbb{R}^2 \\ &u = 0 \quad \text{on} \quad \partial\Omega \qquad\qquad (1.1) \\ &u > 0 \quad \text{in} \quad \Omega, \end{aligned}$$

This work was supported in part by NSF Grant #DMS-8600250.

then

$$\lambda \geq \pi j_0^2/A, \tag{1.2}$$

where j_0 is the first zero of $J_0(x)$ and A is the area of Ω. It is assumed that Ω is a bounded region with Lipschitz boundary. This inequality is isoperimetric since the equality sign holds if Ω is a disc. Inequality (1.2), referred to in the literature as the Faber-Krahn inequality ([10], [21]) is one of the few isoperimetric inequalities (in the extended sense) that was established prior to the appearance of the book of Hardy, Littlewood, and Pólya.

Isoperimetric inequalities are of considerable mathematical interest in themselves, but they also serve a number of practical purposes. As pointed out in [42], they often permit us to obtain sharp bounds for difficult to compute domain functionals in terms of more easily computed quantities, often in terms of geometric quantities such as the volume, surface area, maximum diameter, or the like. However, they also frequently lead to explicit uniqueness or stability criteria, to optimal a priori bounds and to explicit decay rates, explicit existence criteria, and comparison theorems.

We give one simple illustration: Let Ω be a bounded region in $\mathbb{R}^2$ with smooth convex boundary $\partial\Omega$. If $u \in C^2(\Omega) \cup C^0(\bar{\Omega})$ then $\exists$ positive constants k_1 and k_2 such that

$$\left[\int_\Omega u^2 dx\right]^{1/2} \leq k_1\left[\int_{\partial\Omega} u^2 ds\right]^{1/2} + k_2\left[\int_\Omega (\Delta u)^2 dx\right]^{1/2} \tag{1.3}$$

The optimal k_2 is λ^{-1} where λ is given by (1.1), while the optimal k_1 is the square root of the so-called first Dirichlet eigenvalue (see [46]). Using the inequality (1.2) to replace λ^{-1} and an analogous isoperimetric inequality for the Dirichlet eigenvalue in terms of K_m (the minimum value of the curvature of $\partial\Omega$) we have an explicit a priori inequality which is sharp with easily computable constants. This inequality can then be used to approximate the solution of a Dirichlet boundary value problem

$$\begin{aligned} \Delta v &= F \quad \text{in } \Omega \\ v &= f \quad \text{on } \partial\Omega \end{aligned} \tag{1.4}$$

via the Rayleigh-Ritz technique.

The starting point in the proof of most isoperimetric inequalities in mathematical physics is either a variational characterization of the functional under consideration, or a maximum principle for the solution of the underlying boundary value problem. Often a rearrangement is combined with the variational characterization of the functional in establishing the result. The basic ideas of rearrangements are contained in the book of Hardy, Littlewood, and Pólya, but of course these concepts have been expanded over the years (see e.g. [3], [19], [49]).

As mentioned earlier Pólya and Szegö were largely responsible for generating a wide interest in the study of isoperimetric inequalities with their early work which led to their book [42] in 1950 and with the considerable research in the 1950's by themselves and their colleagues and students at Stanford. They and their book had a great influence on the work of Joseph Hersch and his students at the Technical University in Zürich, on that of Weinberger, Payne and their colleagues and students at the University of Maryland in the 1950's and on many other researchers in this area. By the mid-1960's many of the basic isoperimetric inequalities had been established and except for the continued investigations of Hersch and his students, work on isoperimetric inequalities began to slack off somewhat - possibly because most of the easier or more obvious inequalities had been established by that time.

In a one-hour report one cannot hope to give a comprehensive treatment of the post-1934 work on isoperimetric inequalities. It is possible at best to mention only a few results and indicate a few unproved conjectures. For a more complete picture of what has been done over the past 50 years we refer to the book of Pólya and Szegö [42] in 1951, the book of C. Bandle [2] in 1980 and the book of J. Mossino [22] in 1984 as well as the survey papers by Payne [24] in 1967 and Osserman [23] in 1978.

Due to time consideration we shall restrict our attention to some isoperimetric inequalities that arise in modeling a few simple problems from classical continuum physics. We shall not discuss the many purely geometric results, the numerous results for manifolds or tori or the various isoperimetric inequalities which can be established in various modern nonlinear continuum settings. Isoperimetric results for these simple classical problems do, however, often prove useful in the study of related nonlinear problems.

II. Some Model Problems

Isoperimetric results have been derived for physically interesting quantities in the various areas of the physical and life sciences, in fact they appear in almost all contexts that are amenable to mathematical modeling. As indicated above we can deal with only a few examples and we, therefore, restrict our attention to some of the most familiar and easily understandable problems from the literature. These also happen to be some of the most widely studied examples from mathematical physics, whose isoperimetric results are applicable in a variety of contexts. The problems we will consider are now listed.

A. The Clamped Membrane Problem

The eigenvalues and eigenfunctions of the clamped membrane are respectively the solutions λ and u of (1.1) now defined in $\mathbb{R}^N$, i.e., they are solutions of

$$\begin{aligned} \Delta u + \lambda u &= 0 \quad \text{in} \quad \Omega \subset \mathbb{R}^N \\ u &= 0 \quad \partial\Omega. \end{aligned} \tag{2.1}$$

It is well known that (2.1) possesses an infinite set of positive eigenvalues λ_i which may be ordered as

$$0 < \lambda_1 < \lambda_2 \leq \lambda_3 \cdots \leq \lambda_n < \cdots \to \infty. \tag{2.2}$$

In practice it is probably a sharp lower bound for λ_1 that is most often required, but one is also frequently interested in bounds for the first few eigenvalues.

B. The Free Membrane Problem

Of concern here are the eigenvalues μ_k and eigenfunctions v_k of the Laplace operator under homogeneous Neumann boundary conditions: i.e.,

$$\begin{aligned} \Delta v + \mu v &= 0 \quad \text{in} \quad \Omega \subset \mathbb{R}^N \\ \frac{\partial v}{\partial n} &= 0 \quad \text{on} \quad \partial\Omega. \end{aligned} \tag{2.3}$$

In this case the first eigenvalue μ_1 is zero and corresponds to a constant eigenfunction. Clearly then all eigenfunctions corresponding to nonzero eigenvalues have mean value zero over Ω. We again order the eigenvalues as

$$0 = \mu_1 < \mu_2 \leq \mu_3 \cdots \to \infty. \tag{2.4}$$

In treating Neumann type boundary value problems for second order elliptic operators one may well require a lower bound for μ_2. In other contexts one might require bounds for the first few eigenvalues.

C. Problem of Elastic Torsion

Here the governing equation for the stress function Ψ is

$$\begin{aligned} \Delta\Psi + 2 &= 0 \quad \text{in} \quad \Omega \subset \mathbb{R}^2 \\ \Psi &= 0 \quad \text{on} \quad \partial\Omega, \end{aligned} \tag{2.5}$$

where in this case Ω is assumed to be a bounded simply connected region in $\mathbb{R}^2$. The analogous mathematical problem in $\mathbb{R}^N$ of course makes perfectly good sense and inequalities for functionals of Ψ in higher dimensions are useful in analyzing related problems. Of interest are the Dirichlet integral (torsional rigidity) and the maximum value of τ defined as

$$\tau^2 = |\text{grad } \Psi|^2. \tag{2.6}$$

The point at which τ takes its maximum value is known to occur on $\partial\Omega$ and the maximum value of τ is proportional to the maximum stress. It is of interest to find the point on the boundary at which the maximum stress occurs as well as the location of the interior point (or points) at which the stress vanishes, i.e., the point (or points) at which $\tau = 0$.

D. Buckling Problem for the Clamped Plate

The classical model for the deflection $\Phi(x_1,x_2)$ of an elastic clamped plate subjected to uniform compressive loading in the plane of the plate is

$$\Delta^2\phi + \Lambda\Delta\phi = 0 \quad \text{in} \quad \Omega \subset \mathbb{R}^2$$
$$\phi = \frac{\partial\phi}{\partial n} = 0 \qquad \text{on} \quad \partial\Omega. \tag{2.7}$$

The critical buckling load for the plate problem is proportional to the first eigenvalue Λ_1 of (2.7). Here again the eigenvalues are ordered as follows:

$$0 < \Lambda_1 \leq \Lambda_2 \leq \Lambda_3 \leq \cdots \to \infty. \tag{2.8}$$

The mathematical problem makes sense of course in any number of dimensions and for some comparison results we shall make use of the solution of the problem in $\mathbb{R}^3$.

E. <u>Vibration of an Incompressible Elastic Medium</u>

We mention this problem because of its usefulness in the study of the solution of boundary value problems for the Navier-Stokes equations. Of interest here are the eigenvectors (u_i^1, u_i^2, u_i^3) and eigenvalues ν_i of the system

$$\Delta u^\alpha + \nu u^\alpha = \frac{\partial p}{\partial x_\alpha} \quad \text{in} \quad \Omega \subset \mathbb{R}^N,\ N = 2,3$$
$$\sum_{\alpha=1}^{N} \frac{\partial u^\alpha}{\partial x_\alpha} = 0 \quad \text{in} \quad \Omega \tag{2.9}$$
$$u^\alpha = 0 \qquad \text{on} \quad \partial\Omega.$$

Here the "pressure" term p is not known a priori. This eigenvalue problem also arises naturally in the study of the slow Stokes flow of an incompressible viscous fluid.

There are of course many other interesting boundary value problems and eigenvalue problems in which isoperimetric inequalities have been or can be obtained, e.g. the problems of electrostatic capacity, virtual mass, polarization, and pure plastic torsion as well as membrane and plate eigenvalue problems subjected to different types of loading or boundary conditions, Stekloff and Dirichlet eigenvalue problems and a host of one dimensional problems. It is also possible to obtain isoperimetric inequalities for optimal Sobolev constants, for optimal Korn constants in $\mathbb{R}^2$, etc. However we shall not have time to discuss any of the interesting results in these areas.

3. Various Isoperimetric Inequalities

The proof of the Faber-Krahn inequality (1.2) and its analogue in N-dimensions can today be done in various ways, e.g. using symmetrization arguments together with the variational characterization of λ, or using differential inequalities for $\Omega(u)$ (where $\Omega(\tilde{u})$ is the measure of the set on which $u > \tilde{u}$) together with the classical inequality relating the N volume of a body to its surface area.

The classical result is stated as

Theorem 1: The first eigenvalue of (2.1) is not smaller than that for the N-ball whose volume is the same as that of Ω.

The symmetrization arguments carry over to smooth positive functions defined on a region Ω and vanishing on $\partial\Omega$. Under Steiner or spherical symmetrization (see [42]) the Dirichlet integral does not increase while $\int_\Omega F(u)dx = \int_{\Omega^*} F(u^*)dx$ for any integrable function F. Here Ω^* is the symmetrized region and u^* is the symmetrized function on Ω^*.

An interesting extension of the Faber-Krahn inequality was obtained by Payne and Weinberger [37], i.e.,

Theorem 2: If $\Omega \subset \mathbb{R}^2$ lies interior to a wedge of angle π/α for $\alpha \geq 1$ then

$$\lambda_1 \geq \left[\frac{\pi}{4} \frac{K_\alpha^{-1}}{\alpha(\alpha+1)}\right]^{1/\alpha+1} j_\alpha^2 , \tag{3.1}$$

where

$$K_\alpha = \int_\Omega r^{2\alpha} \sin^2\alpha\, \theta \, dx.$$

Equality holds iff Ω is a circular sector.

Here (r,θ) are polar coordinates with the origin taken at the apex of the wedge, and j_α is the first zero of $J_\alpha(x)$. Since any finite region can be made to lie inside a wedge of arbitrary angle by simply choosing the origin appropriately, Theorem 2 permits considerable freedom in the choice of α. To prove this result the authors had to derive a generalization of the classical isoperimetric inequality relating volume and surface area.

Another classical result for λ_1 is given for $N = 2$ by the result of Pólya and Szegö [42, pp.97-98].

Theorem 3: The first eigenvalue λ_1 of problem (2.1), ($N = 2$), is not greater than that for the circle of the same maximum inner radius, $\dot{r}$, i.e.

$$\lambda_1 \leq \left[\frac{j_0}{\dot{r}}\right]^2 . \tag{3.2}$$

The proof of this theorem was effected by use of conformal mapping techniques, together with the variational characterization of the eigenvalue λ_1. Szegö [48] also used conformal mapping to prove the conjecture of Kornhauser and Stakgold [20] that for $N = 2$, μ_2 is not greater than the μ_2 for the circle of the same area. The following was subsequently proved by Weinberger [51]:

Theorem 4: Of all N-dimensional regions of given N-volume, the N-ball yields the largest value for the μ_2 of (1.3).

Weinberger's proof was based on a simple rearrangement argument coupled with the variational characterization of μ_2.

Another upper bound for λ_1 ($N = 2$) was derived by Payne and Weinberger [39], i.e.,

Theorem 5: For a simply connected region Ω in $\mathbb{R}^2$ the eigenvalue λ_1 is not greater than that for the annular domain (concentric circular boundaries) of the same area fixed on the outer boundary whose perimeter is equal to that of $\partial\Omega$ and free along the inner boundary.

The method used here and in complementary work of Hersch [13], [14] involved a judicious choice for a trial function in the Rayleigh quotient used for bounding λ_1. Another isoperimetric inequality due to Hersch [15] is the following:

Theorem 6: *For any convex domains $\Omega \subset \mathbb{R}^N$*

$$\lambda_1 \geq \pi^2/4\delta^2 \tag{3.3}$$

where δ is the radius of the largest inscribed ball.

The equality sign is never achieved for any bounded region, but as Ω approaches the infinite slab $\lambda_1\delta^2 \to \pi^2/4$. Theorem 6 can be proved either from the variational characterization of λ_1 or from a maximum principle.

Many other isoperimetric results involving λ_1 and/or u_1 have been derived in the literature. We mention only two of those which involve the eigenfunction u_1.

Theorem 7: *For $N = 2$ the quotient*

$$Q :\equiv \frac{(\int_\Omega u_1 dx)^2}{\lambda_1 \int_\Omega u_1^2 dx} \tag{3.4}$$

takes its minimum value for the circle.

This result is due to Payne and Rayner [35] and the N dimensional analogue is due to Jobin-Kohler [18]. This result (3.4) again follows from a differential inequality for $\Omega(u)$, while the proof of Jobin-Kohler rests on a new type of symmetrization.

Theorem 8: *For all convex domains in $\mathbb{R}^N$*

$$\int_\Omega u_1 dx/MV < 2/\pi. \tag{3.5}$$

Here V designates the measure of Ω and M is the maximum value of u_1 in Ω. Again the equality sign holds in the limit for an infinite slab. The proof does not actually require the convexity of Ω but rather that the average curvature be non-negative on $\partial\Omega$. This inequality was established by Payne and Stakgold [36] using a differential inequality for $\Omega(u)$ combined with a maximum principle.

A lower bound for μ_2 was derived by Payne and Weinberger [38] using a device of domain cutting combined with the variational characterization of μ_2. It is as follows:

Theorem 9: The first nonzero eigenvalue μ_2, for a convex domain Ω satisfies the inequality

$$\mu_2 \geq \pi^2 D^{-2} \tag{3.6}$$

where D is the diameter of Ω.

Here again the equality sign holds in the limit for the infinite slab.

Let us now make a few remarks on the comparison of the eigenvalues of the various problems. The following isoperimetric theorems have been derived in the literature.

Theorem 10: For a region $\Omega \subset \mathbb{R}^N$

$$\lambda_2 \leq \Lambda_1 . \tag{3.7}$$

Theorem 11: If $\Omega \subset \mathbb{R}^N$ is such that the average curvature of $\partial\Omega$ is everywhere positive then

$$\Lambda_1 \leq 4\lambda_1 . \tag{3.8}$$

Theorem 12: For $N = 2$ the following identity holds

$$\nu_1 = \lambda_2, \tag{3.9}$$

and for general N

$$\nu_1 \leq \Lambda_1. \tag{3.10}$$

Theorem 10 was conjectured by Weinstein [53] and proved by Payne and Weinberger [31] using combinations of ϕ_1 and its derivatives as admissible functions in the Rayleigh quotient for λ_2. Theorem 11 was established by Payne [26] using u_1^2 as an admissible function in the Rayleigh quotient for Λ_1. The identity (3.9) was proved by Velte [50] and (3.10) follows from taking appropriate combinations of the derivatives of ϕ as admissible functions in the Rayleigh quotient for ν_1. By (3.7) it is clear that (3.10) is isoperimetric in two space, but it is not known to be isoperimetric for $N \geq 3$.

The basic isoperimetric inequality in the elastic torsion problem (2.5) is that conjectured by St.Venant [44] in 1856 and proved by Pólya [40] in 1948. We define

$$S := \int_\Omega |\mathrm{grad}\ \psi|^2 dx. \tag{3.11}$$

The quantity S is proportional to the torsional rigidity of the elastic beam. The St. Venant conjecture (proved by Pólya) is stated as

Theorem 13: Of all isotropic beams of given simply connected cross sectional area, the circular beam has the highest torsional rigidity.

Stated as an inequality Theorem 13 reads

$$2\pi S \leq A^2 \tag{3.12}$$

where A is the area of Ω. This result was originally established using symmetrization arguments. Payne [27] has in fact derived the following set of inequalities

$$A - [A^2 - 2\pi S]^{1/2} \leq 2\pi\psi_M \leq (2\pi S)^{1/2} \leq A \tag{3.13}$$

with the equality sign throughout iff Ω is the disc. Here $\psi_M :\equiv \max_\Omega \psi$.

For convex regions we have the following theorem:

Theorem 14: If Ω is convex then

$$\max_\Omega \tau^2 < 4\psi_M < 4\delta^2 \tag{3.14}$$

where δ is the radius of the largest inscribed circle.

The following isoperimetric results are also valid:

$$SA^{-1}\delta^{-2} < 4/3 \tag{3.15}$$

and

$$SA^{-3}L^2 > 4/3 \tag{3.16}$$

with the equality sign being approached in the limit as the domain degenerates to an infinite strip. Inequalities (3.14) were derived by

Payne [25] using a maximum principle, as was (3.15). The inequality (3.16) is due to Pólya [41a]. As was pointed out in [29] another interpretation of the inequality (3.14) states that the point at which Ψ assumes its maximum value must occur at a distance greater than A/L from the boundary.

We mention finally two isoperimetric inequalities which relate functionals in the elastic torsion problem with functionals in the clamped membrane problem. For instance it was conjectured by Pólya and Szegö [42] and proved by Jobin-Kohler [17] that

Theorem 15: The quantity $S\lambda_1^2$ takes its minimum value when Ω is a disc,

and Payne [28] has shown that

Theorem 16: If Ω is convex then the quantity $\lambda_1\Psi_M$ takes its minimum value in the limit as Ω tends to the infinite strip.

Numerous other isoperimetric inequalities for functionals arising in the five problems we have considered appear in the by-now extensive literature. For additional results we refer the reader to the papers and books cited in the bibliography.

4. Conjectures and Open Questions

In this section we list a few as yet unproved conjectures regarding functionals or solutions of the five problems we have introduced.

Conjecture 1: The first eigenvalue, Λ_1, in the clamped plate problem is not less than that of the disc of the same area.

This conjecture was made by Pólya and Szegö [42] and was proved by Szegö [48a] for the special situation in which the first eigenfunction does not change sign in Ω. It is an open question as to whether it is possible to characterize geometrically those domains for which the first eigenfunction does not change sign in Ω.

Conjecture 2: For $N = 2$ the kth eigenvalue λ_k satisfies the inequality

$$\lambda_k \geq 4\pi k A^{-1}. \tag{4.1}$$

This conjecture is due to Pólya [41] and has been proved by him for plane covering (tiling) domains. The inequality is isoperimetric in the sense that

$$\lim_{k\to\infty} \lambda_k A k^{-1} = 4\pi. \tag{4.2}$$

<u>Conjecture 3</u>: <u>For $N = 2$ the kth eigenvalue μ_k satisfies the inequality</u>

$$\mu_k \leq 4\pi(k-1)A^{-1}. \tag{4.3}$$

This conjecture is also due to Pólya [41] and has been proved by him for a class of tiling domains. It is again isoperimetric in the sense described above. For more information on the status of Conjectures 2 and 3 see Protter [43].

The analogues of Conjectures 2 and 3 in higher dimensions follow from Pólya's arguments and suggest the weaker conjecture that $\mu_{k+1} < \lambda_k$ for arbitrary k and for arbitrary Ω in $\mathbb{R}^N$. This result is clearly true for tiling domains and for domains with boundaries of non-negative average curvature (see Aviles [1] and Levine and Weinberger [21a]).

<u>Conjecture 4</u>: <u>For any n and arbitrary N the ratio λ_{n+1}/λ_n is not greater than the ratio λ_2/λ_1 for the N-ball.</u>

This conjecture is due to Payne, Pólya, and Weinberger [34] who showed that

$$\lambda_{n+1}/\lambda_n \leq (1 + \frac{4}{N}). \tag{4.4}$$

For $N = 2$ the ratio λ_2/λ_1 for the circle is

$$\lambda_2/\lambda_1 \approx 2.539. \tag{4.5}$$

The bound for λ_{n+1}/λ_n has been improved by various authors (see e.g. Brands [6], de Vries [9], Hile and Protter [43], Chiti [8]), the sharpest bound today being that of Chiti

$$\lambda_2/\lambda_1 \leq 2.586$$

which was obtained by use of rearrangement methods. For related results see [43].

Conjecture 5: The eigenfunction u_2 cannot have a closed interior nodal surface.

We mention this conjecture here even though it is not in the nature of an isoperimetric inequality since certain isoperimetric results could follow from the conjectured behavior of u_2. This conjecture is stated in [24] and has been proved for special geometries (N = 2) by Payne [30] and Lin [21b]. In [24] there is also listed a conjecture that convex domains must have convex level surfaces. This conjecture was proved by Brascamp and Lieb [7] in 1974.

In addition to the above conjectures (and there are many more indicated in [42], some of which have been subsequently proved) there are also questions that need to be answered, e.g.

Question 1: Is it possible to characterize geometrically the boundary point at which the stress τ in problem C takes its maximum value?

It is known (see [29], [32]) that the maximum stress cannot occur at a point on the boundary at which the curvature K is greater than $L/2A$, where L is the perimeter of $\partial\Omega$ and A the area of Ω. Other results for special geometries have been obtained by Kawohl [19a].

Question 2: For $N = 3$ is $\nu_1 \leq \lambda_2$?

As indicated in the text this result is certainly true for $N = 2$ since Velte [50] has shown in this case that $\nu_1 = \lambda_2$.

Question 3: Is it possible to find a sharp explicit lower bound for $\min_{\partial\Omega}|\text{grad } u_1|$ (Problem A) in terms of the geometry of the domain?

Such a lower bound would be of considerable theoretical interest, but it would also lead to improved isoperimetric inequalities (see e.g. [36]). Here u_1 is assumed to be normalized.

Question 4: Is it possible to characterize geometrically those domains for which ϕ_1 does not change sign in Ω?

We have discussed this question earlier as well as a possible application of the result.

Question 5: If Ω is convex is it possible to find an isoperimetric upper bound for the ratio μ_{n+1}/μ_n for $n \geq 2$?

This question was posed recently by Protter [43]. It seems clear that without some restriction on the geometry the ratio could be made larger than any prescribed number M for any given finite value of n.

The following question is also of considerable interest.

Question 6: Of all domains of given area which domain provides the largest value for $\lambda_2 - \lambda_1$? The gap between eigenvalues for the Schrödinger equation has received considerable attention in the recent literature and the above question might be thought of in this context for the case in which the potential is zero.

We have listed only a few of the multitude of interesting questions that could be asked about solutions (or functionals of solutions) of the five problems listed in this paper.

There are also related classical questions for overdetermined problems of the following type:

Question 7: (Pompeiu's Problem) If for some region Ω, Problem B has a solution v_k (corresponding to some eigenvalue μ_k) which has the property that v_k = constant on $\partial\Omega$, is Ω the N-ball?

or

Question 8: If for some region Ω Problem A has a solution u_k (corresponding to some eigenvalue λ_k) which has the property that $\partial u_k/\partial n$ = constant on $\partial\Omega$, is Ω the N-ball?

These questions have been discussed in the literature [1], [4] but as of the present moment the answer is not known. It is known that if the boundary $\partial\Omega$ contains a segment of a spherical surface which has finite measure then the answer to both of questions 7 and 8 is yes. The analogous result for Problem C was answered in the affirmative by Serrin [45] and Weinberger [52]. One could of course ask related questions in Problem 4 and 5.

Let us list as a final question one posed by Van den Berg, i.e.,

Question 9: Is it possible to obtain for convex Ω an explicit bound of the following type for u_M, the maximum value in Ω of the first eigenfunction in Problem A?

$$u_M \le F(V, \int_\Omega u^2 dx)?$$

One looks for an explicit F which is independent of the eigenvalue or other geometrical quantities. Here V is of course the volume of Ω.

Again let us refer to the books and survey articles mentioned in the bibliography for references to many additional isoperimetric inequalities and conjectures.

Bibliography

1. Aviles, P., Symmetry theorems related to Pompeiu's problem, Amer. J. Math. 108 (1986), pp.347-350.

2. Bandle, C., Isoperimetric inequalities and applications, Pitman Lecture Series #7 (1980).

3. Bandle, C. and Mossino, J., Rearrangement in variational inequalities, Ann. Mat. Pura Appl., 138 (1984), pp.1-14.

4. Berenstein, C. O., The problem of Pompeiu, Proceedings of the Ninth Brazilian Colloquium, 1 (1973), pp.31-37.

5. Blaschke, W., Kreis und Kugel, Leipzig (1916).

6. Brands, J. J. A. M., Bounds for the ratios of the first three membrane eigenvalues, Arch. Rat. Mech. Anal., 16 (1964), pp.265-268.

7. Brascamp, H. J. and Lieb, E. H., Some inequalities of gaussian measure and a long-range order of the one dimensional plasma, in Functional Integration and its Applications (ed. A. Arthur) Oxford (1975), pp.1-14.

8. Chiti, G., A bound for the ratio of the first two eigenvalues of a membrane, SIAM J. Math. Anal., 14 (1983), pp.1163-1167.

9. de Vries, H. L., On the upper bound for the ratio of the first two membrane eigenvalues, Z. Naturforsch, 22a (1967), pp.152-153.

10. Faber, G., Beweis dass unter allen homogenen membranen von gleicher Fläche und gleicher Spannung die kreisförmige den tiefsten Grundton gibt, Sitz. Bayer. Akad. Wiss. (1923), pp.169-172.

11. Hadwiger, H., Vorlesungen über Inhalt, Oberfläche und Isoperimetrie, Springer Verlag, Berlin, Heidelberg and New York (1957).

12. Hardy, G. H., Littlewood, J. E. and Pólya, G., Inequalities, Cambridge University Press (1934).

13. Hersch, J., Contribution to the method of interior parallels applied to vibrating membranes, Studies in Mathematical Analysis and Related Topics: Essays in honor of G. Pólya, Stanford University Press, Stanford, CA (1962), pp.132-139.

14. Hersch, J., The method of interior parallels applied to polygonal or multiply connected membranes, Pacific J. Math., 13 (1963), pp.1229-1238.

15. Hersch, J., Sur la frequence fondamentale d'une membrane vibrante: Evaluations par defaut et principe de maximum, ZAMP, 11 (1960), pp.387-415.

16. Hile, G. N. and Protter, M. H., Inequalities for eigenvalues of the Laplacian, Indiana Univ. Math. J., 29 (1980), pp.523-538.

17. Jobin-Kohler, M. T., Une methode de comparaison isopérimétrique de fonctionelles de domaines de la physique mathématique. 1 Une démonstration de la conjecture isoperimetrique $P\lambda^2 \geq \pi j_0^4/2$ de Pólya et Szegö, ZAMP, 24 (1978), pp.757-766.

18. Jobin-Kohler, M. T., Sur le première fonction propre d'une membrane: une extension a N dimensions de ℓ'inègalité isoperimétrique de Payne-Rayner, ZAMP, 28 (1977), pp.1137-1140.

19. Kawohl, B., Rearrangements and convexity of level sets in P.D.E., Lecture Notes in Math. #1150, Springer Verlag, Berlin (1985).

19a. Kawohl, B., On the location of the maxima of the gradient of solutions to quasilinear elliptic problems and a problem of St. Venant, J. Elasticity, 17 (1987), pp.195-206.

20. Kornhauser, E. T. and Stakgold, I., A variational theorem for $\nabla^2 u + \lambda u = 0$ and its applications, J. Math. and Phys., 31 (1952), pp.45-54.

21. Krahn, E., Über eine von Rayleigh formulierte Minimaleigenschaft des Kreises, Math. Ann., 94 (1924), pp.97-100. See also Über Minimaleigenschaften der Kugel in drei und mehr Dimensionen, Acta Comm. Univ. Tartu (Dorpat), Ag (1926), pp.1-44.

21a. Levine, H. A. and Weinberger, H. F., Inequalities between Dirichlet and Neumann eigenvalues, Archive Rat. Mech. Anal., 94 (1986), pp.193-208.

21b. Lin, Chang-Sou, On the second eigenfunctions of the Laplacian in $\mathbb{R}^2$, preprint.

22. Mossino, J., Inégalites Isopérimétriques et Applications en Physique, Hermann, Paris (1984).

23. Osserman, R., The isoperimetric inequality, Bull. Amer. Math. Soc., 84 (1978), pp.1182-1238.

24. Payne, L. E., Isoperimetric inequalities and their applications, SIAM Review, 9 (1967), pp.453-488.

25. Payne, L. E., Bounds for the maximum stress in the St. Venant torsion problem, Indian J. Mech. and Math., Special issue dedicated to B. Sen (1968), pp.51-59.

26. Payne, L. E., A note on inequalities for plate eigenvalues, J. Math. and Phys., 39 (1960), pp.155-159.

27. Payne, L. E., Some isoperimetric inequalities in the torsion problem for multiply connected regions, Studies in Mathematical Analysis and Related Topics: Essays in honor of G. Pólya, Stanford University Press, Stanford, CA (1962), pp.270-280.

28. Payne, L. E., Bounds for solutions of a class of quasilinear elliptic boundary value problems in terms of the torsion function, Proc. Roy. Soc. Edinburgh, 88A (1981), pp.251-256.

29. Payne, L. E., Some applications of "best possible" maximum principles in elliptic boundary value problems, Partial Differential Equations and Dynamical Systems, Pitman Research Notes #101 (1984), pp.286-313.

30. Payne, L. E., On two conjectures in the fixed membrane eigenvalue problem, ZAMP, 24 (1970), pp.721-729.

31. Payne, L. E., Inequalities for eigenvalues of membranes and plates, J. Rat. Mech. Anal., 4 (1955), pp.517-528.

32. Payne, L. E. and Philippin, G. A., Isoperimetric inequalities in the torsion and clamped membrane problems for convex plane domains, SIAM J. Math. Anal., 14 (1983), pp.1154-1162.

33. Payne, L. E. and Philippin, G. A., Some remarks on the problems of elastic torsion and of torsional creep, Some Aspects of Mechanics of Continua, Sen Memorial Committee (1977), pp.32-40.

34. Payne, L. E., Pólya, G., and Weinberger, H. F., On the ratio of two consecutive eigenvalues, J. of Math. and Phys., 33 (1956), pp.289-298.

35. Payne, L. E. and Rayner, M., An isoperimetric inequality for the first eigenfunction in the fixed membrane problem, ZAMP, 23 (1972), pp.13-15.

36. Payne, L. E. and Stakgold, I., On the mean value of the fundamental mode in the fixed membrane problem, Appl. Analy., 3 (1973), pp.295-303.

37. Payne, L. E. and Weinberger, H. F., A Faber-Krahn inequality for wedge-like membranes, J. Math. and Phys., 39 (1960), pp.182-188.

38. Payne, L. E. and Weinberger, H. F., An optimal Poincaré inequality for convex domains, Arch. Rat. Mech. Anal., 5 (1960), pp.182-188.

39. Payne, L. E. and Weinberger, H. F., Some isoperimetric inequalities for membrane frequencies and torsional rigidity, J. Math. Anal. Appl., 2 (1961), pp.210-216.

40. Pólya, G., Torsional rigidity, principal frequency, electrostatic capacity and symmetrization, Quart. Appl. Math., 6 (1948), pp.267-277.

41. Pólya, G., On the eigenvalues of vibrating membranes, Proc. Lond. Math. Soc., 11 (1961), pp.419-433.

41a. Pólya, G., Two more inequalities between physical and geometrical quantities, J. Ind. Math. Soc., 24 (1960), pp.413-419.

42. Pólya, G. and Szegö, G., Isoperimetric inequalities in mathematical physics, Annals of Math. Studies #27, Princeton Univ. Press, Princeton, N.J. (1951).

43. Protter, M. H., Can one hear the shape of a drum? revisited, SIAM Review, 29 (1987), pp.185-197.

44. St. Venant, B. de, Mémoire sur la torsion des prismes, pres., divers. savants Acad. Sci., 14 (1856), pp.233-560.

45. Serrin, J., A symmetry problem in potential theory, Arch. Rat. Mech. Anal., 43 (1971), pp.304-318.

46. Sigilitto, V., Explicit a prior inequalities with applications to boundary value problems, Research Notes in Math. #13, Pitman Publishing, London (1977).

47. Steiner, J., Einfache Beweise der isoperimetrischen Hauptsätze, Ges. Werke II, Berlin (1882), pp.75-91.

48. Szegö, G., Inequalities for certain eigenvalues of a membrane of given area, J. Rat. Mech. Anal., 3 (1954), pp.343-356.

48a. Szegö, G., On membranes and plates, Proc. Natl. Acad. Sci. U.S.A., 36 (1950), pp.210-216, see also Note to my paper "On membranes and plates", Proc. Natl. Acad. Sci. U.S.A., 44 (1958), pp.314-316.

49. Talenti, G., Elliptic equations and rearrangements, Ann. Scuola Norm. Sup. Pisa Cl. Sci., 3 (1976), pp.697-718.

50. Velte, W., Über ein Stabilitatskriterium in Hydrodynamik, Arch. Rat. Mech. Anal., 9 (1962), pp.9-20.

51. Weinberger, H. F., An isoperimetric inequality for the N-dimensional free membrane problem, J. Rat. Mech. Anal., 5 (1956), pp.633-636.

52. Weinberger, H. F., Remark on the preceding paper of Serrin, Arch. Rat. Mech. Anal., 43 (1971), pp.319-320.

53. Weinstein, A., Étude des spectres des équations aux derivees partielles de la theorie des plaques élastiques, Mem. Sci. Math. #88 (1937).

10 Operator Inequalities and Applications

Johann Schröder Universität Köln, Cologne, Federal Republic of Germany

This paper is concerned with inequalities which involve abstract operators, briefly: *operator inequalities*. We will mainly consider operator inequalities which originate from *operator equations*

$$Mu = r \quad , \tag{0.1}$$

where $M : R \to S$ is a given (linear or nonlinear) operator, $r \in S$, and $u \in R$ denotes the solution to be determined. The relation $Mu \leq r$ (in an ordered space S) is a very simple example of such an inequality.

One of our main objects is to obtain information on a solution of an operator equation (0.1) by solving operator inequalities related to this equation. The theory, however, allows also other applications, for instance, to eigenvalue theory.

Sections 1, 2, and 3 deal with (a priori) *inclusion statements*. The various statements considered can all be written in the form

$$Mu \in C \quad \Rightarrow \quad u \in K \quad , \tag{0.2}$$

where "$u \in K$" represents the inclusion statement desired, and C is a set described by inequalities involving M, i.e., operator inequalities. A statement like (0.2) will also be called a *range-domain implication* (for M). A simple example is the statement

$$M\varphi \leq Mu \leq M\psi \quad \Rightarrow \quad \varphi \leq u \leq \psi \quad ,$$

which holds for *inverse-monotone* operators M in ordered spaces.

By applying (0.2) to the operator equation (0.1), one obtains the following statement:

$$r \in C \quad \Rightarrow \quad u \in K \text{ for each solution of } Mu = r \quad . \tag{0.3}$$

Section 4 treats *existence and inclusion statements* of a corresponding form:

$$r \in C \quad \Rightarrow \quad \text{there is a solution } u \in K \text{ of } Mu = r \quad . \tag{0.4}$$

If a statement of the latter form holds, one can, in theory, carry out an existence proof by verifying the inequalities which the premise "$r \in C$" represents.

In section 5 we will show that, for certain problems, the verification of all assumptions required for an existence proof can be reduced to the construction of two algorithms: an algorithm for calculating approximate solutions, and an algorithm for calculating (constant) bounds for certain elements. In addition, we will briefly report on a corresponding program for two-point boundary problems.

The theories in sections 1 through 4 contain large parts of the theories on *differential inequalities, integral inequalities, matrix inequalities*, etc. In this sense our paper represents a survey. It is not our intention, however, to discuss systematically all theories which lead to statements of the form (0.3) or (0.4). Instead, we will describe *one particular approach*, which is very simple, and consistent in its terms and arguments. For example, the theory of existence and inclusion statements (0.4) will be based on the theory of the corresponding range-domain implications (0.2).

Although, naturally, the abstract theory will not provide "all" results which have been proved for more special operators (such as differential operators, etc.), the abstract approach has some advantages. The unified abstract theory more clearly exposes some features common to the more special theories. For instance, it becomes rather obvious why certain assumptions are made, certain restrictions have to be imposed, etc. Also, this understanding of the underlying ideas makes it comparibly easy to obtain new results. For example, certain results on functional-integro-differential inequalities can be derived from the abstract theorems almost without any effort. We will need such a generalization, for instance, in section 5.22.

The publication of the book of Hardy, Littlewood and Polya [24] has stimulated the development of the theory of inequalities in general. There seems to be no direct relation, however, between the content of the book and the theory of operator inequalities described here. As explained above, we present here an approach to incorporate various special theories into a unified abstract theory. The historical development of the ideas and methods described has mostly taken place in the special fields and theories from which the abstract theory originated. The most important ones are:

(1) matrix inequalities and related subjects;

(2) differential inequalities (a priori inclusion by two-sided bounds);

(3) comparison theories for differential equations (existence, and inclusion by two-sided bounds);

(4) the Perron-Frobenius theory of positive operators;

(5) the theory of iterative procedures;

(6) invariance theory, and shape-invariant bounds.

It is certainly not possible to describe here the history of all these fields, and we will make no attempt to do this. Instead we will, in most cases, mention some important early papers, some survey papers, and a few books where further references can be found.

To (1): Ostrowski [49,50] introduced the notion of an M-matrix and made important contributions to this field. Systematical treatments and surveys are due to Fan [18] and Fiedler and Ptàk [19]. See also the book of Berman and Plemmons [10].

To (2), *initial value problems*: This field has a long history, the early part of which is marked by names such as Kamke [29], Müller [43,44], Nagumo [45], Szarski [74], Wazewski [81]. A detailed historical survey can be found in the book of Walter [80]. See also the books of Szarski [75], Lakshmikantham and Leela [37].

To (2), *boundary value problems*: This theory is closely related to the boundary maximum principle. See Hopf [27], and the book of Protter and Weinberger [54]. Much work in this field has been done by Adams, Redheffer, Spreuer (see, e.g., [2], [55], [56].)

A first abstract formulation of the operator properties used in (1) and (2) is due to Collatz [13], who introduced the concept of an *operator of monotonic kind*, and discussed a series of applications to matrices, differential operators, etc. See also the books of Collatz [14,15], and that of Beckenbach and Bellman [9]. The development of the operator theory described in this paper started in [66,67].

To (3): In this paper we report only on results derived with the method of modification, emphasizing the general procedure more than the results. Modified problems of one or the other form have been used in the existence theories for two-point boundary value problems by Erbe [17], Hartman [25,26], Jackson and Schrader [28], Knobloch [31], Schmitt [63], Schrader [65], and others.

To (4): This theory, which essentially started with the papers of Perron [51], and Frobenius [20], is treated here only marginally. Corresponding operator theories are provided by Krasnosel'skii [32], Krein and Rutman [33], Sawashima [61], and Schaefer [62].

To (5): The theory of iterative procedures is closely related to that of inverse-positive matrices, and, more generally, inverse-positive operators. See Ostrowski [50] for early work in this field. A few remarks in our paper will make this close relation apparent. For more details one may consult, for example, the books of Bohl [11], Collatz [15], Varga [76].

To (6): Many results on invariance and shape-invariant bounds have been obtained more recently. Some comments on relevant literature are made in section 3.

That inverse-positivity and inverse-monotonicity (or monotonic type) are appropriate tools for deriving a posteriori error estimates was shown by Collatz [13]. We cannot possibly attempt to survey the abundance of examples where such methods were applied in the meantime. In section 5 we report on possibilities for carrying out rigorous existence proofs with algorithms based on the theories in sections 1 through 4. Various other attempts have been made for obtaining verified bounds by numerical means. For approaches using iterative techniques, fixed-point theorems, and other means, combined with interval arithmetic, see Adams and Ames [1], Kaucher and Miranker [30] and Nickel [48], to give just some references as typical examples.

Many more references on the subjects (1) through (5) can be found in [69]. This book contains some of the basic results reported on in the present paper, e.g., the Monotonicity Theorem, and its generalization, Theorem 2.1. (Proofs which can be found in [69] are not repeated in this paper.)

In this paper, however, we emphasize even more strongly the description of important properties of operators by positive linear functions ℓ, and certain characteristic terms P_ℓ and Q_ℓ related to these functionals. In this way, many basic results on inequalites for matrices, differential operators, etc. can be derived from the abstract results by a kind of (rather simple) translation process: one chooses the (point-)functionals ℓ properly, determines the quantities P_ℓ and Q_ℓ, and applies the corresponding abstract theorem . — We also like to mention that the theory of $\mathbf{Z}$-operators and its applications have partly been reformulated, and extended.

There is practically no difference in the way the results on a priori inclusion are applied to ordinary differential equations (of the first or second order), and to elliptic-parabolic partial differential equations (of the second order). Naturally, essential differences come up, whenever existence statements are needed.

A series of further theories and results related to the subject of this paper cannot be discussed here, for example: the theory of weak differential inequalities, special results for differential operators of divergence form, methods for treating differential operators on unbounded domains, the theory of super- and subfunctions (for existence proofs), the use of degree theory for existence proofs, inclusion and existence by monotonic iterations, interval analysis.

Notation. We will use some simple terms and definitions of order theory. Most of them are standard. Therefore, we will explain here only a few of them.

All (abstract or concrete) linear spaces considered in this paper are supposed to be real linear spaces.

Let $(R, \leq)$ be an *ordered linear space* with *order cone* $K_R = \{u \in R : u \geq 0\}$. (Some authors use the term *partially* ordered linear space.) Such a space is called *Archimedian* if the intersection of any line with K_R is closed in the line topology, or, equivalently, the following is true:

if $u \geq \frac{1}{n} v$ for some $u, v \in R$ and all $n \in \mathbb{N}$, then $u \geq 0$.

An essential feature of the theories below is the simultaneous use of several different linear orders (linear order relations), denoted by

$\leq$ (the given order),

$\preceq$ (the corresponding *strict order*),

$\overset{\backprime}{\leq}$ (a *strong order*, i.e., an order stronger than $\leq$) .

Elements $u \geq 0$ (i.e., $0 \leq u$) are called *positive*. The signs $\geq$ and $>$ are related to each other as follows:

$$u > 0 \quad \Leftrightarrow \quad (u \geq 0 \ , \ \text{and } u \neq 0) \quad ; \quad u \geq 0 \quad \Leftrightarrow \quad (u > 0 \ , \ \text{or } u = 0) \quad .$$

Corresponding relations hold for $\succeq$ and $\succ$, as well as for $\overset{\backprime}{\geq}$ and $\overset{\backprime}{>}$.

The *strictly positive* elements $u \succ 0$ are defined by the property that to each $v \in R$ there is an $n \in \mathbb{N}$ with $nu \geq v$. A linear order $\overset{\backprime}{\leq}$ is called *stronger* than the given order $\leq$ (or, simply, a *strong order* in $(R, \leq)$), if $u \overset{\backprime}{>} 0$ for at least one $u \in R$, and

$$u \overset{\backprime}{\geq} 0 \quad \Rightarrow \quad u \geq 0 \quad ; \quad (u \overset{\backprime}{>} 0 \ , \ v \geq 0) \quad \Rightarrow \quad u + v \overset{\backprime}{>} 0 \quad . \tag{0.5}$$

Elements $u \overset{\backprime}{>} 0$ are called *strongly positive*.

Example 0.1 The properties (0.5) hold in the following two cases:

$$\text{(i)} \quad u \overset{\backprime}{>} 0 \quad :\Leftrightarrow \quad u > 0 \quad , \quad \text{(ii)} \quad u \overset{\backprime}{>} 0 \quad :\Leftrightarrow \quad u \succ 0 \quad .$$

When considering operators $M : R \to S$ we will often require that the following assumption on the spaces R and S is satisfied (even if not all of the properties mentioned are needed for each of the results).

Assumption (**S**).

$(R, \leq)$ and $(S, \leq)$ are ordered linear spaces,

$(R, \leq)$ is Archimedian,

$\overset{\backprime}{\leq}$ denotes a strong order in $(S, \leq)$ with order cone W.

As in this assumption we will use the sign $\leq$ for order relations in different spaces, even if these spaces may have the same elements. The notation $(R, \leq) \subset (S, \leq)$ indicates that $(R, \leq)$ is an ordered linear subspace of $(S, \leq)$, i.e., R is a linear subspace of S, and the order in R is induced by the order in S.

A linear operator A which maps an ordered linear space $(R, \leq)$ into an ordered linear space $(S, \leq)$ is called *positive*, $A \geq 0$, if $u \geq 0 \Rightarrow Au \geq 0$ for all $u \in R$. In particular, this notation will be used for linear functionals $\ell : R \to \mathbb{R}$.

Whenever in the following a linear space of functions with values in $\mathbb{R}^n$ is considered, the sign $\leq$ will denote the *natural* (component- and pointwise) order.

Example 0.2 Let Ω be a bounded domain in IR^m (with boundary $\partial\Omega$ and closure $\overline{\Omega}$),

$S = \mathrm{IR}(\overline{\Omega})$ the space of real-valued functions on $\overline{\Omega}$,

$R = C_0(\overline{\Omega})$ the subspace of continuous functions.

Then the strictly positive elements $u \in R$ are described by

$$u \succ 0 \quad \Leftrightarrow \quad u(x) > 0 \quad \text{for all } x \in \overline{\Omega} \quad .$$

(The same is true for $R = C_2(\overline{\Omega})$, $R = C_0(\overline{\Omega}) \cap C_2(\Omega)$, etc., where the lower indices indicate differentiability properties.)

The space $(S, \leq)$ does not contain any strictly positive element. A strong order in S is given by

$$u \mathrel{\grave{>}} 0 \quad :\Leftrightarrow \quad u(x) > 0 \quad \text{for all } x \in \overline{\Omega} \quad .$$

Other strong orders are given by (i) in Example (0.1), and by

$$u \mathrel{\grave{>}} 0 \quad :\Leftrightarrow \quad (u \geq 0 \ , \ u(x) > 0 \text{ for all } x \in \partial\Omega) \quad .$$

Different choices of the strong order lead to different results.

By

$$\mathrm{IR}^n(\overline{\Omega}) \quad , \quad C_0^n(\overline{\Omega}) \quad , \quad C_2^n(\Omega) \quad ,\ldots$$

we denote the linear spaces of functions with values $u(x) = (u_i(x))$ in IR^n such that all components u_i $(i = 1, 2, \ldots, n)$ belong to, respectively,

$$\mathrm{IR}(\overline{\Omega}), \quad , \quad C_0(\overline{\Omega}) \quad , \quad C_2(\Omega) \quad ,\ldots$$

Acknowledgement. I would like to thank Mrs. Marion Adam for carefully preparing this manuscript on TEX, and Michael Göhlen for his technical assistance (in constructing new symbols for TEX, etc.). I also thank Michael Plum for fruitful discussions on various subjects.

1. Inverse-positive linear operators

Throughout this section we assume that Assumption **(S)** is satisfied.

1.1 The Monotonicity Theorem

Let $M : R \to S$ denote a linear operator. M is called *inverse-positive* (on R), if for each $u \in R$

$$Mu \geq 0 \quad \Rightarrow \quad u \geq 0 \quad .$$

This implication holds *if and only if M has a positive inverse*, i.e., M^{-1} exists, and $M^{-1}U \geq 0$ for all $U \geq 0$ in MR. The following theorem is the basic result on inverse-positivity.

Theorem 1.1 (*Monotonicity Theorem*) *Let M have the following two properties.*

I. *M is pre-inverse-positive, i.e., for each $u \in R$*

$$(u \geq 0 \ , \ Mu \mathrel{\grave{>}} 0) \quad \Rightarrow \quad u \succ 0 \quad .$$

II. *M has a majorizing element z, i.e., there is a $z \in R$ with $z \geq 0$, $Mz \mathrel{\grave{>}} 0$.*

Then for each $u \in R$,

$$Mu \geq 0 \quad \Rightarrow \quad u \geq 0 \quad , \text{ and } \quad Mu \mathrel{\grave{\geq}} 0 \quad \Rightarrow \quad u \succeq 0 \quad . \tag{1.1}$$

The terms introduced in I, II depend on the given strong order with order cone W. Sometimes we will indicate this by using the term W-*pre-inverse-positive*. We will, in particular, say that M is *weakly pre-inverse-positive* if $(u \geq 0 \ , \ Mu \succ 0) \Rightarrow u \succ 0$, and that M is *strictly pre-inverse-positive* if $(u \geq 0 \ , \ Mu > 0) \Rightarrow u \succ 0$. Moreover, M is called *strictly inverse-positive*, if $Mu \geq 0 \Rightarrow u \succeq 0$. This property is equivalent to the second property in (1.1) for $\mathrel{\grave{\leq}}$ identical with $\leq$.

If the range of M contains an element $\succ 0$, *then condition* II *is necessary for inverse-positivity.*

Example 1.2 Let $R = S = \mathbb{R}^n$. An $n \times n$-matrix $M = (m_{ik})$ is called a Z-*matrix* if $m_{ik} \le 0$ for $i \ne k$ (see [19]). **Z** denotes the class of all Z-matrices. *Each* Z-*matrix is weakly pre-inverse-positive; each irreducible* Z-*matrix is strictly pre-inverse-positive.* An inverse-positive Z-matrix is called an M-*matrix*. *An irreducible* M-*matrix is strictly pre-inverse-positive*, i.e., $\mu_{ik} > 0$ for all elements of $M^{-1} = (\mu_{ik})$. *A* Z-*matrix (an irreducible* Z-*matrix) is an* M-*matrix if and only if* $z \ge 0$, $Mz \succ 0$ $(z \ge 0\ ,\ Mz > 0)$ *for some* $z \in \mathbb{R}^n$.

The two assumptions of the monotonicity theorem are of a quite different character. These assumptions are not only means for proving inverse-positivity, but in many cases of interest themselves. For instance, majorizing elements can be used for numerical purposes in various ways.

Pre-inverse-positivity is usually proved for classes of operators, such as Z-matrices in the above example. For some operator classes (and suitable strong orders) this proof is very easy, for others it is very complicated (if the property holds at all). Pre-inverse-positivity (with respect to a suitable strong order) is easily proved for

Z-matrices ,
certain elliptic-parabolic differential operators of the second order,
certain integral operators , (1.2)
...

It turns out that for these operators further results can be proved which are related to inverse-positivity or pre-inverse-positivity. For instance, one obtains results on eigenvalue theory, and on iterative procedures for solving equations (0.1).

Our next object is to define classes (**Q** and **Z**) of abstract pre-inverse-positive operators such that

a) the operators in (1.2) are special cases,

b) the further results mentioned (on eigenvalue theory, iterative procedures, etc.) hold also for these abstract operators.

For preparing the definition of these operator classes we study the proof of a further simple example.

Example 1.3 Let $R = C_2[0,1]$, $S = \mathbb{R}[0,1]$,

$$\begin{aligned} Mu(x) &= -u''(x) + b(x)u'(x) + c(x)u(x) \qquad \text{for } x \in (0,1)\ , \\ Mu(0) &= -u'(0) + \gamma_0 u(0)\ ,\quad Mu(1) = u'(1) + \gamma_1 u(1) \end{aligned} \tag{1.3}$$

and define $U \succ 0 :\Leftrightarrow U(x) > 0\ (0 \le x \le 1)$.

M is pre-inverse-positive if and only if for each $u \in R$

$$(u \ge 0\ ,\ u \not> 0) \quad \Rightarrow \quad Mu \not\succ 0\ .$$

This property holds if for each $u \in R$ and each $\xi \in [0,1]$

$$(u \ge 0\ ,\ u(\xi) = 0) \quad \Rightarrow \quad Mu(\xi) \le 0\ . \tag{1.4}$$

This sufficient condition is here satisfied, since

$$\left.\begin{array}{lllll} \text{for } 0 < \xi < 1: & (u \ge 0, u(\xi) = 0) & \Rightarrow & (u'(\xi) = 0\ ,\ u''(\xi) \ge 0) \\ \text{for } \xi = 0: & (u \ge 0, u(\xi) = 0) & \Rightarrow & u'(\xi) \ge 0 \\ \text{for } \xi = 1: & (u \ge 0, u(\xi) = 0) & \Rightarrow & -u'(\xi) \ge 0 \end{array}\right\} \tag{1.5}$$

For the proof it is essential, that $Mu(\xi)$ depends only on the quantities which occur in the corresponding implication (1.5), and that $Mu(\xi)$ is an antitone (non-increasing) function of those terms for which only inequality is stated in (1.5).

These considerations show also that M remains pre-inverse-positive if one adds to $Mu(x)$ in (1.3) negative functional terms like

$$-\int_0^1 K(x,t)u(t)dt \quad \text{with} \quad K(x,t) \geq 0 \quad , \quad \text{or} \ -d(x)u(1-x) \quad \text{with} \quad d(x) \geq 0 \quad .$$

(Observe that the inequality $u \geq 0$ occurs in each of the implications (1.5).)

Using the linear functional ℓ defined by $\ell u = u(\xi)$ we can write (1.4) in the form $(u \geq 0, \ell u = 0) \Rightarrow \ell Mu \leq 0$. This formulation of (1.4) and a corresponding formulation of (1.5) will lead us to the definition of the operator classes $\mathbf{Q}(\mathcal{L})$, in the next section.

1.2 A class of abstract pre-inverse-positive operators

There is a series of operator classes for which pre-inverse-positivity can be shown with arguments analogous to those used in Example 1.3. We will define a class of abstract pre-inverse-positive operators which contains most of these special operator classes. This makes it possible to treat large parts of the inverse-positivity theory of matrices, scalar and vector-valued ordinary and partial differential operators, and integral operators in a unified way. The treatment becomes very simple after one has introduced the abstract terms. Moreover, we will see, that for this class of operators a theory can be developed which is much richer than the theory of pre-inverse-positive operators in general.

The formulation of (1.4) and (1.5) with linear functionals $\ell_\xi u = u(\xi)$ leads us to the concepts and results described in the following. Now we require

Assumption **(L)**

(a) Let $\mathcal{L}$ denote a set of linear functionals on S such that for each $U \in S$

$$U \geq 0 \quad \Leftrightarrow \quad \ell U \geq 0 \quad \text{for all} \quad \ell \in \mathcal{L} \quad , \tag{1.6}$$

and that there is a $U \in S$ with $\ell U > 0$ for all $\ell \in \mathcal{L}$.

(b) Suppose that $(R, \leq) \subset (S, \leq)$, and that for each $u \in R$

$$\begin{aligned} u \geq 0 \quad &\Leftrightarrow \quad \ell u \geq 0 \quad \text{for all} \quad \ell \in \mathcal{L} \quad , \\ u \succ 0 \quad &\Leftrightarrow \quad \ell u > 0 \quad \text{for all} \quad \ell \in \mathcal{L} \quad . \end{aligned} \tag{1.7}$$

The *strong order induced by* $\mathcal{L}$ is then defined by

$$U \gtrdot 0 \quad :\Leftrightarrow \quad \ell U > 0 \quad \text{for all} \quad \ell \in \mathcal{L} \quad . \tag{1.8}$$

The corresponding order cone will be denoted by $W(\mathcal{L})$.

The first line in (1.7) could be dropped, since the property described is contained in (1.6). In some cases, however, we will only assume that $\mathcal{L}$ is a set of linear functionals on R such that (1.7) holds for each $u \in R$. Then one can *define* $(S, \leq)$ in such a way that **(L)** is satisfied. One chooses $S = \mathbb{R}(\mathcal{L})$, the space of real-valued functions $U : \ell \to U_\ell$ on $\mathcal{L}$ (in other terms: $S = \mathbb{R}^{\mathcal{L}}$, $U = (U_\ell)$), and defines $\ell U = U_\ell$ for $U \in S$, $\ell \in \mathcal{L}$.

Definition 1.4 A linear operator $M : R \to S$ is called a Q($\mathcal{L}$)-*operator*, if for each $u \in R$ and $\ell \in \mathcal{L}$

$$(u \geq 0\ , \ell u = 0) \quad \Rightarrow \quad \ell M u \leq 0 \quad .$$

Q($\mathcal{L}$) denotes the class of all Q($\mathcal{L}$)-operators.

Lemma 1.5 *Each* Q($\mathcal{L}$)*-operator* $M : R \to S$ *is pre-inverse-positive with respect to the strong order induced by* $\mathcal{L}$.

Lemma 1.6 *If* $A : R \to S$ *is a* Q($\mathcal{L}$)*-operator,* $B : R \to S$ *a positive linear operator, and* $c \in \mathbb{R}$*, then* $A - B$ *and* $A + cI$ *are* Q($\mathcal{L}$)*-operators.*

For proving that a given operator M is a Q($\mathcal{L}$)-operator one may try to derive from $(u \geq 0, \ell u = 0)$ certain linear equations $Q_\ell u = 0$ and linear inequalities $P_\ell u \geq 0$, and then show that $\ell M u$ is a function of the four variables $\ell u, Q_\ell u, P_\ell u, u$ which is antitone with respect to the last two of these variables.

This leads to the formulation of the following property **(B)** of the space $(R, \leq)$, which involves relations holding on the boundary of the order cone.

(B) For each $\ell \in \mathcal{L}$, there are linear operators P_ℓ and Q_ℓ such that P_ℓ maps R into an ordered linear space $(X_\ell, \leq)$, Q_ℓ maps R into a linear space Y_ℓ, and for $u \in R$

$$(u \geq 0\ ,\ \ell u = 0) \quad \Rightarrow \quad (P_\ell u \geq 0\ ,\ Q_\ell u = 0) \quad . \tag{1.9}$$

Lemma 1.7 $M : R \to S$ *is a* Q($\mathcal{L}$)*-operator, if for each* $\ell \in \mathcal{L}$ *and* $u \in R$ *the term* $\ell M u$ *can be written as*

$$\ell M u = -a_\ell P_\ell u + b_\ell Q_\ell u + c_\ell \ell u - d_\ell u \quad , \tag{1.10}$$

where

(i) P_ℓ, Q_ℓ *are as described in* **(B)**,

(ii) a_ℓ, b_ℓ, c_ℓ, *and* d_ℓ *are linear functionals on, respectively,* $X_\ell, Y_\ell, \mathbb{R}$, *and* R (*so that* $c_\ell \in \mathbb{R}$),

(iii) *the functionals* a_ℓ *and* d_ℓ *are positive.*

The operators P_ℓ and Q_ℓ in **(B)** are not uniquely determined. In particular, the terms u and ℓu could be incorporated into, respectively, $P_\ell u$ and $Q_\ell u$, so that we could write $\ell M u = -a_\ell P_\ell u + b_\ell Q_\ell u$ instead of (1.10). Such a formula was used in [69] (with P_ℓ and Q_ℓ exchanged). For reasons which will become apparent later, we prefer here the form (1.10).

Example 1.8 (*Z-matrices*) Let $R = S = \mathbb{R}^n$, $u = (u_i)$, and $\mathcal{L}$ be the set of functionals ℓ_i $(i = 1, 2, \ldots, n)$ defined by $\ell_i u = u_i$. A Z-matrix $M = (m_{ik})$ determines a Q($\mathcal{L}$)-operator, since

$$\ell_i M u = c_i \ell_i u - d_i u \quad \text{with} \quad c_i = m_{ii}\ ,\ d_i u = \sum_{k \neq i} |m_{ik}| u_k \ . \tag{1.11}$$

Example 1.9 (*Scalar elliptic functional differential operators*) Let $\Omega \subset \mathbb{R}^m$ denote a bounded domain of points $x = (x_i)$, $R = C_1(\bar{\Omega}) \cap C_2(\Omega)$, $S = \mathbb{R}(\bar{\Omega})$, $\mathcal{L}$ the set of all point-functionals ℓ_x $(x \in \bar{\Omega})$ defined by $\ell_x U = U(x)$. Moreover, let $M : R \to S$ be given by

$$\begin{aligned} Mu(x) &= -a(x) \cdot u_{xx}(x) + b(x) \cdot u_x(x) + c(x)u(x) - d(x)u \qquad &\text{for} \quad x \in \Omega \quad , \\ Mu(x) &= -\alpha(x)u_\nu(x) + \gamma(x)u(x) - \delta(x)u &\text{for} \quad x \in \partial\Omega \quad , \end{aligned}$$

where u_x denotes the gradient, u_{xx} the matrix of the second derivatives, $a(x) = (a_{ik}(x))$ a symmetric m×m-matrix, $a \cdot u_{xx} = \sum_{i,k} a_{ik} \partial^2 u / \partial x_i \partial x_k$, $b(x) = (b_i(x))$ an m-vector, $b \cdot u_x = \sum_i b_i \partial u / \partial x_i$,

$c(x) \in \mathbb{R}$, $d(x)$ a linear functional on R; $u_\nu(x)$ the inner normal derivative at $x \in \partial\Omega$ (or any other directional derivative at this point corresponding to a vector ν pointing into Ω), $\alpha(x) \in \mathbb{R}$, $\gamma(x) \in \mathbb{R}$, $\delta(x)$ a linear functional on R.

M is a Q($\mathcal{L}$)-operator, if

$$\begin{array}{llll} a(x) \geq_d 0 \quad , & d(x) \geq 0 & \textit{for} \quad x \in \Omega & , \\ \alpha(x) \geq 0 \quad , & \delta(x) \geq 0 & \textit{for} \quad x \in \partial\Omega & , \end{array} \qquad (1.12)$$

where $\geq_d$ indicates positive semi-definiteness.

For the proof observe that for $\xi \in \Omega$

$$(u \geq 0, u(\xi) = 0) \quad \Rightarrow \quad (u_x(\xi) = 0 \,, u_{xx}(\xi) \geq_d 0) \quad ,$$

and that $a(\xi) \cdot u_{xx}(\xi) \geq 0$ for positive semi-definite $a(\xi)$ and $u_{xx}(\xi)$.

According to the Monotonicity Theorem and Lemma 1.5, the operator M is inverse-positive, if (1.12) holds, and if M has a majorizing function z. Concerning the construction of such functions see [66]. For instance, a majorizing function exists, if a is uniformly positive definite, b is bounded, $c \geq 0$, $d = 0$ on Ω, and if α is bounded, $\gamma \geq$ const. > 0, $\delta = 0$ on $\partial\Omega$.

Example 1.10 (*Scalar parabolic functional differential operators*) Now we consider functions $u(x,t)$ for $x \in \overline{\Omega}$, $0 \leq t \leq T$, where Ω denotes a domain as in Example 1.9, and $T > 0$. Let R denote the set of all continuous functions on $\overline{\Omega} \times [0,T]$ such that u_x is continuous on $\overline{\Omega} \times (0,T]$, and u_t and u_{xx} are continuous on $\Omega \times (0,T]$; let $S = \mathbb{R}(\overline{\Omega} \times [0,T])$, and $\mathcal{L}$ be the set of all point-functionals $\ell_{x,t}$ ($x \in \overline{\Omega}, 0 \leq t \leq T$). Moreover, let $M : R \to S$ be an operator of the form

$$\begin{array}{llll} Mu(x,t) = (u_t - a \cdot u_{xx} + b \cdot u_x + cu)(x,t) - d(x,t)u & \text{for} \quad x \in \Omega & , \quad 0 < t \leq T & , \\ Mu(x,t) = (-\alpha u_\nu + \gamma u)(x,t) - \delta(x,t)u & \text{for} \quad x \in \partial\Omega & , \quad 0 < t \leq T & , \\ Mu(x,0) = u(x,0) & \text{for} \quad x \in \overline{\Omega} & , & \end{array}$$

where $a, b, ..$ are quantities as in Example 1.9, which may now depend on x and t.

M is a Q($\mathcal{L}$)-operator, if

$$\begin{array}{llll} a(x,t) \geq_d 0 \quad , & d(x,t) \geq 0 & \textit{for} \quad x \in \Omega \quad , & 0 < t \leq T \quad , \\ \alpha(x,t) \geq 0 \quad , & \delta(x,t) \geq 0 & \textit{for} \quad x \in \partial\Omega \quad , & 0 < t \leq T \quad . \end{array}$$

For this operator M one may try to construct a majorizing function of the form $z = \exp Nt$ with constant N. Such a majorizing function exists, for instance, if c is bounded below, $d = 0$ on $\Omega \times (0,T]$, and if $\gamma > 0$, $\delta = 0$ on $\partial\Omega \times (0,T]$.

The next example reveals some typical difficulties arizing for vector-valued operators. For simplicity, we restrict ourselves to ordinary differential operators. The results can immediately be carried over to partial differential operators.

Example 1.11 (*Vector-valued functional differential operators*) Let $S = \mathbb{R}^n[0,1]$, the space of functions $U = (U_i) : [0,1] \to \mathbb{R}^n$, $R = C_2^n[0,1]$, $\mathcal{L}$ the set of all $\ell_{i,x}$ ($i = 1,2,\ldots,n; x \in [0,1]$) defined by $\ell_{i,x}U = U_i(x)$. Moreover, let $M : R \to S$ have the form

$$\begin{array}{l} Mu(x) = -a(x)u''(x) + b(x)u'(x) + c(x)u(x) - d(x)u \qquad \text{for} \quad 0 < x < 1 \,, \\ Mu(0) = -a^0u'(0) + c^0u(0) - d^0u \quad , \quad Mu(1) = a^1u'(1) + c^1u(1) - d^1u \quad , \end{array}$$

with n×n-matrices $a(x), b(x), c(x), a^0, c^0, a^1, c^1$, and linear functionals $d(x), d^0, d^1$ on R.

M is a Q($\mathcal{L}$)*-operator if*

(i) *the matrices* $a(x), b(x), a^0, a^1$ *are diagonal,*

(ii) *the matrices* $a(x), a^0, a^1$ *are positive (componentwise* ≥ 0 *),*

(iii) $c(x), c^0, c^1$ *are Z-matrices,*

(iv) *the functionals* $d(x), d^0, d^1$ *are positive.*

Observe that we have the same generality, if we assume $c(x), c^0, c^1$ to be diagonal, since the non-diagonal terms can be incorporated into the functional terms occuring (compare (1.11)).

The above restrictions (i) and (iii) are due to the fact that for $\ell = \ell_{i,\xi}$ the relations $(u \geq 0\ ,\ \ell u = 0)$ do not imply any relations for derivatives of $u_k(\xi)$ with $k \neq i$, and only inequality $u_k(\xi) \geq 0$ for $k \neq i$.

In other words, we have the restriction of

weak coupling : $(Mu)_i(x)$ does not depend on any derivative of any $u_k(x)$ with $k \neq i$;

quasi-antitonicity : $(Mu)_i(x)$ is an antitone function of $u_k(x)$ for $k \neq i$.

Conditions of this type are also to be required for nonlinear inverse-monotone operators. Obviously, these conditions severely restrict the *immediate* applicability of the theory of inverse-monotone operators to vector-valued problems. (Concerning the term *quasi-antitone* see the comments at the end of section 2.)

The following remarks will show that the above concept of a Q($\mathcal{L}$)-operator includes more special cases than it may seem at first sight.

A problem $Mu = r$ for $u \in R$ with an operator as in (1.10) and some $r \in S$ can be written in the form

$$m_\ell u := -a_\ell P_\ell u + b_\ell Q_\ell u + c_\ell \ell u - d_\ell u = r_\ell \in \mathsf{IR} \quad (\ell \in \mathcal{L}) \quad , \tag{1.13}$$

without explicit reference to the space S and the operator M . Indeed, if $\mathcal{L}$ denotes a set of functionals defined only on R such that (1.7) and (**B**) hold, then all terms in (1.13) are meaningful. In order to make the operator theory applicable, one may then choose $S = \mathsf{IR}(\mathcal{L})$, and define an operator $M : R \to S$ by $(Mu)_\ell = m_\ell u$ for $\ell \in \mathcal{L}$. This operator belongs to **Q**($\mathcal{L}$), if properties (ii), (iii) in Lemma 1.7 hold; and problem (1.13) is equivalent to $Mu = r$ with $r = (r_\ell) \in S$.

Example 1.12 Let $R = \{u \in C_2[0,1] : u(0) = u(1) = 0\}$, $L[u](x) = -u''(x) + b(x)u'(x) + c(x)u(x)$ for $0 \leq x \leq 1$. One has $z \succ 0 \Leftrightarrow (z(x) > 0 \text{ for } 0 < x < 1\ ,\ z'(0) > 0\ ,\ z'(1) < 0)$. Thus (1.7) holds for $\mathcal{L}$ consisting of all ℓ_x $(0 < x < 1)$, ℓ'_0, and ℓ'_1, where $\ell_x u = u(x)$, $\ell'_0 u = u'(0)$, $\ell'_1 u = -u'(1)$. Here, we obtain a Q($\mathcal{L}$)-operator $M : R \to S = \mathsf{IR}(\mathcal{L})$ by defining $(Mu)_\ell = L[u](x)$ for $\ell = \ell_x$, $(Mu)_\ell = L[u](0)$ for $\ell = \ell'_0$, $(Mu)_\ell = L[u](1)$ for $\ell = \ell'_1$.

1.3 Further results related to the Monotonicity Theorem

1.31 Connected sets of inverse-positive operators

Interesting characterizations of inverse-positive operators can be obtained by using continuity arguments. The idea is to consider sets $\mathcal{M}$ of operators, and to prove that *each* operator $M \in \mathcal{M}$ has a majorizing element, if *one* operator $M_0 \in \mathcal{M}$ has such an element. We assume here that $(R, \leq)$ contains a strictly positive element ζ, and define a norm on R by

$$\|u\| = \|u\|_\zeta = \inf\{\alpha \in \mathsf{IR} : \ -\alpha\zeta \leq u \leq \alpha\zeta\} \quad . \tag{1.14}$$

Topological terms such as continuity are to be understood with respect to this norm.

Theorem 1.13 *Suppose that $\mathcal{M}$ is a set of linear operators $M : R \to S$ with the following properties:*

(i) *all $M \in \mathcal{M}$ are pre-inverse-positive ;*

(ii) *all $M \in \mathcal{M}$ are invertible ;*

(iii) *the ranges of all $M \in \mathcal{M}$ have at least one element $r \succ 0$ in common ;*

(iv) *$\mathcal{M}$ is connected in the following sense : for each pair M_0, M_1 of operators in $\mathcal{M}$ there is a family $\{M(t) : 0 \leq t \leq 1\} \subset \mathcal{M}$, such that $M(0) = M_0$, $M(1) = M_1$, and $z(t) = M^{-1}(t)r$ is continuous on $[0,1]$.*

Under these assumptions, all operators $M \in \mathcal{M}$ are inverse-positive, if one operator $M_0 \in \mathcal{M}$ is inverse-positive.

From this result one can, for instance, directly derive that

a symmetric Z-matrix is inverse-positive if and only if it is positive definite .

These statements describe interesting relations between two different types of orders. Analogous statements can also be proved for differential operators. Such statements, however, are also contained in the theory of section 1.4.

1.32 Non-invertible operators

For the application to eigenvalue theory, we want to consider also non-invertible operators.

Theorem 1.14 *Suppose that $M : R \to S$ is a linear operator such that*

I. $(u \in R\ , u > 0\ ,\ Mu \geq 0) \Rightarrow u \succ 0$; II. $z > 0\ ,\ Mz \geq 0$ *for some* $z \in R$.

Then,

if $Mz > 0$, the operator M is inverse-positive ,

if $Mz = 0$, the operator M has the simple eigenvalue $\lambda = 0$, corresponding to the strictly positive eigenelement z.

1.33 Stronger types of inverse-positivity, strong boundary maximum principle

For many Q($\mathcal{L}$)-operators $M : R \to S$, the inequality $Mu \geq 0$ implies positivity properties of u stronger than $u \geq 0$. For (second order) differential operators these properties are closely related to the strong boundary maximum principle. We will describe a corresponding theory in abstract terms.

In the following we assume that $M : R \to S$ is a linear operator, that (**L**) holds, and that

$$\mathcal{L} = \mathcal{L}^b \cup \mathcal{L}^i \quad \text{with} \quad \mathcal{L}^b \cap \mathcal{L}^i = \emptyset \quad .$$

Theorem 1.15 *Suppose that the following conditions are satisfied.*

I.(a) $(u \in R,\ u > 0,\ \ell Mu \geq 0$ *for all* $\ell \in \mathcal{L}^i) \;\Rightarrow\; \ell u > 0$ *for all* $\ell \in \mathcal{L}^i$;

(b) $(u \in R,\ u \geq 0,\ \ell u = 0$ *for some* $\ell \in \mathcal{L}^b) \;\Rightarrow\; \ell Mu \leq 0$.

II. *There is a $z \in R$ such that $z \geq 0$, $Mz > 0$, $\ell Mz > 0$ for all $\ell \in \mathcal{L}^b$.*

Then for $u \in R$,

$$Mu \geq 0 \quad \Rightarrow \quad \begin{cases} u \geq 0 \quad , \text{ and} \\ \text{either} \quad u = 0 \quad , \text{ or} \quad \ell u > 0 \text{ for all } \ell \in \mathcal{L}^i \end{cases} \quad . \tag{1.15}$$

Proof The inverse-positivity of M follows from the Monotonicity Theorem. One verifies that the assumptions of this theorem hold for $U \succ 0$ defined by : $U > 0$, $\ell U > 0$ for all $\ell \in \mathcal{L}^b$. The second statement in (1.15) then is a consequence of I(a).

Remark: I(b) holds for each Q($\mathcal{L}$)-operator M. In the case $\mathcal{L} = \mathcal{L}^b$ assumption I is satisfied if and only if M is a Q($\mathcal{L}$)-operator. In the case $\mathcal{L} = \mathcal{L}^i$ assumption I is equivalent to : $(u \in R\ ,\ u > 0\ ,\ Mu \geq 0) \Rightarrow u \succ 0$. This property was required in Theorem 1.14. Here, we are interested in cases, where $\mathcal{L}^b \neq \emptyset$ and $\mathcal{L}^i \neq \emptyset$.

Corollary 1.16 (*$\mathcal{L}^b$-maximum principle*) *Suppose that*

$$\ell M u = \ell u \quad \text{for} \quad \ell \in \mathcal{L}^b \quad ,$$

and that assumptions I(a) *and* II *of Theorem* 1.15 *are satisfied. If* $u \in R\ ,\mu \in \mathbb{R}$ *and*

$$\ell u \leq \mu \ell z \quad \text{for all} \quad \ell \in \mathcal{L}^b \quad , \quad \mu \geq 0 \quad , \quad \ell M u \leq 0 \quad \text{for all} \quad \ell \in \mathcal{L}^i \quad ,$$

then

$$\text{either} \quad u = \mu z \quad , \quad \text{or} \quad \ell u < \mu \ell z \quad \text{for all} \quad \ell \in \mathcal{L}^i \quad .$$

Proof One applies the statement of Theorem 1.15 to $v = \mu z - u$ instead of u. Observe that here I(b) is trivial.

The following examples deal with Q($\mathcal{L}$)-operators as considered in Section 1.2. Since here I(b) always holds, we are mainly interested in sufficient conditions for I(a).

Example 1.17 (*Scalar elliptic differential operators*) Let R, S, and $\mathcal{L}$ be defined as in Example 1.9 and consider the following special case of the operator M considered there :

$$\begin{aligned} Mu(x) &= -a(x)\cdot u_{xx}(x) + b(x)\cdot u_x(x) + c(x)u(x) \ \text{ with } \ a(x) \geq_d 0 && \text{for} \quad x \in \Omega \quad , \\ Mu(x) &= \ u(x) && \text{for} \quad x \in \partial\Omega \quad . \end{aligned}$$

Then choose $\mathcal{L}^i = \{\ell_x : x \in \Omega\}$, $\mathcal{L}^b = \{\ell_x : x \in \partial\Omega\}$.

Here, I(a) *holds if and only if for* $u \in R$

$$\left.\begin{array}{l} u \geq 0 \quad , \quad u(\xi) = 0 \text{ for some } \xi \in \Omega \\ Mu(x) \geq 0 \quad \text{for all } x \in \Omega \end{array}\right\} \quad \Rightarrow \quad u \equiv 0 \quad , \tag{1.16}$$

and the statement of Theorem 1.15 says, that for $u \in R$

$$Mu \geq 0 \text{ implies that either } u \equiv 0, \text{ or } u(x) > 0 \text{ for all } x \in \Omega \quad . \tag{1.17}$$

By applying the technique used by Hopf [27] for deriving the strong boundary maximum principle one proves that

(1.16) *holds for each* $u \in R$, *if* $a(x)$ *is uniformly positive definite on* Ω, *and* $b(x), c(x)$ *are bounded on* Ω.

The usual strong boundary maximum principle for the elliptic operator considered is equivalent to the corresponding statement of Corollary 1.16 with $z(x) \equiv 1$. Observe that in this case condition II requires $c(x) \geq 0$ for $x \in \Omega$.

The above theorem and corollary can also be applied to vector-valued elliptic operators. For simplicity, we restrict ourselves again to ordinary differential operators, in the following example.

Example 1.18 (*Vector-valued differential operators*) Let R, S and $\mathcal{L}$ be defined as in Example 1.11, and consider the following special case of the operator M considered there:

$$\begin{aligned} Mu(x) &= -u''(x) + b(x)u'(x) + c(x)u(x) \quad \text{for} \quad 0 < x < 1 \quad , \\ Mu(0) &= u(0) \quad , \quad Mu(1) = u(1) \quad . \end{aligned}$$

Then choose $\mathcal{L}^i = \{\ell_{i,x} : 0 < x < 1; i = 1, 2, \ldots, n\}$, $\mathcal{L}^b = \mathcal{L} \setminus \mathcal{L}^i$.

Here *assumption* I(a) *of Theorem* 1.15 *is equivalent to the following property:*

I'.(a) *If* $u \in R$, $u \geq 0$, $(Mu)_i(x) \geq 0$ *for all indices* i *and all* $x \in (0,1)$, *and* $u_j(\xi) = 0$ *for some index* j *and some* $\xi \in (0,1)$, *then* $u_i(x) = 0$ *for all indices* i *and all* $x \in (0,1)$.

M has this property, if

(i) *the coefficients* b, c *are continuous on* $[0,1]$,

(ii) $c(x)$ *is an irreducible Z-matrix for each* $x \in (0,1)$.

The irreducibility of $c(x)$ is used to show that the premise in I'(a) implies $u_i(\xi) = 0$ for all indices i. Suppose that $u_k(\xi) > 0$ for some index k. Then, due to the irreducibility of $c(\xi)$, there are indices p, q such that $c_{pq}(\xi) < 0$, $u_p(\xi) = 0$, and $u_q(\xi) > 0$. The latter inequality and $u \geq 0$ imply $(-u_p'' + b_p u_p' + c_{pp} u_p)(\xi) > 0$, which contradicts $u_p(\xi) = 0$, since M is a Q($\mathcal{L}$)-operator.

Using $u_i(\xi) = 0$ one derives $u_i \equiv 0$ from (1.16) with $\Omega = (0,1)$, u replaced by u_i, and Mu replaced by $-u_i'' + b_i u_i' + c_{ii} u_i$.

For the operator M considered here assumption II holds for a constant vector z if

$$z \succ 0 \quad , \quad c(x)z \geq 0 \qquad (0 < x < 1) \quad .$$

Under this assumption, the maximum principle of Corollary 1.16 says:

if $Mu(x) \leq 0$ *for* $0 < x < 1$,

then either $u = \mu z$, *or* $u_i(x) < \mu z_i$ *for* $0 < x < 1$, *and* $i = 1, 2, \ldots, n$,
where $\mu = \max\{0, \mu_1, \mu_2, \ldots, \mu_n\}$, $\mu_i = \max\{u_i(0)/z_i, u_i(1)/z_i\}$.

1.34 Order cones with empty interior

The Monotonicity Theorem can be generalized, so that the existence of a strictly positive element in $(R, \leq)$ need not be required. Suppose that

$p > 0$ denotes a fixed element in R,

$u \succ_p 0 : \Leftrightarrow u \geq \alpha p$ for some $\alpha > 0$,

B_p is the set of all $u \in R$ such that $u \geq -\gamma p$ for some $\gamma \in \mathbb{R}$.

Theorem 1.19 *Let a linear operator* $M : R \to S$ *have the following two properties:*

I. $u \in B_p$, $u \geq 0$, $Mu \succ 0) \quad \Rightarrow \quad u \succ_p 0$.

II. *There is a* $z \in R$ *with* $z \geq 0$, $Mz \succ 0$.

Then for each $u \in B_p$,

$$Mu \geq 0 \quad \Rightarrow \quad u \geq 0 \quad , \quad \text{and} \quad Mu \succ 0 \quad \Rightarrow \quad u \succ_p 0 \quad .$$

1.4 Z-operators, and related concepts

In this section we assume, in addition to **(S)**, that $(R,\leq) \subset (S,\leq)$, and that R contains an element $\zeta \succ 0$. Then we define $\|u\| = \|u\|_\zeta$ as in (1.14).

1.41 Definition of certain operator classes

The theory of Z-matrices is related to the theory of eigenvalue problems, positive matrices, iterative procedures etc. Many of the corresponding results can be carried over to abstract operators. We will here be interested in the theory of eigenvalue problems

$$(M+\lambda N)\varphi = 0 \tag{1.18}$$

with linear operators

$$M : R \to S \quad , \quad N : S \to S \quad , \quad N \text{ positive} \quad . \tag{1.19}$$

For generalizing the notion of a Z-matrix and an irreducible Z-matrix we observe that for quadratic matrices M the following statements are true:

(i) *M is a Z-matrix if and only if $M+\lambda I$ is weakly pre-inverse-positive for all sufficiently large λ.*

(ii) *M is an irreducible Z-matrix if and only if $M+\lambda I$ is strictly pre-inverse-positive for all sufficiently large λ.*

These equivalences lead us to the following

Definition 1.20 $\mathbf{Z}_P(W)$ denotes the class of all pairs (M,N) of linear operators with property (1.19) such that for all sufficiently large λ the operator $M+\lambda N$ is W-pre-inverse-positive, i.e.

$$(u \in R\,,\ u \geq 0\,,\ (M+\lambda N)u \overset{W}{>} 0) \quad \Rightarrow \quad u \succ 0 \quad . \tag{1.20}$$

In the case $(R,\leq) = (S,\leq)$, $\mathbf{Z}(W)$ denotes the class of all operators $M : R \to R$ such that $(M,I) \in \mathbf{Z}_P(W)$. In particular,

$$\mathbf{Z} := \mathbf{Z}(W) \quad \text{for} \quad W = \{u \in S \,:\, u \succeq 0\} \quad .$$

An operator $M \in \mathbf{Z}$ is called a *Z-operator*, and an operator $M \in \mathbf{Z}(W)$ for $W = \{u \in S : u \geq 0\}$ is called an *irreducible Z-operator.*

Thus, by definition, the above statements (i) and (ii) hold also for operators instead of matrices.

Proposition 1.21 *If $(M,N) \in \mathbf{Z}_P(W)$, then (1.20) holds for all $\lambda \in \mathbb{R}$.*

Theorem 1.22 *Suppose that **(L)** holds, and that*

$$M \in \mathbf{Q}(\mathcal{L}) \quad , \quad \ell N u = \gamma_\ell \ell u \quad \textit{for} \quad u \in S\,,\ \ell \in \mathcal{L}$$

with constants $\gamma_\ell \geq 0$. Then

$$(M,N) \in \mathbf{Z}_P(W) \quad \textit{for} \quad W = W(\mathcal{L}) \quad .$$

In particular,

$$\mathbf{Q}(\mathcal{L}) \subset \mathbf{Z}(W) \quad \textit{for} \quad W = W(\mathcal{L}) \quad .$$

1.42 On inverse-positivity related to spectral properties, and positive eigenelements.

Now suppose that $(M,N) \in \mathbf{Z}_P(W)$. Assume, moreover, that

$$(M+\mu_0 N)z \succ 0 \ , \ z \geq 0 \quad \text{for some } z \in R \text{ and some } \mu_0 \in \mathbb{R} \quad , \tag{1.21}$$

so that $M + \lambda N$ is inverse-positive for all $\lambda \geq \mu_0$. We will derive results on the inverse-positivity of the operators $M + \lambda N$ as a function of λ and use these results for proving the existence of a positive (or strictly positive) eigenelement of the eigenvalue problem (1.18). The assumptions to be made will be related to the invertibility of $M + \lambda N$, the range of $M + \lambda N$, the boundedness of $(M + \lambda N)^{-1}N$, and the compactness of certain operators involved.

For including a series of different cases in our general theory we introduce sets J and Λ_0 of the following type. Suppose that

$$\mathrm{J} = (\omega, \infty) \quad \text{with some} \quad \omega \in [-\infty, \mu_0) \quad ,$$

and that Λ_0 is a subset of $\mathbb{R}$ which consists of all (real) eigenvalues of problem (1.18) and possibly some further λ for which $M + \lambda N$ is not inverse-positive. (In the simplest case, Λ_0 is the set of all eigenvalues, and $\mathrm{J} = (-\infty, \infty)$.)

Then define $\Lambda_1 = \mathrm{J} \setminus \Lambda_0$,

$$\kappa = \sup(\mathrm{J} \cap \Lambda_0) \quad \text{if} \quad \mathrm{J} \cap \Lambda_0 \neq \emptyset \quad ; \quad \kappa = \omega \quad \text{if} \quad \mathrm{J} \cap \Lambda_0 = \emptyset \quad .$$

Obviously, $\omega \leq \kappa < \mu_0$.

In the theory of this section we will use one or several of the following properties.

(P1) There is an $r \succ 0$ in S such that for each $\lambda \in \Lambda_1$

$$(M + \lambda N)z = r \quad \text{has a solution} \quad z = z_\lambda \in R \quad . \tag{1.22}$$

(P2) For each $\lambda \in \Lambda_1$ there is a constant C_λ such that

$$\left.\begin{array}{l}(M+\lambda N)u = Nv \\ u, v \in R\end{array}\right\} \quad \Rightarrow \quad \|u\| \leq C_\lambda \|v\| \quad . \tag{1.23}$$

(P3) If $\kappa > -\infty$, and if $\{u_n\}$ is a bounded sequence in R, and $\{\lambda_n\}$, $\{\alpha_n\}$ are convergent sequences in $\mathbb{R}$ such that

$$(M + \lambda_n N)u_n = \alpha_n r \quad , \quad \lambda_n > \kappa \quad (n = 1, 2, \ldots) \quad ; \quad \lambda_n \to \kappa \quad \text{for} \quad n \to \infty \quad ,$$

then a subsequence $\{u_{n_i}\}$ and a $u_0 \in R$ exist such that

$$u_{n_i} \to u_o \quad \text{for} \quad i \to \infty \quad , \quad (M + \kappa N)u_0 = \alpha_0 r \quad \text{with} \quad \alpha_0 = \lim \alpha_n \quad .$$

(Q1) To each $u \in R$ exists an $m \in \mathbb{N}$ such that

$$Nu \prec mr \quad .$$

(Q2) There are $v \in R$, $s \in S$, $n \in \mathbb{N}$ such that

$$(M + \mu_0 N)v = s > 0 \quad , \quad \text{and} \quad nNv \geq s \quad . \tag{1.24}$$

(P1) is a condition on the range of $M + \lambda N$, and **(P2)** a condition related to the boundedness of $(M + \lambda N)^{-1}N$. **(P3)** may be verified in different ways by exploiting compactness properties of certain operators, for instance, $(M + \mu_0 N)^{-1}$, or M.

The above form of the conditions is sufficiently weak, so that they can be verified rather easily in many cases, e.g., for certain elliptic differential operators. (One could try to reduce the general case of an eigenvalue problem (1.18) to the special case $N = I$ by introducing $T = (M + \mu_0 N)^{-1}N$. The approach described here, however, is more suitable for our purposes.)

Condition **(Q1)** is satisfied, if r is a strictly positive element of $(S, \leq)$. Moreover, condition **(Q2)** holds, if $u \succ 0 \Rightarrow Nu \succ 0$ for $u \in R$. (Observe that for $s = r$, the solution z of $(M + \mu_0 N)z = r$ satisfies $z \succ 0$.) We will, however, have to consider other cases, too.

Theorem 1.23 *If* **(P1)** *and* **(Q1)** *hold, then there is a* $\lambda_0 \in [\omega, \infty)$ *such that for each* $\lambda \in \mathrm{J}$ *the following three properties are equivalent :*

(i) $M + \lambda N$ *is inverse-positive ;*

(ii) $z \geq 0$, $(M + \lambda N)z \succ 0$ *for some* $z \in R$;

(iii) $\lambda_0 < \lambda$.

Theorem 1.24 *If* **(P1)**, **(Q1)** *and* **(P2)** *are satisfied, then the equivalence statement of Theorem* 1.23 *holds with*

$$\lambda_0 = \kappa \quad ; \quad \textit{and either} \quad \kappa = \omega \quad , \quad \textit{or} \quad \kappa \in \Lambda_0 \quad .$$

Corollary 1.25 *If* $\mathrm{J} = (-\infty, \infty)$, *and* **(P1)**, **(Q1)**, **(P2)** *and* **(Q2)** *hold, then* $\kappa > -\infty$.

Theorem 1.26 *If* **(P1)**, **(Q1)**, **(P2)** *and* **(P3)** *hold, and* $\kappa > \omega$, *then there is a* $\varphi \in R$ *such that*

$$\varphi > 0 \quad , \quad (M + \kappa N)\varphi = 0 \quad . \tag{1.25}$$

Theorem 1.27 *Suppose that* $\kappa > \omega$, $(M, N) \in \mathbf{Z}_P(W)$ *for*

$$W = \{u \in S \ : \ u \geq 0\} \quad , \quad u > 0 \Rightarrow Nu > 0 \quad \textit{for } u \in R \quad , \tag{1.26}$$

and that (1.25) *holds for some* $\varphi \in R$.

Then κ *is a simple eigenvalue of problem* (1.18) *and*

$$\varphi \succ 0 \quad , \quad (M + \kappa N)\varphi = 0 \quad , \tag{1.27}$$

Moreover,

$$\kappa = \min\{\lambda : \ (M + \lambda N)z \geq 0 \ , \ z > 0 \textit{ for some } z \in R\} \quad ,$$
$$\kappa = \max\{\lambda : \ (M + \lambda N)z \leq 0 \ , \ z > 0 \textit{ for some } z \in R\} \quad .$$

Remark The latter formulas for κ can be used to calculate upper and lower bounds for this eigenvalue by constructing suitable elements z.

Proofs. Let Λ denote the set of all $\lambda \in \mathrm{J}$ such that $M + \lambda N$ is inverse-positive. Obviously, $\mu_0 \in \Lambda \subset \Lambda_1$.

1) Because of the Monotonicity Theorem, condition (ii) in Theorem 1.23 implies (i) for each $\lambda \in \mathrm{J}$.

2) Let **(P1)** be satisfied. Then Λ is the set of all $\lambda \in \mathrm{J}$ such that

$$(M + \lambda N)z = r \succ 0 \quad , \quad z \geq 0 \quad \text{for some} \quad z \in R \quad , \tag{1.28}$$

and, consequently, (i) $\Rightarrow$ (ii). To show this, observe that (1.22) holds for $\lambda \in \Lambda_1$, and that $z \geq 0$ for $\lambda \in \Lambda \subset \Lambda_1$.

Since a majorizing element for $M + \lambda N$ is also a majorizing element for each $M + \lambda' N$ with $\lambda' \geq \lambda$, we have

$$\begin{array}{llll} \text{either} & \Lambda = (\lambda_0, \infty) & \text{with some} & \lambda_0 \in [\omega, \infty) \quad , \\ \text{or} & \Lambda = [\lambda_0, \infty) & \text{with some} & \lambda_0 \in (\omega, \infty) \quad . \end{array} \tag{1.29}$$

Moreover, $\lambda_0 \geq \kappa$. To prove this latter property, suppose that $\lambda_0 < \kappa$. Then, due to the definition of κ, a μ exists such that $\lambda_0 < \mu \leq \kappa$, and $\mu \in \mathrm{J} \cap \Lambda_0$. On the other hand, $\mu \in \Lambda \subset \Lambda_1$, because of (1.29).

3) If **(P1)** and **(Q1)** hold, then the second case in (1.29) cannot occur. For suppose that $\lambda_0 \in \Lambda$, that z satisfies (1.28) with $\lambda = \lambda_0$, and that $Nz \not> mr$. Then z is a majorizing element for $M + \lambda_1 N$ with $\lambda_1 = \lambda_0 - m^{-1}$, so that $M + \lambda_1 N$, too, is inverse-positive, contrary to the definition of λ_0.

4) Let **(P1)**, **(Q1)** and **(P2)** hold. Then $\lambda_0 = \kappa$ and thus

$$\Lambda = (\kappa, \infty) \quad . \tag{1.30}$$

For proving this it suffices to show that $(\kappa, \infty) \subset \Lambda$. Define $\mathcal{M}$ to be the set of all operators $M + \lambda N$ with $\kappa < \lambda < \infty$. For this set all assumptions of Theorem 1.13 are satisfied. In particular, the solutions $z = z_\lambda$ in (1.28) depend continuously on λ, because of **(P2)**. Thus, $(\kappa, \infty) \subset \Lambda$, since $\mu_0 \in \Lambda \cap (\kappa, \infty)$.

Moreover, $\kappa \in \Lambda_0$ if $\kappa > \omega$. To prove this, suppose that $\omega < \kappa \in \Lambda_1$ and apply Theorem 1.13 to the set $\mathcal{M}$ of all $M + \lambda N$ with $\lambda \in [\kappa, \infty)$. Then it follows that $[\kappa, \infty) \subset \Lambda$, contrary to (1.30).

5) If $\omega = -\infty$ and **(P1)**, **(Q1)**, **(P2)**, and **(Q2)** hold, then $\kappa > -\infty$, since $M + \nu N$ with $\nu = \mu_0 - n$ is not inverse positive. For suppose that $M + \nu N$ is inverse-positive. Then $(M + \nu N)(-v) = nNv - s \geq 0$ implies $-v \geq 0$, while $(M + \mu_0 N)v > 0$ yields $v > 0$.

6) Now suppose that **(P1)**, **(Q1)**, **(P2)**, and **(P3)** hold, and that $\kappa > \omega$. Let $\{\lambda_n\}$ be a sequence of values $\lambda_n > \kappa$, converging towards κ, and let $z_n \in R$ satisfy

$$(M + \lambda_n N) z_n = r \quad , \quad z_n \geq 0 \qquad (n = 1, 2, \ldots) \quad .$$

First suppose that the sequence $\{z_n\}$ is bounded. Then **(P3)** yields the existence of a subsequence $\{z_{n_i}\}$ with a limit $z_0 \in R$ such that

$$(M + \kappa N) z_0 = r \succ 0 \quad , \quad z_0 \geq 0 \quad .$$

Consequently, $\kappa \in \Lambda$, which contradicts (1.30).

Thus, we may assume that $\|z_n\| \to \infty$ for $n \to \infty$. The elements $\zeta_n = \|z_n\|^{-1} z_n$ satisfy

$$(M + \lambda_n N) \zeta_n = \|z_n\|^{-1} r \quad , \quad \zeta_n \geq 0 \quad , \quad \|\zeta_n\| = 1 \quad .$$

By again using **(P3)** we conclude that there is a subsequence $\{\zeta_{n_i}\}$ with a limit $\varphi \in R$ such that (1.25) holds.

7) If (1.25) and (1.26) hold, then we have for $u \in R$ and $\lambda' > \lambda \in \mathbb{R}$

$$(u > 0 \, , \, (M + \lambda N) u \geq 0) \quad \Rightarrow \quad (u \geq 0 \, , \, (M + \lambda' N) u > 0) \quad \Rightarrow \quad u \succ 0 \quad .$$

Hence, φ in (1.25) satisfies $\varphi \succ 0$.

Applying Theorem 1.14 to $M + \kappa N$ instead of M one sees that κ is a simple eigenvalue.

The formulas for κ in Theorem 1.27 are an easy consequence of the statements proved.

1.43 Elliptic differential operators

The results of the preceeding section can be applied to elliptic differential operators using the existence theory in $H_{j,p}$-spaces. We will consider eigenvalue problems of the form

$$\begin{aligned} L[\varphi](x) + \lambda g(x) \varphi(x) &= 0 \quad \text{for} \quad x \in \Omega \quad , \\ \varphi(x) &= 0 \quad \text{for} \quad x \in \partial\Omega \end{aligned} \tag{1.31}$$

for $\varphi \in C_2(\overline{\Omega})$, where Ω is a bounded domain in $\mathbb{R}^m$ with $m \geq 2$,

$$L[u](x) = -a(x) \cdot u_{xx}(x) + b(x) \cdot u_x(x) + c(x)u(x)$$

with a, b, c as in Example 1.9, a continuous and positive definite on $\overline{\Omega}$, b and c continuous on $\overline{\Omega}$, and

$$g \in C_1(\overline{\Omega}) \quad , \quad g(x) > 0 \quad \text{for} \quad x \in \overline{\Omega} \quad .$$

We will not formulate further conditions on the coefficients of L and on Ω in detail. Instead, we simply assume that some of the basic statements of the existence theory are valid for the case considered here.

Let $H_{j,p} = H_{j,p}(\Omega)$ denote the Sobolev spaces with norm $\| \ \|_{j,p}$, and $C_{j+\alpha} = C_{j+\alpha}(\overline{\Omega})$ the Hölder spaces with norm $| \ |_{j+\alpha}$, so that, in particular, $| \ |_0 = \| \ \|_\infty$.

We assume that $p \in \mathbb{R}$ with

$$p > m$$

and that for

$$L_\lambda[u] = L[u] + \lambda g u \quad \text{and each} \quad \lambda \in \mathbb{R}$$

the following statements hold.

(E1) If

$$(u \in C_1(\overline{\Omega}) \cap C_2(\Omega), L_\lambda[u] = 0 \ \text{ on } \Omega \ , u = 0 \ \text{ on } \partial\Omega) \quad \Rightarrow \quad u = 0 \quad , \tag{1.32}$$

then for $k \in \{0, 1\}$ and each $v \in H_{k,p}$ there is a unique

$$u \in H_{2+k,p} \cap \overset{\circ}{H}_{1,p} \quad \text{with} \quad L_\lambda[u] = v \quad ;$$

moreover,

$$\|u\|_{2+k,p} \leq C_\lambda \|v\|_{k,p}$$

with a constant C_λ independent of v.

(E2) If $\alpha \in (0, 1 - \frac{m}{p})$, then for $k \in \{0, 1\}$

$$H_{2+k,p} \subset C_{1+k+\alpha}$$

and

$$|u|_{1+k+\alpha} \leq C \|u\|_{2+k,p} \quad \text{for } u \in H_{2+k,p}$$

with a constant C independent of u.

Remark In the existence theory in Sobolev spaces one usually requires, instead of (1.32), that

$$(u \in H_{2,p} \cap \overset{\circ}{H}_{1,p} \ , \ L_\lambda[u] = 0) \quad \Rightarrow \quad u = 0 \quad .$$

This property, however, follows from (1.32) under suitable conditions. One exploits that the above implication holds for at least one $\lambda \in \mathbb{R}$.

(Concerning **(E1)**, **(E2)**, and other results on partial differential equations used in this paper, see, for instance, Gilbarg and Trudinger [22], Ladyzhenskaya and Ural'tseva [36], Simader [72].)

Now define

$$R = C_2(\overline{\Omega})\ ;\ \zeta(x) \equiv 1\ ,\ \text{so that}\ \ \|u\| = \|u\|_\zeta = \|u\|_\infty\ \ ;$$
$$S = \mathbb{R}(\overline{\Omega})\ ;\ U \succ 0 \Leftrightarrow U(x) > 0\ \ \text{for all}\ x \in \overline{\Omega}\ \ ;$$

$$\begin{array}{llll} Mu(x) = L[u](x) & ,\ Nu(x) = g(x)u(x) & \text{for}\ x \in \Omega & , \\ Mu(x) = u(x) & ,\ Nu(x) = 0 & \text{for}\ x \in \partial\Omega & . \end{array} \tag{1.33}$$

Then problem (1.18) is equivalent to the given eigenvalue problem (1.31), $(M, N) \in \mathbf{Z}_P(W)$ for W corresponding to the strong order defined above, and (1.21) is satisfied for $z(x) \equiv 1$ and each sufficiently large μ_0.

Choose, moreover, $\omega = -\infty$, and Λ_0 to be the set of all real eigenvalues of problem (1.31), so that κ is the supremum of these eigenvalues, if $\Lambda_0 \neq \emptyset$.

Because of **(E1)** and **(E2)**, properties **(P1)**, **(P2)**, and **(P3)** hold with $r(x) \equiv 1$. As an example, we will verify **(P3)** by proving:

(P'3) If $\{u_n\}$ is a bounded sequence in R, and $\{\alpha_n\}$, $\{\beta_n\}$ are convergent sequences in $\mathbb{R}$ such that

$$(M + \mu_0 N)u_n = \alpha_n r + \beta_n N u_n \quad (n = 1, 2, \ldots)$$

then a subsequence $\{u_{n_i}\}$ and a $u_0 \in R$ exist such that, uniformly on $\overline{\Omega}$,

$$u_{n_i} \to u_0 \quad , \quad Mu_{n_i} \to Mu_0 \quad , \quad Nu_{n_i} \to Nu_0 \quad \text{for}\ i \to \infty \quad .$$

The equations for the elements u_n are here equivalent to

$$L[u_n] + \mu_0 g u_n = \alpha_n + \beta_n g u_n \quad \text{on} \quad \Omega \quad , \quad u_n = 0 \quad \text{on} \quad \partial\Omega \quad .$$

If the sequence $\{u_n\}$ is bounded with respect to the norm $\|\ \| = \|\ \|_\infty$, the functions $v_n = \alpha_n + \beta_n g u_n$ on the right hand side of the differential equations constitute a bounded sequence in C_0. Therefore, **(E1)** and **(E2)** yield for $k = 0$:

$$|\!|\!| u_n |\!|\!|_{1+\alpha} \le c_1 \|u_n\|_{2,p} \le c_2 \|v_n\|_{0,p} \le c_3 \|v_n\|_\infty \le c_4$$

with constants c_j independent of n. Therefore the functions v_n constitute a bounded sequence in $H_{1,p}$. Applying then **(E1)** and **(E2)** for $k = 1$ in a similar way, one shows that the sequence $\{u_n\}$ is bounded with respect to $|\!|\!|\ |\!|\!|_{2+\alpha}$, and hence contains a subsequence $\{u_{n_i}\}$ which converges in the norm $|\!|\!|\ |\!|\!|_2$ towards some $u_0 \in C_2$. Consequently, $Mu_{n_i} \to Mu_o$ and $Nu_{n_i} \to Nu_0$ for $i \to \infty$ in the norm $\|\ \|_\infty$.

Properties **(Q1)** and **(Q2)** hold also. For proving **(Q2)** choose $s \in C_1(\overline{\Omega})$ such that $s \ge 0$, $s \not\equiv 0$, and s has compact support in Ω. According to (1.17) the solution v of $(M + \mu_0 N)v = s$ satisfies $v(x) > 0$ for all $x \in \Omega$, so that there is a number n with $nNv = ngv > s$.

Thus we can apply all results of the preceeding section. In particular, the following theorem is obtained.

<u>Theorem 1.28</u> *The eigenvalue problem* (1.31) *has a largest real eigenvalue* κ. *This eigenvalue is simple and corresponds to an eigenfunction* $\varphi \in C_2(\overline{\Omega})$ *such that* $\varphi(x) > 0$ *for all* $x \in \Omega$.

The operator $M + \lambda N$ *with* M, N *defined in* (1.33) *and* $\lambda \in \mathbb{R}$ *is inverse-positive on* $C_2(\overline{\Omega})$ *if and only if* $\lambda > \kappa$.

1.44 Positive operators

Suppose now that $(R, \leq) = (S, \leq)$, that R is a Banach space with respect to the norm (1.14), and that the given strong order $\lneq$ is identical with the strict order $\prec$. Let $T : R \to R$ denote a given positive linear operator with spectral radius $\rho(T)$ (T is bounded, since T is positive).

We consider the family of operators $\lambda I - T$ $(-\infty < \lambda < \infty)$ and, in particular, the eigenvalue problem

$$T\varphi = \lambda\varphi \quad . \tag{1.34}$$

There are various possibilities of applying the results of section 1.42 to this case. As examples we prove Theorems 1.29 and 1.30 below. One shows that

$$\alpha I - T \in \mathbf{Z} \quad \text{for each } \alpha \in \mathbb{R} \quad .$$

We will say that *the positive operator T is irreducible*, if $I - T$ is irreducible in the sense of Definition 1.20.

Theorem 1.29 *For each $\lambda > \rho(T)$, the operator $\lambda I - T$ is inverse-positive, and there is a $z \in R$ such that $z \geq 0$, $\lambda z - Tz \succ 0$.*

Remark The inequalities for z in this theorem imply that

$$z \succ 0 \quad , \quad \text{and} \quad \|T\|_z < \lambda \quad ,$$

where $\|T\|_z$ denotes the operator norm corresponding to the norm $\|u\|_z$ in R. Consequently,

$$\rho(T) = \inf\{\|T\|_z \ : \ z \in R \ , \ z \succ 0\} \quad . \tag{1.35}$$

Thus, one can use the contraction mapping principle in order to show that *the iterative procedure* $u_{n+1} = Tu_n + r$ $(n = 0, 1, \ldots)$ *is convergent* (for arbitrary $u_0, r \in R$) *if* $\rho(T) < 1$.

Theorem 1.30 *Suppose that T is compact and $\rho(T) > 0$. Then the following three properties are equivalent for each $\lambda \in (-\infty, \infty)$:*

(i) *$\lambda I - T$ is inverse-positive ;*

(ii) *$z \geq 0$, $(\lambda I - T)z \succ 0$ for some $z \in R$;*

(iii) *$\rho(T) < \lambda$.*

Moreover, there is a $\varphi \in R$ such that

$$T\varphi = \rho(T)\varphi \quad , \quad \varphi > 0 \quad .$$

If, in addition, T is irreducible, then $\varphi \succ 0$, $\rho(T)$ is a simple eigenvalue of T, and

$$\rho(T) = \min\{\lambda \ : \ (\lambda I - T)z \geq 0 \ , \ z > 0 \text{ for some } z \in R\} \quad ,$$
$$\rho(T) = \max\{\lambda \ : \ (\lambda I - T)z \leq 0 \ , \ z > 0 \text{ for some } z \in R\} \quad .$$

For proving these theorems we use the results of the preceeding section with $M = -T$, $N = I$ and observe the following facts:

the inequalities (1.21) hold for $z = \zeta \succ 0$ and each sufficiently large $\mu_0 > \rho(T)$;

(P1) and **(P2)** hold with $r = \zeta$, if Λ_1 contains only points of the residual set of T, as in the cases considered below (choose $C_\lambda = \|(\lambda I - T)^{-1}\|$);

(Q1) and **(Q2)** hold with $r = \zeta$ and $s = \zeta$, respectively.

Theorem 1.29 follows from Theorem 1.24 by choosing $\omega = \rho(T)$, and Λ_0 to be the set of all real eigenvalues.

For proving Theorem 1.30 we choose $\omega = -\infty$ and define Λ_0 to be the real spectrum of T. Then, for each $\lambda \in \Lambda_0$ the operator $\lambda I - T$ is not inverse-positive. (If $\lambda = 0 \in \Lambda_0$ and $-T$ were inverse-positive, then $\zeta \succ 0 \Rightarrow T\zeta \geq 0 \Rightarrow (-T)(-\zeta) \geq 0 \Rightarrow -\zeta \geq 0$; contradiction.) Clearly, $\kappa = \sup\Lambda_0 \leq \rho(T)$. Moreover, $\kappa > -\infty$ according to Corollary 1.25. Theorem 1.24 yields the equivalence of the properties (i), (ii) in Theorem 1.30, and $\lambda > \kappa$. In particular, we have $\rho(T) \leq \|T\|_z < \lambda$ for $\lambda > \kappa$ and z in (ii), and hence $\rho(T) = \kappa$ (see the remark following Theorem 1.29.)

The remaining statements of Theorem 1.30 follow from Theorems 1.26 and 1.27, since the compactness of T implies **(P3)**.

2. Nonlinear operators, two-sided bounds.

Suppose that Assumption **(S)** is satisfied, and that $M : R \to S$ denotes an operator which may be nonlinear.

2.1 Inverse-monotone operators

Here we are interested in statements of the form

$$Mv \leq Mw \quad \Rightarrow \quad v \leq w \quad . \tag{2.1}$$

If this implication holds for all $v, w \in D \subset R$, M is called *inverse-monotone on* D. In the following we assume that $v, w \in R$ are fixed elements.

Theorem 2.1 *Suppose there is a* $z \succ 0$ *in* R *such that the following two conditions are satisfied for* $\lambda > 0$.

I. *The relations*

$$v \leq w + \lambda z \quad , \quad v \not\prec w + \lambda z \tag{2.2}$$

imply

$$Mv \not\prec' M(w + \lambda z) \quad .$$

II. $Mw \prec' M(w + \lambda z)$, *if* (2.2) *holds.*

Then $\quad Mv \leq Mw \Rightarrow v \leq w$.

Corollary 2.2 *Theorem* 2.1 *remains true if the relations* $Mv \not\prec' M(w + \lambda z)$, *and* $Mw \prec' M(w + \lambda z)$ *are replaced by, respectively,* $M(v - \lambda z) \not\prec' Mw$, *and* $M(v - \lambda z) \prec' Mv$.

Remark: While inequalities (2.2) are essential in I, they are only side conditions in II, which are useful in treating certain nonlinear operators.

If $(u \leq \tilde{u}\ ,\ u \not\prec \tilde{u}) \Rightarrow Mu \not\prec' M\tilde{u}$ for arbitrary $u, \tilde{u} \in R$, M is called *pre-inverse-monotone on* R. We will, however, also call I a property of pre-inverse-monotonicity (which depends on the elements involved).

Suppose now, that **(L)** and **(B)** hold and that the strong order is induced by $\mathcal{L}$. Assume, furthermore, that for each $\ell \in \mathcal{L}$ and $u \in R$

$$\ell Mu = F_\ell(P_\ell u, Q_\ell u, \ell u, u) \tag{2.3}$$

with a real-valued function F_ℓ on $X_\ell \times Y_\ell \times \mathbb{R} \times R$. For such an operator M the following result can be derived.

Theorem 2.3 *Suppose there is a* $z \succ 0$ *in* R *such that for each* $\ell \in \mathcal{L}$ *and each* $\lambda > 0$ *with*

$$\ell v = \ell(w + \lambda z) \quad , \quad Q_\ell v = Q_\ell(w + \lambda z) \tag{2.4}$$

the following two conditions hold:

I. $F_\ell(P_\ell v, Q_\ell v, \ell v, v) \geq F_\ell(P_\ell v + q, Q_\ell v, \ell v, v + r)$ *for* $q \geq 0$ *in* X_ℓ, $r \geq 0$ *in* R,

II. $\ell M w < \ell M(w + \lambda z)$.

Then $Mv \leq Mw \Rightarrow v \leq w$.

Remark: The relations (2.4), which correspond to (2.2), serve only as side conditions in *both* assumptions, I and II, of this theorem. Assumption I need only be satisfied for $r = w - v + \lambda z$, $q = P_\ell r$.

Definition 2.4 Let **(L)** and **(B)** hold. An operator $M : R \to S$ is called $\mathcal{L}$*-quasilinear*, if for $\ell \in \mathcal{L}$, $u \in R$

$$\ell M u = -a_\ell(Q_\ell u, \ell u) P_\ell u + f_\ell(Q_\ell u, \ell u) - d_\ell(Q_\ell u, \ell u) u \tag{2.5}$$

with linear functionals a_ℓ, f_ℓ and d_ℓ (on X_ℓ, $\mathbb{R}$, and R, respectively) which depend on $Q_\ell u$ and ℓu. (In other words: The term $\ell M u$ depends linearly on those quantitites, $P_\ell u$ and u, for which only inequality is stated in (1.9).)

Lemma 2.5 *If* M *is* $\mathcal{L}$*-quasilinear, then assumption* I *of Theorem* 2.3 *is satisfied, if*

$$a_\ell(Q_\ell v, \ell v) \geq 0 \quad , \quad d_\ell(Q_\ell v, \ell v) \geq 0 \quad \text{for } \ell \in \mathcal{L} \quad . \tag{2.6}$$

An example for operators M of the type (2.5), (2.6) are *quasilinear elliptic differential operators* of the form

$$\begin{aligned} Mu(x) &= -a(x, u(x), u_x(x)) \cdot u_{xx}(x) + f(x, u(x), u_x(x)) && \text{for } x \in \Omega \, , \\ Mu(x) &= -\alpha(x, u(x)) u_\nu(x) + \tilde{f}(x, u(x)) && \text{for } x \in \partial\Omega \end{aligned} \tag{2.7}$$

with $a(x, v(x), v_x(x)) \geq_d 0$, $\alpha(x, v(x)) \geq 0$. (Here, we have used the notation of Example 1.9.)

For *vector-valued nonlinear differential operators*, Eq.(2.3) and assumption I require *weak coupling*, and a certain condition of *quasi-antitonicity*.

2.2 Splitting of operators

The property of inverse-monotonicity can be used to obtain two-sided bounds $\varphi \leq u \leq \psi$ for a solution of an equation $Mu = r$: one applies (2.1) separately to $(v, w) = (\varphi, u)$, and $(v, w) = (u, \psi)$. By treating the two inequalities in $\varphi \leq u \leq \psi$ simultaneoulsy one can develop a more general theory described in the following. In applying this theory to vector-valued problems the restriction of quasi-antitonicity is not needed. This has to be paid for by requiring a stronger condition than $M\varphi \leq r \leq M\psi$. (Let us remark that corresponding results on inclusion and existence statements in section 4 will not be derived from Theorem 2.7 below, but from Theorem 2.1.)

Suppose that $(R, \leq) \subset (S, \leq)$, and that

$$Mu = H(u, u) \tag{2.8}$$

with $H(u, v) \in S$ defined for all $u \in R$, $v \in S$, and that $\varphi, \psi \in R$ with $\varphi \leq \psi$. Define $[\Phi, \Psi] = \{U \in S : \Phi \leq U \leq \Psi\}$ for $\Phi, \Psi \in S$. We are interested in the property

$$H(\varphi,[\varphi,\psi]) \leq Mu \leq H(\psi,[\varphi,\psi]) \quad \Rightarrow \quad \varphi \leq u \leq \psi \quad , \tag{2.9}$$

i.e., $H(\varphi,h) \leq Mu \leq H(\psi,h)$ for all $h \in [\varphi,\psi] \Rightarrow \varphi \leq u \leq \psi$.

The splitting (2.8) will be so chosen that $H(u,v)$ as a function of u is pre-inverse-monotone as described in conditions 1.I and 2.I below. For proving this property one may use the results of the preceeding section.

Example 2.6 Consider an operator $M : C_2^n[0,1] \to \mathrm{IR}^n[0,1]$ with components of the form

$$(Mu)_i(x) = \begin{cases} -u_i''(x) + f_i(x, u_i(x), u_i'(x), u) & \text{for } 0 < x < 1 \\ -u_i'(0) + f_i^0(u_i(0), u) & \text{for } x = 0 \\ u_i'(1) + f_i^1(u_i(1), u) & \text{for } x = 1 \end{cases} \quad .$$

Because of the occurance of the functional term u as the last variable of each of the functions f_i, f_i^0, f_i^1, these functions may, for instance, also depend on $u_k(x)$ with $k \neq i$.

Here we choose $H(u,v) = (H_i(u,v))$ such that $H_i(u,v)(x)$ is obtained, when in the formula for $(Mu)_i(x)$ the functional term u mentioned is replaced by v.

In the following theorem we assume, in addition, that $(S,\leq)$ is a linear lattice, so that the absolute value $|U| \in S$, and the truncation $U^\# = \sup\{\varphi, \inf\{U,\psi\}\} \in S$ are defined for $U \in S$.

Theorem 2.7 *The implication* (2.9) *holds for a (fixed)* $u \in R$, *if there is an element* $z \succ 0$ *in* R *such that the following conditions are satisfied for each* $\lambda > 0$.

1) *If* $\varphi - \lambda z \leq u \leq \psi + \lambda z$ *and* $u \not< \psi + \lambda z$, *then*
   - I. $H(u,u) \not\lessdot H(\psi + \lambda z, u)$,
   - II. $H(\psi, u^\#) \lessdot H(\psi + \lambda z, u)$.
2) *If* $\varphi - \lambda z \leq u \leq \psi + \lambda z$ *and* $\varphi - \lambda z \not< u$, *then*
   - I. $H(\varphi - \lambda z, u) \not\lessdot H(u,u)$,
   - II. $H(\varphi - \lambda z, u) \lessdot H(\varphi, u^\#)$.

Corollary 2.8 *Let* $M = A - B$ *with a linear operator* $A : R \to S$ *and a nonlinear operator* $B : S \to S$, *and define* $H(u,v) = Au - Bv$. *Then assumptions* 1), *and* 2) *of the above theorem are satisfied if*

- I. A *is pre-inverse-positive,*
- II. $\lambda Az - |Bu - Bu^\#| \gtrdot 0$ *for* $\lambda > 0$, $u \in [\varphi - \lambda z, \psi + \lambda z]$, *and the truncation* $u^\# \in [\varphi,\psi]$.

If (**L**) *holds and* $\ell \sup\{u,v\} = \max\{\ell u, \ell v\}$ *for* $u, v \in R$, $\ell \in \mathcal{L}$, *then condition* II *is satisfied, if there is a positive linear operator* $C : S \to S$ *such that*

$$|Bu - Bv| \leq C|u - v| \quad \textit{for } u,v \in S \quad ; \quad (A - C)z \gtrdot 0 \quad .$$

The latter inequality implies that $A - C$ is inverse-positive.

Again we may consider operators M described by functionals as in (2.3). We will not discuss this in detail, but only mention, that here antitonicity with respect to the last variable u in (2.3) need not be required. One defines $H_\ell(u,v) = F_\ell(P_\ell u, Q_\ell u, \ell u, v)$, and $H(u,v) = (H_\ell(u,v)) \in \mathrm{IR}(\mathcal{L})$.

2.3 A different method for initial value problems

2.31 General remarks

The theorems of sections 2.1 and 2.2 can be applied to boundary value problems as well as to initial value problems, such as problems with parabolic differential equations. But for deriving a priori inclusion statements for initial value problems, often a different proof is used in the literature. One first derives statements which involve only strong inequality signs (pointwise $<$), and then uses additional *local* assumptions for proving corresponding statements with the sign $\leq$. These additional assumptions often have the form of estimates by *uniqueness functions* (see Walter [80]). For parabolic problems the results with the $<$-sign are known as the Nagumo-Westphal Lemma ([46], [73], [83], [80], p. 187).

We will here describe an approach which is closely related to the method just mentioned, but yields results of a form similar to the theorems in sections 2.1 and 2.2. For nonlinear problems these results require weaker assumptions than the theorems above: conditions such as assumptions II in Theorems 2.1 and 2.7, which are required to hold for all $\lambda > 0$, are replaced by similar *local* conditions, which involve only small $\lambda > 0$. It must also be said, however, that this advantage disappears, if the results on a priori inclusion statements are used for proving existence and inclusion statements by the method of modification described in section 4.

We will consider problems of the form

$$(M_t u)(t) = r(t) \quad \text{for} \quad 0 < t \leq T \quad , \quad u(0) = r^0 \tag{2.10}$$

for elements u in a certain ordered linear space $\widetilde{R}$. The elements $u \in \widetilde{R}$ are supposed to be functions of $t \in [0,T]$ $(T > 0)$ such that for $t \in (0,T]$, $u(t)$ belongs to some ordered linear space R_t. For $t \in (0,T]$, M_t denotes an operator on $\widetilde{R}$ such that $(M_t u)(t)$ belongs to some ordered linear space S_t.

The given problem can be written in the form $Mu = r$ with an operator M on $\widetilde{R}$ defined by

$$(Mu)(t) = (M_t u)(t) \quad \text{for} \quad 0 < t \leq T \quad , \quad (Mu)(0) = u(0) \quad . \tag{2.11}$$

One may try to apply the results of sections 2.1 and 2.2 to this operator in order to obtain a priori inclusions by two-sided bounds. But for such problems one may also use the different approach mentioned.

For simplicity, we describe this approach in the following only for a special case where, in particular, all spaces R_t $(0 < t \leq T)$ are equal, and all spaces S_t $(0 < t \leq T)$ are equal.

2.32 Inverse-monotonicity

We assume that

$(Y, \leq)$ is an ordered linear space, which is Archimedian and contains an element $\zeta \succ 0$, so that a norm $\| \ \| = \| \ \|_\zeta$ is defined on Y (see (1.14));

$(R, \leq) \subset (Y, \leq)$, and $(S, \leq)$ are ordered linear spaces such that **(S)** holds;

$u \geq 0 \quad :\Leftrightarrow \quad u(t) \geq 0 \quad (0 \leq t \leq T), \quad \text{if} \quad u : [0,T] \to Y;$

$\widetilde{R}$ is a linear subspace of $C_0([0,T], Y)$; $\quad v, w \in \widetilde{R}$;

$u(t) \in R, \quad (M_t u)(t) \in S \quad \text{for} \quad u \in \widetilde{R}, 0 < t \leq T.$

We want to prove the implication $Mv \le Mw \Rightarrow v \le w$ for the operator M in (2.11). This implication is equivalent to the statement (2.15) of the following theorem.

Theorem 2.9 *Suppose there is a $z \in \widetilde{R}$ such that the following two conditions are satisfied for each $t \in (0,T]$ and each $\lambda \in (0,\varepsilon]$ (with some $\varepsilon > 0$).*

I. *The relations*

$$v(s) \prec (w+\lambda z)(s) \quad \textit{for} \quad 0 \le s < t \quad , \quad v(t) \not\prec (w+\lambda z)(t) \tag{2.12}$$

imply

$$(M_t v)(t) \not<^{!} (M_t(w+\lambda z))(t) \quad . \tag{2.13}$$

II. $z(0) \succ 0$, *and*

$$(M_t w)(t) <^{!} (M_t(w+\lambda z))(t) \quad \text{if (2.12) holds} \quad . \tag{2.14}$$

Then

$$\left.\begin{array}{lll} (M_t v)(t) & \le (M_t w)(t) & \textit{for} \quad 0 < t \le T \\ v(0) & \le w(0) & \end{array}\right\} \Rightarrow \quad v(t) \le w(t) \ \textit{for} \quad 0 \le t \le T \quad . \tag{2.15}$$

Proof. Suppose that the premise in (2.15) holds, and let $\lambda \in (0,\varepsilon]$ be fixed. First we will prove that

$$v(t) \prec (w+\lambda z)(t) \quad \text{for} \quad 0 \le t \le T \quad . \tag{2.16}$$

Suppose that these inequalities are not true. Then, due to $v(0) \le w(0) \prec (w+\lambda z)(0)$, there is a $t \in (0,T]$ such that (2.12) holds, so that (2.13) follows. On the other hand, by assumption II, $(M_t v)(t) \le (M_t w)(t) <^{!} (M_t(w+\lambda z))(t)$. Because of this contradiction, (2.16) must be true.

The inequalities $v(t) \le w(t)$ for $0 \le t \le T$ follow from (2.16) for $\lambda \to 0$.

Corollary 2.10. *If assumption* I *holds for each $t \in (0,T]$ and $\lambda = 0$, then*

$$\left.\begin{array}{lll} (M_t v)(t) & <^{!} (M_t w)(t) & \textit{for} \quad 0 < t \le T \\ v(0) & \prec w(0) & \end{array}\right\} \Rightarrow \quad v(t) \prec w(t) \quad \textit{for} \quad 0 \le t \le T \quad . \tag{2.17}$$

In this corollary an element z does not occur. The proof is by arguments analogous to those used in the proof of the theorem.

Suppose now, that **(L)** and **(B)** hold and that the strong order in S is induced by $\mathcal{L}$. Then the inequalities (2.12) imply that $u = w - v + \lambda z$ satisfies

$$\begin{array}{l} u(s) \ge 0 \quad \text{for} \quad 0 \le s \le t \quad , \\ \ell u(t) = 0 \quad , \quad P_\ell u(t) \ge 0 \quad , \quad Q_\ell u(t) = 0 \quad \text{for some} \quad \ell \in \mathcal{L} \quad , \\ (d/dt\ \ell u)(t) \le 0, \text{ if this derivative exists.} \end{array} \tag{2.18}$$

We will now consider operators such that $\ell M_t u$ depends only on quantities occuring in the above equations and inequalities. For simplicity, we restrict ourselves to the following case.

Suppose that for each $t \in (0,T]$, $\ell \in \mathcal{L}$, and $u \in \widetilde{R}$

$$(\ell M_t u)(t) = \gamma_{\ell,t}(d/dt\ \ell u)(t) + \ell A_t u(t) \tag{2.19}$$

with constants $\gamma_{\ell,t} \ge 0$ and operators $A_t : R \to S$, and that

$$\ell A_t u(t) = F_{\ell,t}(P_\ell u(t), Q_\ell u(t), \ell u(t), u(t)) \tag{2.20}$$

with a real-valued function $F_{\ell,t}$ on $X_\ell \times Y_\ell \times \mathbb{R} \times R$.

We require now that for ℓ, t with $\gamma_{\ell,t} > 0$, the derivative $(d/dt\ \ell u)(t)$ exists for each $u \in \widetilde{R}$.

Theorem 2.11 *Suppose there is a* $z \in \widetilde{R}$ *such that* $z(0) \succ 0$, *and for all* $\ell \in \mathcal{L}$, $t \in (0,T]$, *and* $\lambda \in (0,\varepsilon]$ *satisfying*

$$\ell v(t) = \ell(w + \lambda z)(t) \quad , \quad Q_\ell v(t) = Q_\ell(w + \lambda z)(t) \tag{2.21}$$

the following two conditions hold:

I. $F_{\ell,t}(P_\ell v(t), Q_\ell v(t), \ell v(t), v(t)) \geq F_{\ell,t}(P_\ell v(t) + q, Q_\ell v(t), \ell v(t), v(t) + r)$ *for* $q \geq 0$ *in* X_ℓ, $r \geq 0$ *in* R.

II. $\lambda \gamma_{\ell,t}(d/dt\ \ell z)(t) + \ell[A_t(w + \lambda z)(t) - A_t w(t)] > 0 \quad .$ (2.22)

Then (2.15) *is true.*

Comparing (2.21) and assumption I of the above theorem with (2.4) and assumption I of Theorem 2.3, respectively, one sees that for each t the operator A_t considered here is essentially of the same type as M in Theorem 2.3. For example, each A_t may be a quasilinear elliptic operator (see (2.7)). Then, M_t is a parabolic operator of the usual form. (One may choose $\gamma_{\ell,t} = 1$ for point functionals ℓ belonging to $x \in \Omega$, and $\gamma_{\ell,t} = 0$ for ℓ belonging to $x \in \partial\Omega$.)

For a function $z(t) = \exp(Nt)z_0$ with $z_0 \succ 0$ condition II assumes the form of a (local, one-sided) Lipschitz condition:

$$\ell A_t(w + \lambda z)(t) - \ell A_t w(t) > -N\gamma_{\ell,t}\lambda \ell z(t) \quad . \tag{2.23}$$

One can also treat operators M_t more general than those described by (2.19), (2.20). For instance, $(\ell M_t u)(t)$ may depend on $(d/dt\ \ell u)(t)$ in a nonlinear way. Moreover, because of (2.18), $(\ell M_t u)(t)$ may also depend on all $u(s)$ with $0 \leq s \leq t$. This indicates, that the theory can be generalized, for instance, in such a way, that it can be applied to *problems with retarded arguments.*

Furthermore, instead of elements $z_\lambda = \lambda z$ one may also use elements z_λ which depend on λ in a nonlinear way. This is related to replacing the Lipschitz-condition (2.23) by some more general (uniqueness) condition, (see [69], p. 253).

2.33 Splitting

Analogous to the procedure in Section 2.2 one can consider splittings of the operators M_t in order to develop a theory on two-sided bounds which can be applied to vector-valued problems which are not quasi-antitone. We will here only formulate a corresponding result for operators M_t of the form (2.19), (2.20), assuming that the general assumptions made there are satisfied. In addition, we assume that $(Y, \leq)$ is a linear lattice, and that the functions $F_{\ell,t}$ are defined on $X_\ell \times Y_\ell \times \mathbb{R} \times Y$.

Let $u, \varphi, \psi \in \widetilde{R}$; $\varphi(t) \leq \psi(t)$ for $0 \leq t \leq T$, and define $u^\#(t) = \sup\{\varphi(t), \inf\{u(t), \psi(t)\}\}$,

$$H_{\ell,t}[v,w] = F_{\ell,t}(P_\ell v(t), Q_\ell v(t), \ell v(t), w(t)) \quad \text{for suitable } v, w \quad .$$

Theorem 2.12. *Suppose there is a* $z \in \widetilde{R}$ *such that* $z(0) \succ 0$, *and the following conditions are satisfied for* $\ell \in \mathcal{L}$, $0 < t \leq T$, $0 < \lambda \leq \varepsilon$.

1. *If* $(\varphi - \lambda z)(t) \leq u(t) \leq (\psi + \lambda z)(t)$ *and* $\ell u(t) = \ell(\psi + \lambda z)(t)$, $Q_\ell u(t) = Q_\ell(\psi + \lambda z)(t)$, *then*

   I. $H_{\ell,t}[u,u] \geq F_{\ell,t}(P_\ell u(t) + q, Q_\ell u(t), \ell u(t), u(t))$ *for* $q \geq 0$ *in* X_ℓ,

   II. $\lambda \gamma_{\ell,t}(d/dt\ \ell z)(t) + H_{\ell,t}[\psi + \lambda z, u] - H_{\ell,t}[\psi, u^\#] > 0$.

2. *If* $(\varphi - \lambda z)(t) \leq u(t) \leq (\psi + \lambda z)(t)$ *and* $\ell(\varphi - \lambda z)(t) = \ell u(t)$, $Q_\ell(\varphi - \lambda z)(t) = Q_\ell u(t)$, *then*

   I. $F_{\ell,t}(P_\ell u(t) - q, Q_\ell u(t), \ell u(t), u(t)) \geq H_{\ell,t}[u,u]$ *for* $q \geq 0$ *in* X_ℓ,

   II. $\lambda \gamma_{\ell,t}(d/dt\ \ell z)(t) + H_{\ell,t}[\varphi, u^\#] - H_{\ell,t}[\varphi - \lambda z, u] > 0$.

Under these assumptions the following statement is true. If $\varphi(0) \le u(0) \le \psi(0)$, *and*

$$\gamma_{\ell,t}(d/dt\ \ell\varphi)(t) + H_{\ell,t}[\varphi,h] \le \ell M_t u(t) \le \gamma_{\ell,t}(d/dt\ \ell\psi)(t) + H_{\ell,t}[\psi,h]$$
$$\text{for all}\quad \ell \in \mathcal{L}\ , t \in (0,T]\ , h \in Y \text{ with}\quad \varphi(t) \le h \le \psi(t)\ ,$$

then $\varphi(t) \le u(t) \le \psi(t)$ *for* $0 \le t \le T$.

Some remarks on notation and literature. An operator T is called *monotone* if $u \le v \Rightarrow Tu \le Tv$ (for u, v in the domain of T), and T is called *antitone* if $u \le v \Rightarrow Tu \ge Tv$. For vector-valued operators $T : \mathbb{R}^n \to \mathbb{R}^n$ Walter introduced the following notation: T is *quasi-monotone* if, for each index i, $(u \le v, u_i = v_i) \Rightarrow (Tu)_i \le (Tv)_i$ (see [80]). This concept is important, for instance, in the theory of differential inequalities with operators of the form $Mu = u' - Tu$. This was already recognized by Müller [43], Kamke [29], Szarski [74], and Wazewski [81]. We have essentially adapted this notation of Walter. Since, however, not all operators M considered here can be written as a difference $M = A - T$ with quasi-monotone T, we use the term *quasi-antitone* (for the operator M). In particular, the letter Q in Definition 1.4 stands for this term.

For abstract operators the term quasi-monotone was defined by Volkmann [77], in considering differential operators $u' - Tu$. The use of linear functionals is a common tool in the theory of operator inequalities. The description of operators by linear functionals ℓ and corresponding quantitites P_ℓ, Q_ℓ extends the work in [68] for differential operators.

Inequalities which involve order intervals were first considered by Müller [44] for initial value problems with ordinary differential equations.

3. More general estimates by transforming a given problem

In this section we describe methods for transforming a given problem such that the theorems of the foregoing sections can be applied to the transformed problem. In this way we obtain estimates other than those by two-sided bounds, and we can also treat operators different from those considered above. (For instance, vector-valued operators may be strongly coupled.)

We will consider differential equations in a Banach space Y. The inclusion statements derived will have the form

$$u(x) \in \psi(x)G \quad \text{for all } x \text{ considered} \quad , \tag{3.1}$$

where G denotes a certain star-shaped subset of Y, and ψ is a real-valued *bound function* (≥ 0) to be determined.

For constant ψ, say $\psi(x) \equiv 1$, the above statement describes an *invariance property*: $u(x)$ stays in G for all x. For variable ψ, we will speak of a ***shape-invariant estimate***, and ***shape-invariant bounds***.

We describe the method of transformation for the simple case of *pointwise norm bounds* $\|u(x)\| \le \psi(x)$, and ordinary differential equations in a Banach space, and then report on some other cases.

3.1 Pointwise norm bounds for a boundary value problem

Let Y be a Banach space with norm $\|\ \|$, S the space of all functions $U : [0,1] \to Y$, $R = C_0([0,1], Y) \cap C_2((0,1), Y)$, $r \in S$, and $M : R \to S$ given by

$$Mu(x) = u(x) \quad \text{for } x \in \{0,1\}\quad ,$$
$$Mu(x) = -u''(x) + B(x)u'(x) + C(x)u(x) + f(x, u(x), u'(x)) \quad \text{for } 0 < x < 1\quad ,$$

where $B(x)$, $C(x)$ denote linear operators mapping Y into Y, and $f : (0,1) \times Y \times Y \to Y$. We consider the equation $Mu = r$.

Assume, in addition, that an inner product $\langle\ ,\ \rangle$ is defined on Y which is continuous with respect to the given norm $\|\ \|$, and write $\|y\| = \langle y, y\rangle^{\frac{1}{2}}$. We are interested in estimates $\|u(x)\| \le \psi(x)$ $(0 \le x \le 1)$ with some $\psi \ge 0$. This estimate can also be written as (3.1) with $G = \{y \in Y : \|y\| \le 1\}$. Let $\Gamma = \{y \in Y : \|y\| = 1\}$.

Now suppose that $u \in R$ is a fixed (though unknown) solution of $Mu = r$. The values $u(x)$ can be written in the form

$$u(x) = \rho(x)\eta(x) \quad \text{with real} \quad \rho(x) \ge 0 \quad , \quad \text{and} \quad \eta(x) \in \Gamma \ . \tag{3.2}$$

$\rho(x)$ is the term to be estimated by $\psi(x)$. Let X denote the set of all $x \in [0,1]$ with $\rho(x) \ne 0$. For $x \in X$ the element $\eta(x)$ is uniquely determined; moreover, ρ and η are twice continuously differentiable on $X \cap (0,1)$. Corresponding properties do not hold for $x \notin X$, but this fact does not cause any difficulties. Observe that the inequality $\rho(x) \le \psi(x)$ is trivial for $x \notin X$.

We will transform M on R into a scalar differential operator $\mathcal{M}$ on $\mathcal{R} = C_0[0,1] \cap C_2((0,1) \cap X)$, and r into a real-valued function $\mathtt{r}$ such that $(Mu = r) \Rightarrow (\mathcal{M}\rho = \mathtt{r})$, and then apply Theorem 2.1 for proving $\mathcal{M}\rho \le \mathcal{M}\psi \Rightarrow \rho \le \psi$.

If $x \notin X$, we simply define $\mathtt{r}(x) = 0$, and $\mathcal{M}\varphi(x) = \varphi(x)$ for $\varphi \in \mathcal{R}$.

Now let $x \in X$. We want to choose $\mathcal{M}$ such that

$$\mathcal{M}\rho(x) = \langle \eta, Mu\rangle(x) \tag{3.3}$$

with u, ρ, η in (3.2). For $x \in \{0,1\}$ this relation yields $\mathcal{M}\rho(x) = \rho(x)$, and for $x \in (0,1)$ one derives

$$\mathcal{M}\rho(x) = F(x, \eta(x), \eta'(x), \rho) \quad , \tag{3.4}$$

where

$$F(x,h,q,\varphi) = (\mathcal{A}[h]\varphi)(x) + c[h,q](x)\varphi(x) + \langle h, f(x, \varphi(x)h, \varphi(x)q + \varphi'(x)h\rangle$$

for $h, q \in Y$ with a linear differential operator $\mathcal{A}[h]$ and a function $c[h,q]$ defined by

$$\begin{aligned}(\mathcal{A}[h]\varphi)(x) &= -\varphi''(x) + \langle h, B(x)h\rangle\varphi'(x) + \langle h, C(x)h\rangle\varphi(x) \quad , \\ c[h,q](x) &= \langle q,q\rangle + \langle h, B(x)q\rangle \quad .\end{aligned}$$

The above relations are derived from (3.3) by some formal calculations using $\langle\eta,\eta\rangle = 1$, $\langle\eta,\eta'\rangle = 0$, and $\langle\eta',\eta'\rangle + \langle\eta,\eta''\rangle = 0$, and eliminating η''.

These considerations lead us to defining $\mathtt{r}(x)$ and $\mathcal{M}\varphi(x)$ for $x \in X$ and $\varphi \in \mathcal{R}$ by

$$\mathtt{r}(x) = \langle \eta, r\rangle(x) \quad ,$$

$$\mathcal{M}\varphi(x) = \varphi(x) \ \text{ if } x \in \{0,1\} \ ; \quad \mathcal{M}\varphi(x) = F(x, \eta(x), \eta'(x), \varphi) \ \text{ if } x \in (0,1) \quad .$$

Now we will apply Theorem 2.1 to $\mathcal{M} : \mathcal{R} \to \mathbb{R}[0,1], \rho, \psi$ instead of $M : R \to S, v, w$, defining $U \succ 0 :\Leftrightarrow U(x) > 0$ for $0 \le x \le 1$. The operator $\mathcal{M}$ depends on the fixed, but unknown quantities η and X. Without knowing η and X, however, one can show that assumption I is satisfied. The remaining assumptions can be replaced by sufficient conditions independent of η and X as in the following result (observe that $\langle\eta(x), \eta'(x)\rangle = 0$ for $x \in X \cap (0,1)$).

Theorem 3.1 *Suppose there are* $\psi, z \in C_0[0,1] \cap C_2(0,1)$ *such that* $\psi(x) \geq 0$, $z(x) > 0$ $(0 \leq x \leq 1)$, *and*

$$(\mathcal{A}[h]z)(x) + c[h,q](x)z(x) + \lambda^{-1}\langle h, f(x,(\psi+\lambda z)(x)h, (\psi+\lambda z)(x)q + (\psi+\lambda z)'(x)h) - f(x,\psi(x)h,\psi(x)q+\psi'(x)h)\rangle > 0 \tag{3.5}$$

for all $\lambda > 0$ *and*

$$x \in (0,1) \quad ; \quad h, q \in Y \quad \text{with} \quad \|h\| = 1 \quad , \quad \langle h, q\rangle = 0 \quad . \tag{3.6}$$

Then, for $u \in R$ *with* $Mu = r$, *the estimate*

$$\|u(x)\| \leq \psi(x) \quad (0 \leq x \leq 1) \tag{3.7}$$

holds, if

$$\begin{aligned} \langle h, r\rangle(x) &\leq \psi(x) \quad \text{for} \quad x \in \{0,1\} \quad ; \\ \langle h, r\rangle(x) &\leq (\mathcal{A}[h]\psi)(x) + c[h,q](x)\psi(x) + \langle h, f(x,\psi(x)h,\psi(x)q+\psi'(x)h\rangle \\ &\quad \text{for} \quad x,h,q \quad \text{in } (3.6) \quad . \end{aligned} \tag{3.8}$$

Remarks 1) In (3.5) and (3.8) the term $c[h,q](x)$ can be replaced by

$$\bar{c}[h](x) = -\tfrac{1}{4}(\|B^*(x)h\|^2 - \langle h, B(x)h\rangle^2) \quad .$$

$\bar{c}[h]$ is a lower bound of $c[h,q]$ under the side conditions (3.6).

2) The differential inequalities in (3.5) and (3.8) need only hold for x, h, q, λ which satisfy also

$$u(x) = (\psi+\lambda z)(x)h \quad , \quad u'(x) = (\psi+\lambda z)'(x)h + (\psi+\lambda z)(x)q \quad .$$

Let us make some comments on these results for the case $Y = \mathrm{IR}^n$, and corresponding results on two-sided bounds which follow from Theorem 2.7.

It is obvious that the above results on pointwise norm bounds can be applied under quite different assumptions on B, C and f, than the results on two-sided bounds. For instance, we need not require here that the differential operator is weakly coupled, i.e., $B(x)$ need not be diagonal, and $f(x,y,p)$ may be coupled with respect to p.

The formulas (3.8) contain two parameters h and q, and therefore seem to be of an essentially different nature than the corresponding formulas for two-sided bounds.

Consider, however, in the theory of two-sided bounds the special case of vector-valued functions $-\varphi(x) = \psi(x) = (\psi_i(x))$ with $\psi_i(x) = \tilde{\psi}(x)$ $(i = 1, 2, \ldots, n)$. Then the differential inequalities required for $\tilde{\psi}(x)$ can be written in such a form that they contain the parameters

$$h \in \mathrm{IR}^n \quad \text{with} \quad \|h\|_\infty = 1 \quad .$$

Moreover, if in this case one does not require a priori that M is weakly coupled, then a second parameter q occurs also in the theory of two-sided bounds. The occurance of this parameter q, however, essentially forces the operator M to be weakly coupled; and q does not occur, if M has this property.

3.2 Pointwise norm bounds for an initial value problem

Now we consider a problem $Mu = r$ with an operator M of the form

$$\begin{aligned} Mu(0) &= u(0) \quad , \\ Mu(x) &= u'(x) + C(x)u(x) + f(x,u(x)) \quad \text{for} \quad 0 < x \leq 1 \quad . \end{aligned} \tag{3.9}$$

We use the same notation as in the preceeding section, except that now $R = C_0([0,1],Y) \cap C_1((0,1),Y)$, $f : (0,1] \times Y \to Y$, $\mathcal{R} = C_0[0,1] \cap C_1((0,1] \cap X)$.

Applying again Theorem 2.1 we obtain here the following result, where now

$$(\mathcal{A}[h]\varphi)(x) = \varphi'(x) + \langle h, C(x)h\rangle\varphi(x) \quad .$$

Theorem 3.2 *Suppose there are* $\psi, z \in C_0[0,1] \cap C_2(0,1]$ *such that* $\psi(x) \geq 0$, $z(x) > 0$ $(0 \leq x \leq 1)$, *and*

$$(\mathcal{A}[h]z)(x) + \lambda^{-1}\langle \eta, f(x,(\psi + \lambda z)(x) - f(x,\psi(x))\rangle > 0 \tag{3.10}$$

for all $\lambda > 0$, *and*

$$x \in (0,1] \quad , \quad h \in Y \quad \textit{with} \quad \|h\| = 1 \quad . \tag{3.11}$$

Then, for $u \in R$ *with* $Mu = r$, *the estimate* (3.7) *holds if*

$$\begin{aligned} &\langle h, r\rangle(0) \leq \psi(0) \quad , \\ &\langle h, r\rangle(x) \leq (\mathcal{A}[h]\psi)(x) + \langle h, f(x,\psi(x)h)\rangle \quad \textit{for} \quad x, h \;\; \textit{in}\; (3.11) \quad . \end{aligned} \tag{3.12}$$

Remarks 1) The differential inequalities in (3.10) and (3.12) need only hold for x, h which satisfy also $u(x) = (\psi + \lambda z)(x)h$.

2) Condition (3.10) need only be required for $\lambda \in (0,\varepsilon]$ with some $\varepsilon > 0$. This follows by applying Theorem 2.9 instead of Theorem 2.1.

This theorem on pointwise norm bounds, too, requires conditions quite different from those in the theory of two-sided bounds.

Consider, for instance, the simple case

$$Y = \mathbb{R}^2 \quad , \quad \langle y, \eta\rangle = y^T\eta \quad , \quad C(x) = C = \begin{pmatrix} \mu & \nu \\ -\nu & \mu \end{pmatrix} \quad \text{for} \quad 0 \leq x \leq 1 \quad .$$

The matrix C has the complex eigenvalues $\mu \pm i\nu$, and

$$\langle h, Ch\rangle = \mu \quad \text{for} \quad \|h\| = 1 \quad , \tag{3.13}$$

so that the imaginary part ν of the eigenvalues drops out. In this case the matrix $C = \begin{pmatrix} \mu & -|\nu| \\ -|\nu| & \mu \end{pmatrix}$ would enter the conditions in the theory of two-sided bounds.

More generally, if $C(x) = C$ is any constant 2×2-matrix with complex eigenvalues $\mu \pm i\nu$, there is an inner product such that (3.13) holds. This indicates that the use of pointwise norm bounds can be advantageous in stability theory.

The above result can also be applied to parabolic operators. Consider, for instance, an operator M such that for $0 < t \leq 1$ (or $0 < t \leq T$)

$$Mu(t) = \frac{d}{dt}u(t) + C(t)u(t) + f(t, u(t)) \quad ,$$

where $u(t) \in C_2^n(\overline{\Omega})$ ($\Omega \subset \mathbb{R}^m$ a bounded domain), $u(t) = 0$ on $\partial\Omega$, all summands in the above formula are also functions in $\mathbb{R}^n(\overline{\Omega})$, and $C(t)$ is a vector-valued elliptic linear differential operator. One may use the above theorem (with t instead of x) for deriving estimates

$$(\langle u(t), u(t) \rangle)^{\frac{1}{2}} \leq \psi(t)$$

with an inner product $\langle \ , \ \rangle$ on $C_0^n(\overline{\Omega})$, for example,

$$\langle y, \eta \rangle = \int_\Omega y^T \eta \, dx \quad .$$

3.3 Various generalizations

Results analogous to those in the foregoing sections can also be proved for

inclusions (3.1) with certain star-shaped sets G,

second order elliptic and parabolic differential equations.

We will make a few comments on these generalizations, and only mention here that the theory can be carried over to

boundary conditions other than Dirichlet conditions,

estimates of the form $u(x) \in \psi_j(x)G_j$ ($j = 1, 2, \ldots, \nu$), which include estimates by two-sided bounds.

3.31 Star-shaped sets G

Suppose that Y is a Banach space with norm $\| \ \|$. Instead of the unit ball described by $\langle y, y \rangle \leq 1$ with a suitable inner product $\langle \ , \ \rangle$, one may consider more general sets G described by a *level function* $W : Y \to \mathbb{R}$ with the following properties:

(i) $G = \{y \in Y : W(y) \leq 1\}$.

(ii) For each $y \in Y$ there is an $\alpha = \alpha(y) \in (0, \infty]$ such that for $t \in [0, \infty)$

$$W(ty) < 1 \quad \text{if } 0 \leq t < \alpha, \quad W(ty) = 1 \quad \text{if } t = \alpha < \infty, \quad W(ty) > 1 \quad \text{if } t > \alpha \ ;$$

(iii) W is twice continuously differentiable at each y with $\alpha(y) < \infty$; and

$$\omega(y) := W'(y)y > 0 \quad \text{for} \quad y \in \Gamma := \{y \in Y : W(y) = 1\} \quad .$$

(In applications to first order differential equations, W need only be once continuously differentiable at each y with $\alpha(y) < \infty$.)

In order to obtain an inclusion statement (3.1) for a solution u of a given problem $Mu = r$ one uses again (3.2) for transforming the given problem into $\mathcal{M}\rho = \mathbf{r}$ with $\mathcal{M}$ and $\mathbf{r}$ dependent on η.

When this procedure is applied to the problems in Sections 3.1 and 3.2, one arrives at results similar to those in these sections. The formulas obtained differ from those above in the following way:

a) the linear functionals $\langle h, \ldots \rangle$ occurring are to be replaced by $(\omega(h))^{-1}W'(h)$,

b) the term $c[h, q]$ in Section 3.1 is to be replaced by

$$c[h, q] = (\omega(h))^{-1}(W''(h)(q, q) + W'(h)Bq) \quad ,$$

where $W''(h)(q, q)$ denotes the quadratic form corresponding to the second derivative of W at h.

In this way, for instance, the following differential inequality is obtained instead of the one in (3.12):

$$W'(h)r(x) \leq \omega(h)\psi'(x) + W'(h)C(x)h\psi(x) + W'(h)f(x,\psi(x)h) \quad \text{for} \quad x \in (0,1] \; , \; h \in \Gamma \; .$$

3.32 Partial differential equations

The theory of inclusion statements (3.1) can be carried over to problems with partial differential equations. As an example, let us briefly consider parabolic problems

$$Mu(x,t) = 0 \quad \text{for} \quad x \in \overline{\Omega} \; , \quad 0 \leq t \leq T \; ,$$

where $\Omega \subset \mathrm{IR}^m$ is a bounded domain, $u(x,t) \in \mathrm{IR}^n$, and $Mu(x,t) \in \mathrm{IR}^n$ has the following form:

$$\begin{aligned} Mu(x,t) &= Au(x,t) + f(x,t,u(x,t),u_x(x,t)) \quad \text{for} \quad x \in \Omega \; , \; 0 < t \leq T \; , \\ Mu(x,t) &= u(x,t) - r(x,t) \quad \text{for} \quad (x,t) \in \beta := (\partial\Omega \times (0,T]) \cup (\overline{\Omega} \times \{0\}) \; , \\ (Au)_i(x,t) &= L[u_i](x,t) \quad (i = 1,2,\ldots,n) \; , \end{aligned} \tag{3.14}$$

with

$$\begin{aligned} L[\varphi](x,t) &= \tfrac{\partial}{\partial t}\varphi(x,t) - a(x,t)\cdot\varphi_{xx}(x,t) \; , \\ a(x,t) &\in \mathrm{IR}^{m,m} \text{ symmetric and positive semi-definite} \; . \end{aligned}$$

(In the above formulas, $u_x(x,t)$ denotes the n×m matrix of the derivatives $\partial u_i/\partial x_k$.)

Let $G \subset \mathrm{IR}^n = Y$ be a set which is described by a level function as explained in Section 3.31. In order to obtain an inclusion statement

$$u(x,t) \in \psi(x,t)G \quad \text{for } x \in \overline{\Omega} \; , \; 0 \leq t \leq T$$

with a suitable function $\psi \geq 0$, one can proceed in essentially the same way as in the cases discussed above. One writes $u(x,t) = \rho(x,t)\eta(x,t)$ with $\rho(x,t) \in \mathrm{IR}$ and $\eta(x,t) \in \Gamma$, transformes $Mu = 0$ into $\mathcal{M}\rho = 0$, and applies results on a priori inclusion statements to the latter equation.

Here we will only formulate the resulting conditions on ψ (which correspond to (3.8) in Theorem 3.1):

$$\begin{aligned} &u(x,t) \in \psi(x,t)G \quad \text{for} \quad (x,t) \in \beta \; , \\ &\omega(h)L[\psi](x,t) + c[h,q](x,t)\psi(x,t) + W'(\eta)f(x,t,\psi(x,t)h,\psi(x,t)q + h\psi_x(x,t)) \geq 0 \end{aligned} \tag{3.15}$$

for

$$x \in \Omega \; , \quad t \in (0,T] \; , \quad h \in \mathrm{IR}^n \; , \quad q \in \mathrm{IR}^{n,m} \quad \text{with} \quad h \in \Gamma \; , \quad W'(h)q = 0 \; , \tag{3.16}$$

where

$$c[h,q] = a(x,t)\cdot q^T W''(h)q \; , \quad W''(h) \in \mathrm{IR}^{n,n} \; .$$

Invariance properties: In the special case $\psi(x,t) \equiv 1$, the above conditions assume the form

$$u(x,t) \in G \quad \text{for} \quad (x,t) \in \beta \; ,$$

$$c[h,q](x,t) + W'(h)f(x,t,h,q) \geq 0 \quad \text{for} \quad x,t,h,q \text{ in (3.16)} \; ; \tag{3.17}$$

and the inclusion statement says that

$$u(x,t) \in G \quad \text{for all} \quad (x,t) \in \overline{\Omega} \times [0,T] \; .$$

If, for instance $u(x,t) = 0$ for $x \in \partial\Omega$, $0 < t \leq T$, then one obtains conditions such that

$$u(x,0) \in G \quad \text{for} \quad x \in \overline{\Omega} \quad \Rightarrow \quad u(x,t) \in G \quad \text{for} \quad x \in \overline{\Omega}, 0 \leq t \leq T \quad .$$

Tangent conditions: If G is convex, then $c[h,q] \geq 0$ for $h \in \Gamma$, $W'(h)q = 0$, so that the term $c[h,q]$ can be dropped in (3.15) and (3.17). Then (3.17) becomes a tangent condition

$$W'(h)f(x,t,h,q) \geq 0 \quad \text{for} \quad x,t,h,q \text{ in } (3.15) \quad . \tag{3.18}$$

Tangent conditions of one or another form occur in many of the papers mentioned below.

In some cases the (unbounded) term q in $W'(h)f(x,t,h,q)$ can be eliminated by using (3.16). In general, however, (3.18) is stronger than (3.17) as a condition on the dependence of f on q.

Equality of the leading coefficients. In (3.14) we assumed that the components of $(Au)_i$ are given by an operator L which does not depend on i, in other words: we assumed that *all components $(Mu)_i$ have the same leading coefficients.* This requirement, in general, is necessary for carrying out the transformation of $Mu = 0$ into $\mathcal{M}\rho = 0$. Such a condition is also required in most of the papers mentioned below which deal with invariance properties of a similar form. (See the following notes.)

Some remarks on the literature. Statements on invariant sets, other than those by two-sided bounds, have been proved by a series of authors. We mention here only some papers with results more or less strongly related to those above. (Further references and more detailed discussions can be found in these papers.)

For elliptic problems see, for example, Amann [6], Lemmert [40,41], Martin [42], Redheffer and Walter [59], Schmitt [64], Schröder [70], and Weinberger [82]; for parabolic problems and abstract differential equations: the papers [6], [39], [70], [82] mentioned above, and Alikakos [4], Bebernes, Chueh, and Fulks [7], Bebernes, and Schmitt [8], Chueh, Conley, and Smoller [12], Conway, Hoff, and Smoller [16], Lakshmikantham and Vaughn [38], Plum [52,53], Redheffer and Walter [58,60], Schröder [71], Volkmann [78,79].

Some of the papers mentioned treat a priori inclusions (by invariant sets) as considered above, others inclusion and existence statements (see [70], [71] for more details).

There is also an extensive literature on invariance statements and other estimates for ordinary differential equations. For references and discussions see, for example, Gaines and Mawhin [21], Schmitt [64]. In particular, Hartmann [25,26] and Lasota and Yorke [39] derived pointwise norm bounds with constant $\psi(x)$ for two-point boundary value problems.

The main results on shape-invariant bounds formulated above can be found in [70, 71]. The proofs given there, however, differ from the ones presented here. In section 3 above we applied results on inverse-monotonicity to a transformed problem; in the papers mentioned the theorems were proved directly.

The requirement that all leading coefficients are equal is an essential restriction often encountered in invariance theory. This restriction is not necessary for two-sided bounds, or for bounds of integral type as described at the end of section 3.2. (See the discussion in [70], p. 448.) Considerable effort has been made to overcome this restriction in other cases, too. Plum [52,53] has proved corresponding results related to the theory above, and given many references on this subject.

4. Inclusion and existence

This section deals with statements of the form (0.4) for solutions of nonlinear equations in linear spaces. We will assume that $r = 0$ (i.e., r has been incorporated into M). Then the statement to be proved assumes the form

$$0 \in C \quad \Rightarrow \quad \text{there is a } u \in K \text{ such that } Mu = 0 \quad . \tag{4.1}$$

4.1 The method of modification

For proving statements of the above form with sets K and C as considered in the foregoing sections, one usually combines arguments from the theory of inequalities with tools of existence theory. As such tools, fixed-point theorems or degree theory have been used, for example. Discussing these theories in detail would exceed the scope of this paper. We will, however , describe one method which is particularly closely related to the theories presented above. Here the theorems of sections 2 and 3 are applied to some modified operator, or modified problem. In this *method of modification* the proof of inclusion, and the proof of existence are separated, so that, for instance, different tools of existence theory may be used.

Described in a very general form, the method of modification for proving (4.1) consists of three steps:

(1) One defines a *modified problem* $M^{\#}u = 0$ with a *modified operator* $M^{\#}$ on R such that

$$(M^{\#}u = 0 \ , \ u \in K) \quad \Rightarrow \quad Mu = 0 \quad .$$

(2) One proves the *inclusion statement*

$$(u \in R \ , \ M^{\#}u = 0 \ , \ 0 \in C) \quad \Rightarrow \quad u \in K \quad . \tag{4.2}$$

(3) One proves the *existence statement*

$$M^{\#}u = 0 \quad \text{for some } u \in R \quad .$$

For carrying out step (2) we will apply one of the theorems of sections 2 and 3 to $M^{\#}$, or to some other modified operator $\widetilde{M}$, which is closely related to $M^{\#}$. This application then determines the set C. Each of the theorems used contains two assumptions, called I and II. We will choose $M^{\#}$ and and $\widetilde{M}$ in such a way that both these assumptions are satisfied under conditions of a rather simple form. Also, in order to make fixed-point theorems applicable, the operator $M^{\#}$ shall be so constructed that it contains only bounded nonlinearities.More generally speaking , the modified operators will be more suitable than the given operators for the application of inclusion theories and existence theories.

4.2 Two-sided bounds

Suppose that $(R, \leq)$ and $(S, \leq)$ are ordered linear spaces, that $(R, \leq) \subset (S, \leq)$, $(R, \leq)$ is Archimedian, and $(S, \leq)$ is a linear lattice. Let $\varphi, \psi \in R$ with $\varphi \leq \psi$, and define

$$K = \{u \in S : \varphi \leq u \leq \psi\} \quad .$$

The modified problems will be constructed using the *truncations*

$$u^{\#} = \sup\{\varphi, \inf\{u, \psi\}\} \quad \text{of} \quad u \in R \quad ,$$

which have the property that $u^{\#} = u$ if and only if $u \in K$.

In order to explain our procedure we discuss three cases. In each case the emphasis will be on steps (1) and (2) of the method of modification. We shall, however, also make some remarks on the existence proof (step (3)), and give some examples.

Case 1 (applicable, e.g., to vector-valued weakly nonlinear ODEs and PDEs). Suppose that assumption **(L)** holds, and

$$\ell \sup\{U,V\} = \max\{\ell U, \ell V\} \quad \text{for} \quad U,V \in S \quad , \quad \ell \in \mathcal{L} \quad . \tag{4.3}$$

Also assume that $M : R \to S$,

$$M = A + B \quad , \quad A : R \to S \text{ linear} \quad , \quad Bu = \mathcal{B}(u,u) \quad ,$$
$$\mathcal{B}(u,v) \in S \quad , \quad \ell\mathcal{B}(u,v) = f_\ell(\ell u, v) \quad \text{for} \quad u,v \in S \quad , \quad \ell \in \mathcal{L}$$

with $f_\ell : \mathsf{IR} \times S \to \mathsf{IR}$. Then define $M^\# : R \to S$ by

$$M^\# u = Au + c(u - u^\#) + Bu^\# \quad \text{with some } c > 0 \quad .$$

Lemma 4.1 *Suppose that*

I. $A \in \mathbf{Q}(\mathcal{L})$,

II. *there is a* $z \in R$ *with* $z \geq 0$, $\ell(A + cI)z > 0$ *for* $\ell \in \mathcal{L}$.

Then, if the modified problem $M^\# u = 0$ *has a solution* $u^* \in R$, *and if*

$$A\varphi + \mathcal{B}(\varphi, h) \leq 0 \leq A\psi + \mathcal{B}(\psi, h) \quad \text{for all } h \in K \quad , \tag{4.4}$$

u^* *is also a solution of the given problem, and* $u^* \in K$.

This lemma says that the implication (4.2) in step (2) holds with "$0 \in C$" equivalent to (4.4), if A is a $\mathbf{Q}(\mathcal{L})$-operator and $A + cI$ has a majorizing element.

For proving the lemma, we define a further operator $\widetilde{M}$ by

$$\widetilde{M}u = Au + c(u - u^\#) + \mathcal{B}(u^\#, h_0) \quad \text{with} \quad h_0 = (u^*)^\# \quad .$$

In order to derive $u^* \leq \psi$ from (4.4) we apply Theorem 2.1 to $\widetilde{M}, u^*, \psi$ instead of M, v, w, using the strong order induced by $\mathcal{L}$. The corresponding condition I holds, since $u \leq \psi + \lambda z$, $\ell u = \ell(\psi + \lambda z)$ imply $\ell(\widetilde{M}(\psi + \lambda z) - \widetilde{M}u) = \ell A(\psi + \lambda z - u) \leq 0$. Condition II is satisfied, since $\widetilde{M}(\psi + \lambda z) - \widetilde{M}\psi = \lambda(A + cI)z$. Observe also that $\widetilde{M}u^* = 0$, and that $0 \in C$ implies $0 \leq \widetilde{M}\psi$. The inequality $\varphi \leq u^*$ is derived by applying Corollary 2.2 to $\widetilde{M}, \varphi, u^*$.

The existence of a solution of the modified problem can often be verified by applying Schauder's fixed-point theorem. We restrict ourselves to some brief remarks. Under the assumptions of Lemma 4.1 the given operator $A + cI$ is inverse-positive. Thus one may try to write the modified problem as a fixed-point equation $u = Tu := (A + cI)^{-1}(cu^\# - Bu^\#)$ in a suitable Banach space X such that $TX \subset X$. The relative compactness of the set TX then often follows from the compactness of the linear operator $(A + cI)^{-1}$. Observe that only the truncated elements $u^\#$ occur in the definition of Tu.

Such a procedure is successful for many boundary and initial value problems with ODEs. For PDE-problems one may use an extension of $(A + cI)^{-1}$ in defining T.

Example 4.2 (*A weakly nonlinear boundary value problem.*) Let $R = C_1(\overline{\Omega}) \cap C_2(\Omega)$, $S = \mathsf{IR}(\overline{\Omega})$, $M = A + B$,

$$Au(x) = L[u](x) := -a(x) \cdot u_{xx} + b(x) \cdot u_x \quad \text{for } x \in \Omega \quad , \quad Au(x) = u(x) \quad \text{for } x \in \partial\Omega$$

with the notation in Example 1.9, and

$$Bu(x) = f(x, u(x)) \quad \text{for } x \in \Omega \quad \text{with } f : \overline{\Omega} \times \mathsf{IR} \to \mathsf{IR} \quad , \quad Bu(x) = 0 \quad \text{for } x \in \partial\Omega \quad .$$

Then $Mu = 0$ is a weakly nonlinear Dirichlet boundary value problem.

We assume that $a(x) \geq_d 0$ for $x \in \Omega$, that properties **(E1)** and **(E2)** hold, and that f is continuously differentiable on $\overline{\Omega} \times \mathsf{IR}$. Then the following statement will be derived from Lemma 4.1.

If there are $\varphi, \psi \in R$ such that $\varphi \le \psi$ and $M\varphi \le 0 \le M\psi$, then there is a $u^* \in R$ such that $Mu^* = 0$, $\varphi \le u^* \le \psi$.

Proof. Let $\mathcal{L}$ be defined as in Example 1.9, so that $u^\#(x) = \max\{\varphi(x), \min\{u(x), \psi(x)\}\}$. For $c = 1$ the modified problem assumes the form

$$L[u](x) + u(x) = F[u](x) := u^\#(x) - f(x, u^\#(x)) \quad \text{for } x \in \Omega \quad , \quad u(x) = 0 \quad \text{for } x \in \partial\Omega \quad . \tag{4.5}$$

The assumptions I and II of Lemma 4.1 are satisfied (see Example 1.9, and choose $z(x) \equiv 1$), and the inequalities $M\varphi \le 0 \le M\psi$ assumed are equivalent to (4.4) with $\mathcal{B}(u,h) = Bu$. Thus it remains to be shown that the modified problem has a solution.

The operator $A + I$ is invertible on R, and hence (1.32) holds for $\lambda = 1$. Let X be the Banach space $C_0(\overline{\Omega})$ with the maximum norm. For $u \in X$, the right hand side $F[u]$ of the differential equation (4.5) is continuous on $\overline{\Omega}$ and thus belongs to L_p for each $p > m$. Due to **(E1)** and **(E2)** with $k = 0$ there is a $v \in H_{2,p} \cap \overset{\circ}{H}_{1,p}$ such that

$$L[v] + v = F[u] \quad , \tag{4.6}$$

$$\| v \|_{1+\alpha} \le c_1 \|v\|_{2,p} \le c_2 \|F[u]\|_p \le c_3 \|F[u]\|_\infty$$

with $\alpha \in (0, 1 - \frac{m}{p})$ and constants c_j independent of u. Moreover, the modified problem is constructed in such a way that

$$\|F[u]\|_\infty \le c_4 \quad \text{for all } u \in X$$

with some constant c_4 independent of u.

Consequently, a continuous operator $T : X \to X$ is defined by $Tu = v$ and (4.6) such that the image TX is relatively compact. Thus, by the fixed-point theorem of Schauder, the operator T has a fixed point u^*. Since $u^* \in C_{1+\alpha}$, we have $(u^*)^\# \in H_{1,p}$ and $F[u^*] \in H_{1,p}$. Applying **(E1)** and **(E2)** with $k = 1$ to (4.6) with $u = v = u^*$, we see that $u^* = v \in H_{3,p} \cap \overset{\circ}{H}_{1,p} \subset C_{2+\alpha} \subset R$, and also $u^* = 0$ on $\partial\Omega$.

Remark Instead of using the theory in $H_{j,p}$-spaces as above, one may proceed somewhat differently, using a result of Amann [5]. See the existence proof in [70, p.454].

Case 2 Suppose that **(L)** and **(B)** hold and (4.3) is satisfied. Then consider the problem of finding $u \in R$ which satisfies the equations

$$-a_\ell(Q_\ell u, \ell u, u) P_\ell u + f_\ell(Q_\ell u, \ell u, u) = 0 \quad \text{for} \quad \ell \in \mathcal{L} \quad , \tag{4.7}$$

where $a_\ell(y, \lambda, v)$ and $f_\ell(y, \lambda, v)$ are linear functionals on X_ℓ and IR, respectively, which depend on $y \in Y_\ell$, $\lambda \in \mathrm{IR}$, $v \in S$. We define a modified problem for $u \in R$ by

$$-a_\ell(Q_\ell u, \ell u^\#, u^\#) P_\ell u + \ell(u - u^\#) + f_\ell(Q_\ell u, \ell u^\#, u^\#) = 0 \quad \text{for} \quad \ell \in \mathcal{L} \quad . \tag{4.8}$$

The given problem and the modified problem can be written as operator equations $Mu = 0$ and $M^\# u = 0$, respectively, with operators mapping R into $\mathrm{IR}(\mathcal{L})$. For instance, $(Mu)_\ell$ is defined to be the term on the left hand side of (4.7). (See the end of Section 1.2 for analogous definitions. Observe also that we deal then with three spaces R, S and $\mathrm{IR}(\mathcal{L})$. The space S is needed, since $f_\ell(y, \lambda, v)$, for example, must not only be defined for $v = u \in R$, but also for $v = u^\# \in S$.) This operator formulation, however, is only needed for theoretical considerations.

Lemma 4.3 *Suppose that*

I. $a_\ell(Q_\ell\varphi, \ell\varphi, h) \geq 0$, $a_\ell(Q_\ell\psi, \ell\psi, h) \geq 0$ *for all* $h \in K$ *and* $\ell \in \mathcal{L}$,

II. *there is a* $z \in R$ *with*

$$z \succ 0 \quad , \quad P_\ell z = 0 \quad , \quad Q_\ell z = 0 \quad \text{for} \quad \ell \in \mathcal{L} \quad . \tag{4.9}$$

Then, if the modified problem has a solution $u^* \in R$, *and if for all* $h \in K$ *and* $\ell \in \mathcal{L}$

$$-a_\ell(Q_\ell\varphi, \ell\varphi, h)P_\ell\varphi + f_\ell(Q_\ell\varphi, \ell\varphi, h) \leq 0 \leq -a_\ell(Q_\ell\psi, \ell\psi, h)P_\ell\psi + f_\ell(Q_\ell\psi, \ell\psi, h) \quad , \tag{4.10}$$

u^* *is also a solution of the given problem, and* $u^* \in K$.

Remark In most applications P_ℓ and Q_ℓ are differential operators such that $P_\ell z = 0$ and $Q_\ell z = 0$ for constant functions z. For non-constant z one has to replace (4.9) by more complicated conditions.

For proving this lemma one shows that (4.1) holds with $u = u^*$, and "$0 \in C$" equivalent to (4.10). This is done in a way similar to that in Case 1. A further operator $\widetilde{M} : R \to \mathrm{IR}(\mathcal{L})$ is defined by

$$(\widetilde{M}u)_\ell = -a_\ell(Q_\ell u, \ell u^\#, h_0)P_\ell u + \ell(u - u^\#) + f_\ell(Q_\ell u, \ell u^\#, h_0) \quad \text{for} \quad \ell \in \mathcal{L}$$

with $h_0 = (u^*)^\#$. Then results on inverse-monotonicity are applied to $\widetilde{M}$ for proving $u^* \leq \psi$ and $\varphi \leq u^*$. Notice that $(\widetilde{M}u)_\ell$ essentially has the form (2.5) with $g_\ell = 0$. The inequality $u^* \leq \psi$ follows by applying Theorem 2.3 to $\widetilde{M} : R \to \mathrm{IR}(\mathcal{L}), u^*, \psi$ instead of $M : R \to S, v, w$. For verifying the corresponding condition I one uses Lemma 2.5. Condition II holds, since $\ell(\widetilde{M}(\psi+\lambda z) - \widetilde{M}\psi) = \lambda \ell z$, due to (4.9) and the construction of $\widetilde{M}$.

The inequality $\varphi \leq u^*$ can be derived analogously.

Example 4.2 can also be treated using Lemma 4.3 instead of Lemma 4.1.

Case 3 Suppose that $M : R \to S$,

$$M = A + B \quad , \quad A : R \to S \text{ linear} \quad , \quad B : S \to S \quad , \tag{4.11}$$

and define

$$M^\# u = Au + Bu^\# \quad .$$

Lemma 4.4 *Suppose that* A *is inverse-positive.*

Then, if the modified problem $M^\# u = 0$ *has a solution* $u^* \in R$, *and if*

$$A\varphi + Bh \leq 0 \leq A\psi + Bh \quad \text{for all } h \in K \quad ,$$

u^* *is also a solution of* $Mu = 0$, *and* $u^* \in K$.

For proving this lemma one observes that the above implication for $h = (u^*)^\#$ yields $A\varphi \leq Au^* \leq A\psi$, and hence $\varphi \leq u^* \leq \psi$.

Example 4.5 The boundary value problem

$$u^{iv}(x) + f(x, u(x)) = 0 \quad (0 < x < 1) \quad , \quad u(0) = u'(0) = u(1) = u'(1) = 0 \tag{4.12}$$

can be written in the form $Mu = 0$ with M as in (4.11) and

$$R = \{u \in C_4[0,1] : u(0) = u'(0) = u(1) = u'(1) = 0\} \quad , \quad S = C_0[0,1]$$

$$Au(x) = u^{iv}(x) + c(x)u(x) \quad , \quad Bu(x) = -c(x)u(x) + f(x, u(x)) \quad ,$$

where $c \in C_0[0,1]$ denotes any function such that

$$-\lambda_0 < c(x) \leq c_0 \quad (0 \leq x \leq 1) \quad ,$$

with λ_0 being the smallest eigenvalue of the operator u^{iv} subject to the above boundary conditions, and $c_0 = 4\kappa^2 \approx 951$, where κ is the smallest solution of $\tan\kappa = \tanh\kappa$ satisfying $\kappa > 0$. Then A is inverse-positive (see [676], p. 100).

Applying Schauders fixed-point theorem to the operator T defined on S by $Tu = -A^{-1}Bu^{\#}$, one sees that the modified problem has a solution $u^* \in R$. (Observe that $TS \subset S$, and that TS is relatively compact with respect to the maximum norm.)

Thus, Lemma 4.4 yields the following result.

If

$$\varphi^{iv} + c\varphi + (-ch + f(x,h)) \leq 0 \leq \psi^{iv} + c\psi + (-ch + f(x,h)) \quad \text{on } [0,1]$$

for the given functions $\varphi, \psi \in R$ *(with* $\varphi \leq \psi$*) and all* $h \in S$ *with* $\varphi \leq h \leq \psi$*, then the given problem* (4.12) *has a solution* $u^* \in R$ *such that* $\varphi \leq u^* \leq \psi$.

4.3 Shape-invariant bounds

In this section we are concerned with proving the existence of solutions of differential equations which satisfy shape-invariant estimates, as considered in section 3. We will describe one way in which the method of modification may here be applied. (Concerning relevant literature see the remarks at the end of section 3.)

Let Y be a Banach space, and G a subset described by a level function W, as explained in section 3.31. Suppose that $\Omega \subset \mathbb{R}^m$ is a bounded domain, R and S are linear spaces of functions on $\overline{\Omega}$, $R \subset S$, $v(x) \in Y$ $(x \in \overline{\Omega})$ for $v \in S$,

$$M = A + B \quad , \quad A : R \to S \text{ linear} \quad , \quad Bu = \mathcal{B}(u,u) \quad , \quad \mathcal{B}(u,v) \in S \quad \text{for } u, v \in S \quad .$$

Such formulas occurred also in case 1 treated above. In general, however, one will here use a splitting $Bu = \mathcal{B}(u,u)$ different from that used for two-sided bounds. (If no "functional terms", such as integrals, occur, one will now simply define $\mathcal{B}(u,v) = Bu$.)

For obtaining a statement of the form (4.1) with

$$K = \{u \in S : u(x) \in \psi(x)G \text{ for } x \in \overline{\Omega}\}$$

and some real-valued $\psi \geq 0$, one may proceed as follows.

Step (1). Each $u \in R$ can be written in the form $u(x) = \rho(x)\eta(x)$ $(u \in \overline{\Omega})$ as in (3.2). Define now the *truncation* $u^{\#}$ of u by

$$u^{\#}(x) = \rho^{\#}(x)\eta(x) \quad \text{for } x \in \overline{\Omega} \quad , \quad \text{with} \quad \rho^{\#}(x) = \min\{\rho(x), \psi(x)\} \quad ,$$

and a modified operator $M^{\#} : R \to S$ by

$$M^{\#} = Au + c(u - u^{\#}) + Bu^{\#}$$

with a sufficiently large constant $c > 0$. (Here we have to assume that S contains all elements of the form $\rho^{\#}\eta$ for $\rho\eta \in R$.)

Step (2). Assume that $u^* \in R$ is a solution of the equation $M^{\#}u = 0$, and define $\widetilde{M} : R \to S$ by

$$\widetilde{M}u = Au + c(u - u^{\#}) + \mathcal{B}(u^{\#}, (u^*)^{\#}) \quad .$$

Observing $\widetilde{M}u^* = 0$, apply then a suitable result on a priori inclusion by shape-invariant bounds to the equation $\widetilde{M}u = 0$, in order to show that $u^* \in K$, if ψ satisfies certain inequalities. These inequalities constitute the premise "$0 \in C$" in (4.1).

Step (3). For proving the existence of a solution of the modified problem one will, in general, use the same methods as in the case of estimates by two-sided bounds.

Example 4.6 Suppose that $Y = \mathbb{R}^n$, G is bounded , $R = C_0^n[0,1] \cap C_2^n(0,1)$, $S = C_0^n[0,1]$,

$$Au(x) = -u''(x) + b(x)u'(x) \quad \text{for} \quad 0 < x < 1 \quad , \quad Au(0) = u(0) \quad , \quad Au(1) = u(1) \quad ;$$
$$\mathcal{B}(u,v)(x) = f(x, u(x), v) \quad \text{for} \quad 0 < x < 1 \quad , \quad Bu(0) = -r^0 \quad , \quad Bu(1) = -r^1$$

with an n×n matrix $b(x)$ depending continously on $x \in [0,1]$, and continuous $f : [0,1] \times \mathbb{R}^n \times S \to \mathbb{R}^n$, $r^0 \in \mathbb{R}^n$, $r^1 \in \mathbb{R}^n$.

Using the procedure explained above, and applying a suitable generalization of Theorem 3.2 (with $z(x) \equiv 1$), we obtain here the following result.

Suppose that $\psi \in C_0[0,1] \cap C_2(0,1)$, $\psi \geq 0$, $r^0 \in \psi(0)G$, $r^1 \in \psi(1)G$, *and*

$$0 \leq -\omega(h)\psi''(x) + W'(h)b(x)h\psi'(x) + [q^T W''(h)q + W'(h)b(x)q]\psi(x) + W'(h)f(x, \psi(x)h, v)$$

for all $x \in (0,1)$, *all* $h \in \Gamma$, $q \in \mathbb{R}^n$ *with* $W'(h)q = 0$, *and all* $v \in K$.

Then there is a $u^* \in R$ *such that*

$$Mu^*(x) = 0 \quad , \quad u^*(x) \in \psi(x)G \quad \textit{for} \quad 0 \leq x \leq 1 \quad .$$

5. Existence proofs by numerical algorithms

For proving the existence of a solution of a given problem and obtaining bounds for this solution with one of the methods described above, one has to construct a *bound function* ψ (or several such functions). In some cases one may be able to describe ψ by explicit formulas, which can be used to calculate values $\psi(x)$. In most cases, however, this will not be possible. One has to calculate ψ numerically (by some approximation procedure). Then the problem arises to verify the differential inequalities which ψ must satisfy. This is a very difficult task.

One would like to have numerical algorithms which yield a (mathematically rigorous) *existence* statement together with *bounds* for a solution, briefly: EB-algorithms.

5.1 A general approach

We will here describe in abstract terms a general method for constructing EB-algorithms based on the theory of two-sided bounds. These algorithms are composed of an approximation procedure **A**, and an estimation procedure **E**, as subalgorithms.

Assume that $(R, \leq)$ and $(S, \leq)$ are ordered linear spaces, $(R, \leq) \subset (S, \leq)$, $(S, \leq)$ is a linear lattice, and that ζ is a strictly positive element in $(S, \leq)$. Let an equation

$$\mathcal{M}U = r \quad \text{with} \quad \mathcal{M} : R \to S \quad , \quad r \in S \tag{5.1}$$

be given, and suppose that an *approximate solution* $\omega \in R$ with *defect* $d = r - \mathcal{M}\omega$ has been calculated in some way. Then the *error* $u = U - \omega$ satisfies

$$Mu = d \quad \text{with} \quad Mu = \mathcal{M}(\omega + u) - \mathcal{M}\omega \quad . \tag{5.2}$$

Let

$$M = A + B \quad \text{with linear} \quad A : R \to S \quad , \tag{5.3}$$

and an operator $B : S \to S$ such that Bu is "small of higher order" for "small" u. (Under suitable conditions one may choose $A = \mathcal{M}'(\omega)$.) For "small" d one can, in general, exspect that u in (5.2) is also "small", so that B has little influence.

Assume now that M is an operator such that the following existence and inclusion statement holds for $\varphi, \psi \in R$:

$$\begin{array}{l} \textit{If} \quad \varphi \leq \psi \quad , \quad \textit{and} \quad M\varphi \leq d \leq M\psi \quad , \\ \textit{then there is a } u \in R \textit{ with} \quad Mu = d \quad , \quad \varphi \leq u \leq \psi \quad . \end{array} \tag{5.4}$$

We will only consider the case $-\varphi = \psi \geq 0$. The inequalities to be solved,

$$-A\psi + B(-\psi) \leq d \leq A\psi + B\psi \quad , \tag{5.5}$$

in general, contain "small" nonlinearities $B(-\psi)$ and $B\psi$. In order to obtain simple formulas, we use rough estimates of these terms.

Let $G : [0, \infty) \to [0, \infty)$ be a continous *majorizing function* for B, i.e.,

$$|By| \leq G(t)\zeta \quad \text{for} \quad y \in S \quad \text{with} \quad |y| \leq t\zeta \quad , \tag{5.6}$$

such that $G(t)$ is small of higher order for small t, for instance,

$$G(t) = O(t^2) \quad \text{for} \quad t \to 0 \quad . \tag{5.7}$$

Then the inequalities

$$\psi \leq \nu\zeta \quad , \quad |d| \leq A\psi - G(\nu)\zeta \tag{5.8}$$

are sufficient for (5.5).

These inequalities are solved by *linearization.* One first calculates an approximate solution $\rho = \psi_0$ of the linear problem $\zeta = A\rho$ and then tries to find a constant c such that (5.8) holds for $\psi = c\psi_0$.

In this way one obtains the following algorithm, where $\delta, \nu_0, \delta_0, \varepsilon_0, \varepsilon$ denote constants ≥ 0.

EB-algorithm:

1) *Solve* (approximately): $\mathcal{M}\omega \approx r$.
2) *Estimate*: $|d| \le \delta\zeta$.
3) *Solve* (approximately): $A\psi_0 \approx \zeta$.
4) *Estimate*: $|\psi_0| \le \nu_0\zeta$, $|\zeta - A\psi_0| \le \delta_0\zeta$, $\psi_0 \ge \varepsilon_0\zeta$.
5) *Verify*: $\varepsilon_0 \ge 0$.
6) *Verify* (with some fixed $\varepsilon > 0$):

$$G((1+\varepsilon)\nu_0\delta) + (1+\varepsilon)\delta\delta_0 \le \varepsilon\delta \quad . \tag{5.9}$$

Statement on existence and bounds *If the* EB-*algorithm can be carried out, then the given problem has a solution* U^* *such that*

$$|U^* - \omega| \le \psi := (1+\varepsilon)\delta\psi_0 \quad .$$

Modifications a) The construction of the majorizing function G may be avoided. The term $G((1+\varepsilon)\nu_0\delta)$ in (5.9) can be replaced by a constant $k \ge 0$ such that

$$|By| \le k\zeta \quad \text{for} \quad y \in S \quad , \quad |y| \le (1+\varepsilon)\nu_0\delta\zeta \quad . \tag{5.10}$$

b) Instead of choosing a fixed ε in step 6, say $\varepsilon = 0.01$, one may consider (5.9) as a condition for a number ε to be calculated. For example, (5.9) holds for $\varepsilon\delta = \tau$, if

$$\tau \ge f(\tau) \quad \text{with} \quad f(t) = G(\nu_0(\delta + t)) + \delta_0(\delta + t) \quad ,$$

and this condition is often satisfied for $\tau = 2f(0)$. Condition (5.10) can be treated analogously.

Discussion of the EB-algorithm a) For carrying out the EB-algorithm one needs an *approximation procedure* **A** for solving the problems in steps 1 and 3 approximately (with sufficient accuracy), and an *estimation procedure* **E** for calculating the bounds in steps 2 and 4 (with sufficient accuracy).

b) The quantities on the left hand side of (5.9), considered as functions of δ and δ_0, are small of higher order, while the term on the right hand side is linear. This suggests that under certain "regularity conditions" step 6 can be carried out, if a solution U^* exists, and if all calculations in steps 1 through 4 are done with sufficient accuracy. (The "regularity conditions" are needed for taking into account that A, B and G depend on ω).

c) An element ψ_0 such that the inequality in step 5 holds and δ_0 is sufficiently small, in general, exists if and only if

$$z \ge 0 \quad , \quad Az \succ 0 \quad \text{for some} \quad z \in R \quad . \tag{5.11}$$

In many cases this latter property is equivalent to A being inverse-positive (see the Monotonicity Theorem).

5.2 On applications to ordinary differential equations

5.21 A boundary value problem of the second order

As a very simple example, we consider here, for illustration, a boundary value problem

$$-U''(x) + F(x, U(x)) = 0 \quad \text{for} \quad 0 \le x \le 1 \quad , \quad U(0) = U(1) = 0 \quad , \tag{5.12}$$

with $F(x,y)$ twice continuously differentiable for $0 \le x \le 1$, $y \in \mathbb{R}$. This problem can be written as an equation (5.1) by defining

$$\mathcal{M}U(x) = -U''(x) + F(x,U(x)) \quad , \quad R = \{u \in C_2[0,1] : u(0) = u(1) = 0\} \quad , \quad S = C_0[0,1] \quad ,$$

and the operator M in (5.2) assumes the form (5.3) with

$$\begin{aligned} &Au(x) = L[u](x) := -u''(x) + c(x)u(x) \quad , \quad Bu(x) = f(x,u(x)) \quad \text{for} \quad 0 \le x \le 1 \quad , \\ &c(x) = F_y(x,\omega(x)) \quad , \quad f(x,y) = 0(y^2) \quad \text{for} \quad y \to 0 \quad , \end{aligned} \tag{5.13}$$

where all terms depend on ω, in particular, $A = \mathcal{M}'(\omega)$.

For $\zeta(x) \equiv 1$, condition (5.6) is satisfied, if

$$|f(x,y)| \le G(|y|) \quad \text{for} \quad 0 \le x \le 1 \quad , \quad y \in \mathbb{R} \quad ,$$

and we may assume that G has property (5.7).

In the present case, (5.11) holds if and only if A is inverse-positive. If the given problem (5.12) has a solution U^*, and if the operator $\mathcal{M}'(U^*)$ is inverse-positive, then (5.11) holds for $A = \mathcal{M}'(\omega)$, if ω is a sufficiently good approximation of U^*. Thus, in the theory of the above EB-algorithm the inverse-positivity of $\mathcal{M}'(U^*)$ plays an important role.

5.22 Producing inverse-positivity by transformation

One would like to have an EB-algorithm also for problems (5.12) such that A in (5.13) is not inverse-positive. To achieve this we will transform the problem

$$Mu = d \quad \text{for} \quad u \in R \quad \text{with} \quad M = A + B \tag{5.14}$$

into a problem

$$\widetilde{M}u = \widetilde{d} \quad \text{for} \quad u \in \widetilde{R} \quad \text{with} \quad \widetilde{M} = \widetilde{A} + \widetilde{B} \tag{5.15}$$

such that $\widetilde{A}$ is inverse-positive on $\widetilde{R}$, and for "small" u, $\widetilde{B}u$ is "much smaller" than $\widetilde{A}u$.

For an operator A of the form in (5.13) there are always values ξ_j $(j = 1,2,\ldots,N)$ with $0 =: \xi_0 < \xi_1 < \ldots < \xi_N < \xi_{N+1} := 1$, such that $\widetilde{A}$ is inverse-positive on $\widetilde{R}$, where

$$\begin{aligned} &\widetilde{A}u(x) &&= Au(x) \quad \text{for all } x \in [0,1] \quad \text{with} \quad x \ne \xi_j \quad (j = 1,2,\ldots,N) \quad , \\ &\widetilde{A}u(\xi_j) &&= u(\xi_j) \qquad\qquad\qquad\qquad\qquad\qquad\quad (j = 1,2,\ldots,N) \quad , \end{aligned} \tag{5.16}$$

and $\widetilde{R}$ denotes the linear space of all functions $u \in C_0[0,1]$ such that the restrictions of u to the subintervals $[\xi_j, \xi_{j+1}]$ $(j = 0,1,\ldots,N)$ are twice continuously differentiable, and $u(0) = u(1) = 0$. The functions $u \in \widetilde{R}$ may have *breaks* (jumps in the derivatives) at the *breakpoints* ξ_j $(j = 1,2,\ldots,N)$.

We will construct a transformed problem (5.15) with an operator $\widetilde{A}$ as above, and

$$\widetilde{B}u(x) = Bu(x) \quad , \quad \widetilde{d}(x) = d(x) \quad \text{for all } x \in [0,1] \quad \text{with} \quad x \ne \xi_j \quad (j = 1,2,\ldots,N) \quad .$$

Suppose that u is a solution of (5.14), ξ one of the breakpoints, and that $g \in C_0[0,1]$ denotes a function such that the restrictions of g to $[0,\xi]$ and $[\xi,1]$ are twice continuously differentiable, and

$$g(0) = g(1) = 0 \quad , \quad \sigma := g'(\xi - 0) - g'(\xi + 0) \ne 0 \quad .$$

By applying partial integration to $\int_0^1 gu''dx$ and using the differential equation for u, one obtains the relation

$$\widetilde{A}u(\xi) + \widetilde{B}u(\xi) = \widetilde{d}(\xi) \tag{5.17}$$

with $\widetilde{A}u(\xi) = u(\xi)$, and

$$\widetilde{B}u(\xi) = \sigma^{-1}\{\textstyle\int_0^1 L[g]u\;dx + \int_0^1 gf(x,u)\;dx\} \quad , \quad \widetilde{d}(\xi) = \sigma^{-1}\int_0^1 gd\;dx \quad . \tag{5.18}$$

Thus *the transformed problem* (5.15) *is obtained, when at each breakpoint the differential equation* $Mu(\xi) = d(\xi)$ *is replaced by the integral equation* (5.17).

Here we will not further discuss the theory of this transformation (existence of g, equivalence statements, etc.), or the construction of a corresponding EB-algorithm, but only make the following remarks.

1) A statement analogous to (5.4) is not valid for the transformed operator, since at a breakpoint ξ, $\widetilde{B}u(\xi)$ contains integrals. We have seen, however, that such functional terms can also be treated with the theories above. In the present case, one obtains the following statement.

If $\varphi, \psi \in \widetilde{R}$, $\varphi \le \psi$, *and*

$$M\varphi(x) \le d(x) \le M\psi(x) \quad \textit{for all} \quad x \in [0,1] \quad \textit{with} \quad x \ne \xi_j \; (j = 1,2,\ldots,N) \quad ,$$

and if for each breakpoint ξ

$$\varphi(\xi) + \widetilde{B}h(\xi) \le \widetilde{d}(\xi) \le \psi(\xi) + \widetilde{B}h(\xi) \quad \textit{for all} \quad h \in C_0[0,1] \quad \textit{with} \quad \varphi \le h \le \psi \quad ,$$

then there is a $u \in \widetilde{R}$ *such that* $\widetilde{M}u = \widetilde{d}$, $\varphi \le u \le \psi$.

2) If for each breakpoint ξ the corresponding *breakpoint function* g is calculated as an approximate solution of $L[u] = 0$ with small $L[g]$, then $\widetilde{B}u(\xi)$ is "much smaller" than $\widetilde{A}u(\xi)$, for small u, as desired.

3) For the transformed problem an EB-algorithm can be developed which again is composed of an approximation procedure **A**, and an estimation procedure **E**, as explained above. Here, one will not exactly proceed is in the algorithm in section 5.1, but use estimates more suitable for this case. For instance, instead of estimating $|\widetilde{d}|$ on the entire interval $[0,1]$ by one constant δ, one may use one constant as bound for the values $|\widetilde{d}(\xi_j)|$ at the breakpoints, and another constant as bound for $|\widetilde{d}(x)| = |d(x)|$ at the remaining points.

4) In the theory of this EB-algorithm with breakpoints, the *invertibility* (and not the inverse-positivity) of $\mathcal{M}'(U^*)$ plays an important role.

5) Considering bound functions φ, ψ such that $\varphi'(\xi+0) \ge \varphi'(\xi-0)$ and $\psi'(\xi+0) \le \psi'(\xi-0)$ at certain values ξ, represents no essential extension of the theory of smooth bound functions. The range of the applicability of the theory is enlarged, however, when functions φ, ψ are considered such that the derivates may have jumps of arbitrary sign. Such functions were treated by Küpper [34], applying the general theory of range-domain implications. Küpper used auxiliary functions similar to the break-point functions explained above.

Example. The problem

$$-U'' - 25U + U^3 = \lambda x(1-x) \quad , \quad U(0) = U(1) = 0$$

with $\lambda \in \mathbb{R}$ has one solution for $|\lambda| > \lambda_0 \approx 129.29$, two solutions for $|\lambda| = \lambda_0$, and three solutions for $|\lambda| < \lambda_0$. The values $(\lambda, U_\lambda(\frac{1}{2}))$ for all these solutions constitute an S-shaped curve, which is pointsymmetric with respect to the origin. It can be decomposed in three branches, an upper branch defined for $\lambda > -\lambda_0$, a lower branch defined for $\lambda < \lambda_0$, and a middle branch defined for $-\lambda_0 \leq \lambda \leq \lambda_0$.

In principle, the EB-algorithm described in sections 5.1 and 5.21 can be carried out for proving the existence of any solution belonging to the upper or lower branch: the operator $\mathcal{M}'(U^*)$ is inverse-positive for each of these solutions U^*.

For a solution U^* belonging to the middle branch and a value λ with $|\lambda| < \lambda_0$, the operator $\mathcal{M}'(U^*)$ is not inverse-positive, but invertible. For such a solution the EB-algorithm mentioned in the third remark above can, in principle, be carried out, if one breakpoint $\xi = \frac{1}{2}$ is used.

For a solution U^* belonging to the middle branch and $\lambda = \lambda_0$ or $\lambda = -\lambda_0$, the operator $\mathcal{M}'(U^*)$ is not invertible, so that neither of the EB-algorithms mentioned can be applied.

Of course, whether for a given case the EB-algorithm can actually be carried out, depends on the computational tools available.

5.23 A computer program

An EB-algorithm similar to those mentioned above has been developed, and a corresponding computer program has been written for scalar boundary value problems of the form

$$-U'' + F(x, U, U') = 0 \quad ,$$
$$-\alpha^0 U'(0) + F^0(U(0)) = 0 \quad , \quad \alpha^1 U'(1) + F^1(U(1)) = 0$$

with sufficiently smooth nonlinear functions F, F^0, and F^1 (see [23]).

This algorithm, too, works with two subalgorithms, **A** and **E**, and the conditions to be verified are of essentially the same form as those in steps 5 and 6, above.

One of the main problems in constructing the subalgorithms is the consideration of the rounding errors occurring. For this purpose some type of interval arithmetic is needed. We used the subroutine package ACRITH (see [35]).

In the basic theory of existence and inclusion for this EB-algorithm for the above problem, some ideas of Nagumo [47] were used, but a *Nagumo condition* of the usual form need not be required.

5.3 Generalizations

The EB-algorithm in section 5.1 is based on the existence and inclusion statement (5.4). One can develop similar general algorithms based on the theory of two-sided bounds with splitting, or on the theory of pointwise norm bounds. This allows one, in principle, to treat also vector-valued boundary value problems with ordinary differential equations of the second order. Such algorithms, however, have not yet been implemented on a computer.

Fourth order problems such as (4.12) can be transformed into a problem with two differential equations of the second order, by introducing $v = -u''$. But the given boundary conditions then don't have a form suitable for the application of the theories above: one has four boundary conditions for u, and none for v. One can, however, replace two of these boundary conditions for u by conditions for v which contain integrals, and then apply the theories described. The method is similar to the one used for obtaining conditions of the form (5.17), (5.18).

References

1. Adams, E., and Ames, W.F.: On contracting interval iteration for nonlinear problems in $\mathbb{R}^N$. Nonlinear Analysis 3, 773-794 (1979).
2. Adams, E., and Spreuer, H.: Uniqueness and stability for boundary value problems with weakly coupled systems of nonlinear integro-differential equations and applications to chemical reactions. J. Math. Anal. Appl. 49, 393-410 (1975).
3. Agmon, S., Douglis, A., and Nirenberg, L.L: Estimates near the boundary for solutions of elliptic partial differential equations satisfying general boundary conditions I. Comm. Pure Appl. Math. 12, 623-727 (1959).
4. Alikakos, N.D.: Remarks on invariance in reaction-diffusion equations. Nonlinear Analysis 5, 593-614 (1981).
5. Amann, H.: Fixed point equations and nonlinear eigenvalue problems in ordered Banach spaces. SIAM Rev. 18, 620-709 (1976).
6. Amann, H.: Invariant sets and existence theorems for semilinear parabolic and elliptic systems. J. Math. Anal. Appl. 65, 432-467 (1978).
7. Bebernes, J.W., Chueh, K.N., and Fulks, W.: Some applications of invariance for parabolic systems. Indiana Univ. Math. J. 28, 269-277 (1979).
8. Bebernes, J.W., and Schmitt, K.: Invariant sets and the Hukuhara-Kneser property for systems of parabolic partial differential equations. Rocky Mountain J. Math. 7, 557-567 (1977).
9. Beckenbach, E.F., and Bellmann, R.: Inequalities. Springer-Verlag 1961.
10. Berman, A., and Plemmons, R.J.: Nonnegative Matrices in the Mathematical Sciences. Academic Press 1979.
11. Bohl, E.: Monotonie: Lösbarkeit und Numerik bei Operatorgleichungen. Springer-Verlag 1974.
12. Chueh, K.N., Conley, C.C., and Smoller, J.A.: Positively invariant regions for systems of nonlinear diffusion equations. Indiana Univ. Math. J. 26, 373-392 (1977).
13. Collatz, L.: Aufgaben monotoner Art. Arch. Math. 3, 366-376 (1952).
14. Collatz, L.: The Numerical Treatment of Differential Equations. Springer-Verlag 1960.
15. Collatz, L.: Funktionalanalysis und Numerische Mathematik. Springer-Verlag 1964.
16. Conway, E., Hoff, D., and Smoller, J.: Large time behavior of solutions of systems of nonlinear reaction-diffusion equations. SIAM J. Appl. Math. 35, 1-16 (1978).
17. Erbe, L.H.: Nonlinear boundary value problems for second order differential equations. J. Differential Equations 7, 459-472 (1970).
18. Fan, K.: Topological proofs for certain theorems on matrices with non-negative elements. Monatsh. Math. 62, 219-237 (1958).
19. Fiedler, M., and Ptàk, V.: On matrices with nonpositive oft-diagonal elements and positive principal minors. Czechoslovak Math. J. 12(87), 382-400 (1966).
20. Frobenius, C.: Über Matrizen aus nichtnegativen Elementen. Sitzungsber. Preuss. Akad. Wiss. Berlin 456-477 (1912).
21. Gaines, R.E., and Mawhin, J.L.: Coincidence Degree, and Nonlinear Differential Equations. Lecture Notes in Mathematics No. 568, Springer-Verlag 1977.

22. Gilbarg, D., and Trudinger, N.S.: Elliptic Partial Differential Equations of Second Order. Springer-Verlag 1983.

23. Göhlen, M., Plum, M., and Schröder, J.: A numerical algorithm for existence proofs for two-point boundary value problems. To appear.

24. Hardy, G.H., Littlewood, J.E., and Polya, G.: Inequalities. Cambridge Univ. Press 1934.

25. Hartman, P.: On boundary value problems for systems of ordinary, nonlinear, second order differential equations. Trans. Amer. Math. Soc. 96, 493-509 (1960).

26. Hartmann, P.: Ordinary Differential Equations. Wiley 1964.

27. Hopf, E.: Elementare Bemerkungen über die Lösung partieller Differentialgleichungen zweiter Ordnung vom elliptischen Typus. Sitzungsber. Preuss. Akad. Wiss. 19, 147-152 (1927).

28. Jackson, L.K., and Schrader, K.W.: Comparison theorems for nonlinear differential equations. J. Differential Equations 3, 248-255 (1967).

29. Kamke, E.: Zur Theorie der Systeme gewöhnlicher Differentialgleichungen II. Acta Math. 58, 57-85 (1932).

30. Kaucher, E.W., and Miranker, W.L.: Self-Validating Numerics for Function Space Problems. Academic Press 1984.

31. Knobloch, H.W.: Eine neue Methode zur Approximation periodischer Lösungen nichtlinearer Differentialgleichungen zweiter Ordnung. Math. Z. 82, 177-197 (1963).

32. Krasnosel'skii, M.A.: Positive Solutions of Operator Equations. Noordhoff 1964.

33. Krein, M.G., and Rutman, M.A.: Linear operators leaving invariant a cone in a Banach space. Usp. Math. Nauk 3, No. 1(23), 3-95 (1948); [Amer. Math. Soc. Transl. 26 (1950)].

34. Küpper, T.: Einschließungsaussagen für gewöhnliche Differentialoperatoren. Numer. Math. 25, 201-214 (1976).

35. Kulisch, U.W., and Miranker, W.L.: Computer Arithmetic in Theory and Practice. Academic Press 1981.

36. Ladyzhenskaya, O.A., and Ural'tseva, N.N.: Linear and Quasilinear Elliptic Equations. Academic Press 1968.

37. Lakshmikantham, V., and Leela, S.: Differential and Integral Inequalities II. Academic Press 1969.

38. Lakshmikantham, V., and Vaughn, R.: Reaction-diffusion equations in cones. J. Math. Anal. Appl. 70, 1-9 (1979).

39. Lasota, A., and Yorke, J.A.: Existence of solutions of two-point boundary value problems for nonlinear systems. J. Differential Equations 11, 509-518 (1972).

40. Lemmert, R.: Über die Invarianz einer konvexen Menge in bezug auf Systeme von gewöhnlichen, parabolischen und elliptischen Differentialgleichungen. Math. Ann. 230, 49-56 (1977).

41. Lemmert, R.: Über die Invarianz konvexer Mengen eines normierten Raumes in bezug auf elliptische Differentialgleichungen. Comm. Partial Differential Equations 3, 297-318 (1978).

42. Martin, R.H. jr.: Nonlinear perturbations of uncoupled systems of elliptic operators. Math. Ann. 211, 155-169 (1974).

43. Müller, M.: Über das Fundamentaltheorem in der Theorie der gewöhnlichen Differentialgleichungen. Math. Z. 26, 619-645 (1927).

44. Müller, M.: Über die Eindeutigkeit der Integrale eines Systems gewöhnlicher Differentialgleichungen und die Konvergenz einer Gattung von Verfahren zur Approximation dieser Integrale. Sitzungsber. Heidelberger Akad. Wiss., Math.-Naturwiss. Kl. 9, 3-38 (1927).

45. Nagumo, M.: Eine hinreichende Bedingung für die Unität der Lösung von Differentialgleichungen erster Ordnung. Japan J. Math. 3, 107-112 (1926).

46. Nagumo, M.: Note in "Kansū-Hōteisiki" No. 15 (1939).

47. Nagumo, M.: Über das Randwertproblem der nicht linearen gewöhnlichen Differentialgleichungen zweiter Ordnung. Proc. Phys.-Math. Soc. Japan (3), 24, 845-851 (1942).

48. Nickel, K.L.: The construction of a priori bounds for the solution of a two-sided boundary value problem with finite elements I. Computing 23, 247-265 (1979).

49. Ostrowski, A.: Über die Determinanten mit überwiegender Hauptdiagonale. Comment. Math. Helv. 10, 69-96 (1937).

50. Ostrowski, A.: Determinanten mit überwiegender Hauptdiagonale und die absolute Konvergenz von Iterationsprozessen. Comment. Math. Helv. 30, 175-210 (1956).

51. Perron, O.: Zur Theorie der Matrices. Math. Ann. 64, 248-263 (1907).

52. Plum, M.: Pointwise bounds for linear reaction-diffusion systems and an extension to nonlinear problems. To appear in J. Math. Anal. Appl.

53. Plum, M.: Shape-invariant bounds for reaction-diffusion systems with unequal diffusion coefficients. To appear in J. Differential equations.

54. Protter, M.H., and Weinberger, H.F.: Maximum Principles in Differential Equations. Prentice Hall 1967.

55. Redheffer, R.M.: An extension of certain maximum principles. Monatsh. Math. 66, 32-42 (1962).

56. Redheffer, R.M.: Eindeutigkeitssätze bei nichtlinearen Differentialgleichungen. J. Reine Angew. Math. 211, 70-77 (1962).

57. Redheffer, R.: Gewöhnliche Differentialungleichungen mit quasimonotonen Funktionen in normierten linearen Räumen. Arch. Rational Mech. Anal. 52, 121-133 (1973).

58. Redheffer, R., and Walter, W.: Invariant sets for systems of partial differential equations, I. Parabolic equations. Arch. Rational Mech. Anal. 67, 41-52 (1978).

59. Redheffer, R., and Walter, W.: Invariant sets for systems of partial differential equations, II. First-order and elliptic equations. Arch. Rational Mech. Anal. 73, 19-29 (1980).

60. Redheffer, R.M., and Walter, W.: Flow-invariant sets and differential inequalities in normed spaces. Applicable Anal. 5, 149-161 (1975).

61. Sawashima, I.: On spectral properties of some positive operators. Nat. Sci. Rep. Ochanomizu Univ. 15, 53-64 (1964).

62. Schaefer, H.H.: Banach Lattices and Positive Operators. Springer-Verlag 1974.

63. Schmitt, K.: A nonlinear boundary value problem. J. Differential Equations 7, 527-537 (1970).

64. Schmitt, K.: Boundary value problems for quasilinear second-order elliptic equations. Nonlinear Anal. 2, 263-309 (1978).

65. Schrader, K.W.: Boundary value problems for second-order ordinary differential equations. J. Differential Equations 3, 403-413 (1967).

66. Schröder, J.: Lineare Operatoren mit positiven Inversen. Arch. Rational Mech. Anal. 8, 408-434 (1961).

67. Schröder, J.: Monotonie-Eigenschaften bei Differentialgleichungen. Arch. Rational Mech. Anal. 14, 38-60 (1963).

68. Schröder, J.: Operator-Ungleichungen und ihre numerische Anwendung bei Randwertaufgaben. Numer. Math. 9, 149-162 (1966).

69. Schröder, J.: Operator Inequalities. Academic Press 1980.

70. Schröder, J.: Shape-invariant bounds and more general estimates for vector-valued elliptic-parabolic problems. J. Differential Equations 45, 431-460 (1982).

71. Schröder, J.: Generalized maximum principles for strongly coupled parabolic systems, in General Inequalities 3 (E.F. Beckenbach and W. Walter, ed.), pp. 439-454 (1983).

72. Simader, Ch.G.: On Dirichlet's Boundary Value Problem. Lecture Notes in Math. Nr. 268, Springer-Verlag 1972.

73. Simoda, S.: Note sur l'inégalité différentielle concernant les équations du type parabolique. Proc. Japan Acad. 27, 536-539 (1951).

74. Szarski, J.: Sur un système d'inéqalités differentielles. Ann. Soc. Polon. Math. 20, 126-134 (1947).

75. Szarski, J.: Differential Inequalities. Monografie Matematyczne, Tom 73, Warszawa, 1965.

76. Varga, R.S.: Matrix Iterative Analysis. Prentice Hall 1962.

77. Volkmann, P.: Gewöhnliche Differentialungleichungen mit quasimonotonen wachsenden Funktionen in topologischen Vektorräumen. Math. Z. 127, 157-164 (1972).

78. Volkmann, P.: Über die positive Invarianz einer abgeschlossenen Teilmenge eines Banachschen Raumes bezüglich der Differentialgleichung $u' = f(t,u)$. J. Reine Angew. Math. 285, 59-65 (1976).

79. Volkmann, P.: Ein Invarianz-Satz für gewöhnliche und parabolische Differentialgleichungen. Arch. Math. 45, 150-157 (1985).

80. Walter, W.: Differential and Integral Inequalities. Springer-Verlag 1970.

81. Wazewski, T.: Systèmes des équations et des inégalités différentielles ordinaires aux deuxièmes membres monotones et leurs applications. Ann. Soc. Polon. Math. 23, 112-166 (1950).

82. Weinberger, H.F.: Invariant sets for weakly coupled parabolic and elliptic systems. Rendiconti di Matematica 8, Serie VI, 295-310 (1975).

83. Westphal, H.: Zur Abschätzung der Lösungen nichtlinearer parabolischer Differentialgleichungen. Math. Z. 51, 690-695 (1949).

11 Rearrangements and PDE

Giorgio G. Talenti Università Degli Studi, Florence, Italy

§1. Introduction.

Popularized by Hardy & Littlewood [11, ch.10], rearrangements of functions have been used by several authors in real and harmonic analysis, in investigations about function spaces and singular integrals. See [12], [19], [20] for instance. Pòlya & Szegö and their followers have demonstrated a good many isoperimetric theorems and inequalities by means of rearrangements. See [21], a source book on this matter. More recent investigations have shown that rearrangements of functions also fit well into the theory of elliptic second-order partial differential equations. See [30], ... [80].

Several types of rearrangements are known - a catalog is in Kawohl [14]. In this paper we limit ourselves to rearrangements à la Hardy & Littlewood. We briefly review chief properties of these rearrangements, and sketch applications to boundary value problems for elliptic second-order partial differential equations.

§2. Definitions and basic properties.

Let G be a measurable subset of euclidean space $\mathbb{R}^n$; let u be a measurable real-valued function, defined in G. Suppose $m(G)$, the Lebesgue measure of G, is not zero.

Def 1. The distribution function μ is the map from $[0,\infty[$ into $[0,\infty]$ such that

$$\mu(t) = m\left\{x \in G : |u(x)| > t\right\} . \qquad (2.1)$$

The following properties are easy:

(i) μ is a decreasing function;

(ii) μ is right-continuous;

(iii) $m\left\{x \in G : |u(x)| = t\right\} = \mu(t-) - \mu(t)$, the jump of μ at t;

(iv) $\text{sprt}\,\mu$, the support of μ, $= [0, \text{ess sup}\,|u|]$;

(v) $\mu(0) = m(\text{sprt}\, u)$.

A typical situation is sketched in fig.1.

Def 2. The decreasing rearrangement of u, u^*, is the distribution function of μ.

As μ decreases and is right-continuous, we have

$$u^*(s) = \sup\left\{t \geq 0 : \mu(t) > s\right\} = \min\left\{t \geq 0 : \mu(t) \leq s\right\} \tag{2.2}$$

for every $s \geq 0$.

Clearly:

(i) u^* is a decreasing map from $[0,\infty[$ into $[0,\infty]$;

(ii) u^* is right-continuous;

(iii) $\operatorname{sprt} u^* = [0, m(\operatorname{sprt} u)]$;

(iv) $u^*(0) = \operatorname{ess\,sup} |u|$.

The following theorem is crucial:

(v) The distribution function of u^* is exactly μ ; in other words, u and u^* are equimeasurable or equidistributed.

Note that properties (i)(ii)(v) characterize u^*. Fig.2 may illustrate relevant geometric aspects. Property (v) has several consequences; for instance:

(vi) $\int_G A(|u(x)|)\,dx = \int_0^\infty A(u^*(s))\,ds$, where A is any continuous map from $[0,\infty]$ into $[0,\infty]$.

Note that (vi) follows from

$$\int_G A(|u(x)|)\,dx = \int_0^\infty A(t)\left[-d\mu(t)\right] ,$$

a form of Cavalieri's principle.

Def 3. The symmetric rearrangement of u, $u^\star$, is the nonnegative function defined in $\mathbb{R}^n$ by

$$u^\star(x) = u^*(C_n|x|^n) , \tag{2.3}$$

where $|x|=(x_1^2+\ldots+x_n^2)^{\frac{1}{2}}$ and

$$C_n = \frac{\pi^{n/2}}{\Gamma(n/2+1)} ,$$

the measure of the unit n-dimensional ball.

Clearly:

(i) $u^\star$ is a radial function - i.e. invariant under rotations about the

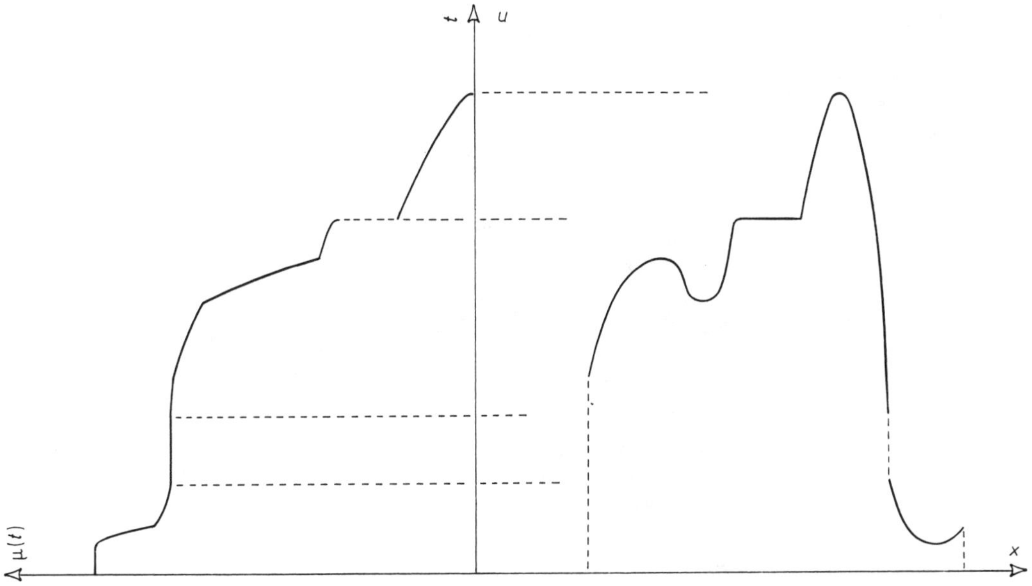

Figure 1

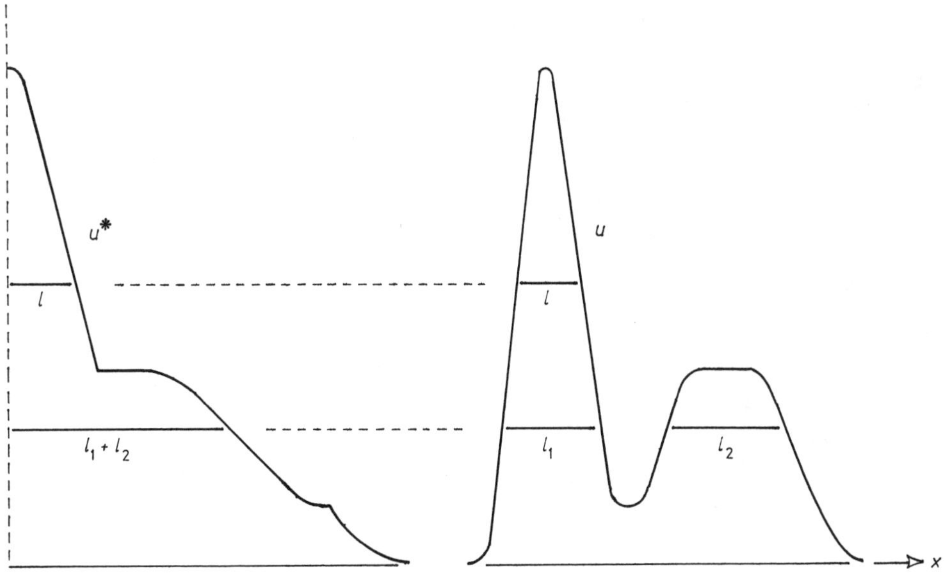

Figure 2

origin - that decreases as the distance from the origin decreases;

(ii) u and $u^\bigstar$ are equidistributed;

(iii) sprt $u^\bigstar \subseteq G^\bigstar$, the ball whose center is the origin and whose measure equals m(G).

Properties (i)(ii) can be summarized this way. For every $t \geq 0$, $\{x \in \mathbb{R}^n : u^\bigstar(x) > t\}$, a level set of $u^\bigstar$, is the ball whose center is the origin and whose measure equals the measure of $\{x \in G : |u(x)| > t\}$, the allied level set of $|u|$. In other words, $u^\bigstar$ is a function whose graph results from a Schwarz symmetrization of the graph of u.

Definitions 2 and 3 can be recast in a more compact form. Recall that any nonnegative integrable function f is the superimposition of the characteristic functions of its level sets, i.e.

$$f = \int_0^\infty \mathbb{1}_{\{x \in \mathbb{R}^n : f(x) > t\}} \, dt ,$$

where $\mathbb{1}$ stands for characteristic function and the integral is à la Bochner. Suppose u is integrable over G. Then we have

$$u^* = \int_0^\infty \mathbb{1}_{[0,\mu(t)]} \, dt , \tag{2.4}$$

$$u^\bigstar = \int_0^\infty \mathbb{1}_{\{x \in \mathbb{R}^n : C_n |x|^n < \mu(t)\}} \, dt . \tag{2.5}$$

§3. Key theorems.

In most applications the main theorems on rearrangements à la Hardy & Littlewood are theorems A,B,C below.

Thm A. Suppose G is a measurable subset of $\mathbb{R}^n$, u and v are nonnegative measurable functions. Then

$$\int_G u(x)\, v(x)\, dx \leq \int_0^{m(G)} u^*(s)\, v^*(s)\, ds . \tag{3.1}$$

Thm B. Suppose f,g,h are nonnegative measurable functions. Then

$$\int_{\mathbb{R}^n} dx \int_{\mathbb{R}^n} f(x) g(y) h(x-y)\, dy \leq \int_{\mathbb{R}^n} dx \int_{\mathbb{R}^n} f^\bigstar(x) g^\bigstar(y) h^\bigstar(x-y)\, dy . \tag{3.2}$$

Thm C. Suppose G is an open subset of $\mathbb{R}^n$, u belongs to Sobolev space $W_o^{1,p}(G)$, $p \geq 1$. Then

$$\int_G |\text{grad } u|^p \, dx \geq \int_{G^\star} |\text{grad } u^\star|^p \, dx \quad . \tag{3.3}$$

Theorem A is by Hardy & Littlewood. A proof is e.g. in [11]. Theorem A is simple,but decisive. One may claim that most theorems from real and harmonic analysis,that can be demonstrated via rearrangements,are consequences of theorem A.

Theorem B is due to F.Riesz. Proofs are in [22] and [11]. Generalizations and improvements are in [3],[8],[17].

Let us sketch a simple application of theorem B. Suppose E is a 3-dimensional homogeneous body, whose volume and density are fixed. Consider the energy of E,i.e. the energy of the gravitational field generated by E. Question: which E renders such an energy a maximum ? We have

$$\text{energy of } E = \int_{\mathbb{R}^3} |\text{grad } u|^2 \, dx \quad ,$$

where u - a potential - is given by

$$u(x) = \int_{\mathbb{R}^3} \mathbb{1}_E(y) |x-y|^{-1} \, dy$$

(apart from a constant factor) and satisfies

$$-\Delta u = 4\pi \, \mathbb{1}_E \quad .$$

Here $\mathbb{1}_E$ stands for the characteristic function of E. Integrations by parts show

$$\text{energy of } E = 4\pi \int_{\mathbb{R}^3} dx \int_{\mathbb{R}^3} \mathbb{1}_E(x) \, \mathbb{1}_E(y) \, |x-y|^{-1} \, dy \quad .$$

Note that $(\mathbb{1}_E)^\star$,the symmetric rearrangement of $\mathbb{1}_E$,is just the characteristic function of $E^\star$,the ball having its center at the origin and the same volume as E. Thus Riesz's theorem tells us immediately

$$\text{energy of } E \leq \text{energy of } E^\star \quad ,$$

i.e. the answer to our question: among all homogeneous bodies,whose volume and density are fixed,the ball generates a gravitational field,whose energy is a maximum.

Theorem C says that the total variation,and Dirichlet type integrals of sufficiently smooth functions that vanish on the boundary,decrease under symmetric rearrangement. Theorem C can be generalized by replacing powers by increasing convex functions of the involved gradients.

Pólya, Szegö & C derived a lot of isoperimetric theorems and isoperimetric properties of eigenvalues (e.g. Faber & Krahn theorem on the principal frequency of a membrane, Poincaré inequality for capacity, St. Venant principle for torsional rigidity) from theorem C, see [21]. Consider for example the following eigenvalue problem:

$$\Delta u + \lambda u = 0 \quad \text{in } G\,, \quad u = 0 \quad \text{on } \partial G \;;$$

where G is a bounded open subset of $\mathbb{R}^n$ and ∂G is the boundary of G. The smallest eigenvalue of this problem, $\lambda(G)$, is the minimum of Rayleigh's quotient

$$\int_G |\text{grad}\, u|^2\, dx \;:\; \int_G u^2\, dx$$

as u runs in Sobolev space $W_o^{1,2}(G)$. A symmetric rearrangement diminishes the numerator - by theorem C - and simultaneously leaves the denominator unchanged - by the equimeasurability of rearrangements. Thus we have

$$\lambda(G) \geq \lambda(G^{\bigstar})\,,$$

the Faber-Krahn theorem [7][15]: among all bounded domains having a fixed measure, the ball renders the first eigenvalue of Laplace operator - coupled with zero Dirichlet boundary condition - a minimum.

Theorem C is also crucial in special questions of functional analysis, e.g. for setting Sobolev type inequalities in a sharp form [1], [19], [28].

The proof of theorem C is based on ideas, that Pòlya and Szegö mastered and ultimately can be found in the papers by Faber and Krahn. However, implementing these ideas needs tools from geometric measure theory, that have been set up in recent times only. Exhaustive proofs of theorem C, and generalizations or variants of it, are offered in [4], [8], [9], [13], [24], [25], [26], [27]. An ingenious proof of the $p=2$ case is due to Lieb [16].

We emphasize that theorem C is basically a consequence of the classic isoperimetric inequality. In loose terms, the isoperimetric inequality in $\mathbb{R}^n$ reads

$$n\, C_n^{1/n} \cdot (n\text{-dim. meas. of } E)^{1-1/n} \leq (n-1)\text{-dim. meas. of } \partial E \;,$$

where E is any sufficiently smooth subset of $\mathbb{R}^n$ having finite measure.

Such inequality is precisely what allows one - when applied to level sets of the relevant function - to compare left and right-hand side of (3.3).

By the way, an isoperimetric inequality also holds on spheres as well as in euclidean spaces equipped with a gaussian measure - Borell [2]. Thus, versions of theorem C are available, where the euclidean space is replaced by a sphere - Sperner [25] - or rearrangements and integrations involve a gaussian measure - Ehrhard [6]. Applications have been made to Sobolev inequalities on spheres [28]; logarithmic Sobolev inequalities [10] could be worked out from this point of view.

As a last remark, let us mention that functions exist, which really differ from their symmetric rearrangements and render (3.3) an equality. Kawohl [14], Friedman & McLeod [8], Brothers & Ziemer [4] have shown that equality in (3.3) implies $u = u^{\star}$ if the set of critical points of u is thin enough.

§4. An algorithm.

In this section we present an algorithm for computing the decreasing rearrangement of a real-valued function of one real variable. Our algorithm has roots in old papers [23] by C. Somigliana - a mathematical physicist who seemingly anticipated the discovery of rearrangements - and is based on the following facts. (i) The decreasing rearrangement of a nonnegative step (i.e. piecewise constant) function of one real variable is still a step function: just the step function which is obtained by assembling the original blocks in order of decreasing height. Thus, rearranging a nonnegative step function, whose steps all have equal length, is exactly the same as listing its values in decreasing order - a mere combinatorial job. (ii) Rearrangements à la Hardy & Littlewood are contractions (i.e. non-expansive mappings) in Lebesgue or, more generally, Orlicz spaces. See Chiti [5]. In other words, if approximations of a function u converge at some rate in a Lebesgue or Orlicz space, then the decreasing, or symmetric, rearrangements of such approximations automatically approach the decreasing, or symmetric, rearrangement of u at the same rate.

Consider a real-valued function u defined in an interval $[a,b]$. Suppose u is smooth enough; for instance, u is absolutely continuous and its derivative u' is in $L^q(a,b)$ for some $q \geq 1$. The decreasing rearrangement of u can be computed in the following way.

Let

$$\ell = b - a \quad , \tag{4.1}$$

the length of interval $[a,b]$. Select a relatively large integer n ,define a mesh size by

$$h = \ell/n \quad , \tag{4.2}$$

then split $[a,b]$ into n subintervals $[x_0, x_1], [x_1, x_2], \ldots [x_{n-1}, x_n]$ by equidistant points

$$x_i = a + i\,h \qquad (i=0,1,\ldots n) \quad . \tag{4.3}$$

For $i=1,2,\ldots n$ sample function u at

$$a+(i-\tfrac{1}{2})h \quad , \tag{4.4}$$

the mid-point of subinterval $[x_{i-1}, x_i]$,and denote by

$$y_i \tag{4.5}$$

the sample value of u.

Consider

$$\text{stpfn} = \sum_{i=1}^{n} y_i \, \mathbb{1}_{[x_{i-1}, x_i[} \quad , \tag{4.6}$$

the step function taking the value y_i at any point of $[x_{i-1}, x_i[$. For any x such that $a \leq x < b$,we have

$$\text{stpfn}(x) = y_i \quad , \tag{4.7a}$$

where i is such that $x_{i-1} \leq x < x_i$,i.e.

$$i = \text{integer part of } 1+(x-a)/h \ . \tag{4.7b}$$

Clearly,

$$(\text{stpfn})^* = \sum_{i=1}^{n} y_i^* \, \mathbb{1}_{[(i-1)h, ih[} \quad , \tag{4.8}$$

the step function whose values are given by:

$$(\text{stpfn})^*(s) = y_i^* \tag{4.9a}$$

$$i = \text{integer part of } 1+s/h \tag{4.9b}$$

for any s such that $0 \leq s < \ell$. Here

$$y_1^*, \ y_2^*, \ldots \ y_n^* \tag{4.10}$$

is the decreasing rearrangement of

$|y_1|$, $|y_2|$,... $|y_n|$.

The decreasing rearrangement of a terminating sequence of real numbers can be produced easily - likewise an alphabetic list of words. An ad hoc procedure is examined in great detail in [29,§4]. Thus, $(\text{stpfn})^*$ is explicitly available. We claim that $(\text{stpfn})^*$ is arbitrarily close to u^* if the number of mesh points, n ,is sufficiently large.

Indeed

$$\|(\text{stpfn})^* - u^*\|_{L^p(0,l)} = O\left(n^{-1-1/p+1/q}\right) .$$

More precisely,

$$\|(\text{stpfn})^* - u^*\|_{L^p(0,l)} \le C(h/2)^{1+1/p-1/q} \|u'\|_{L^q(a,b)} , \qquad (4.11)$$

where p is any exponent $\ge q$,

$$C = \left(1+\frac{q}{p(q-1)}\right)^{1/q}\left(1+p\left(1-\frac{1}{q}\right)\right)^{-1/p}\frac{p}{B(1/p,1-1/q)}$$

and B denotes the Euler beta function. A theorem from [5] ensures that

$$\|(\text{stpfn})^* - u^*\|_{L^p(0,l)} \le \|\text{stpfn} - u\|_{L^p(a,b)} ;$$

a simple argument shows that the r.h.s. of the last inequality cannot exceed the r.h.s. of (4.11). Inequality (4.11) follows.

Our algorithm for computing the decreasing rearrangement of u can be summarized this way: form step function stpfn according to formula (4.6), then compute the authentic decreasing rearrangement of stpfn . See fig.3. Assessments: the output actually is an approximation of u^* ,the accuracy of the output gets better and better as the number of mesh points gets larger and larger.

A BASIC code for implementing the algorithm is offered below.

```
10  A=... : B=...         : REM end points of ground interval
20  N=...                 : REM number of mesh points

45  Clear screen : PRINT "sampling"
50  DIM LS(N)             : REM LS=list of samples
60  H=(B-A)/N             : REM mesh size
70  FOR I=1 TO N
80  X=A+(I-.5)*H          : REM mesh point
130 Y=...                 : REM Define the value of the rele-
vant fnctn at point x. Premise a subroutine,if necessary.
140 LS(I)=ABS(Y)          : REM enroll samples
150 PRINT I;TAB(7)X;TAB(23)Y      : REM display mesh points & samples
160 NEXT I
170 STOP
```

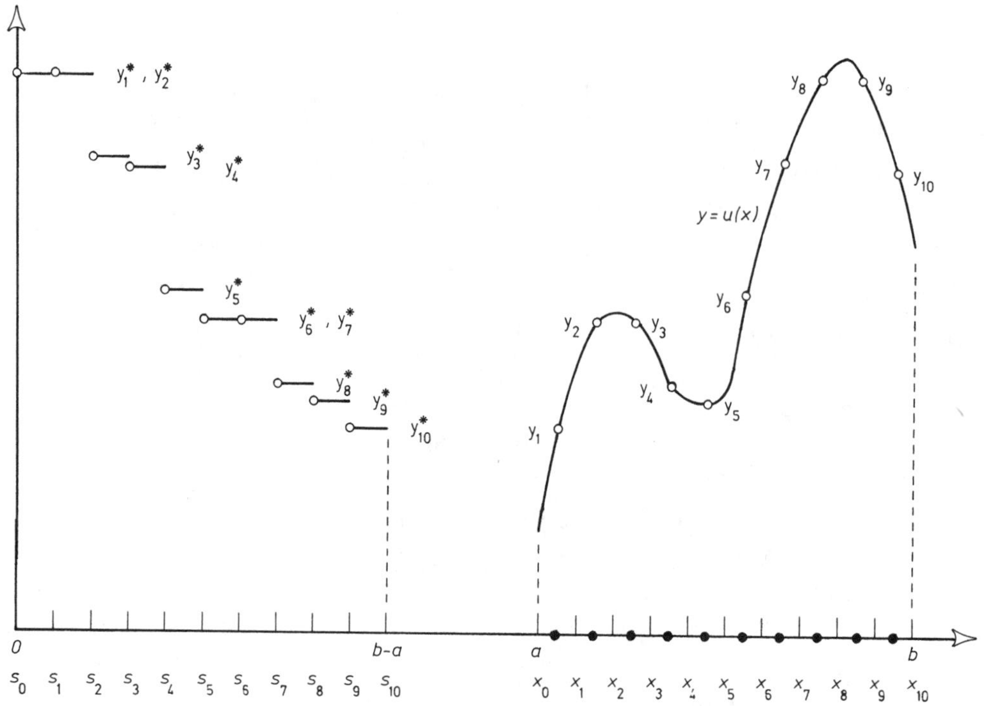

Figure 3

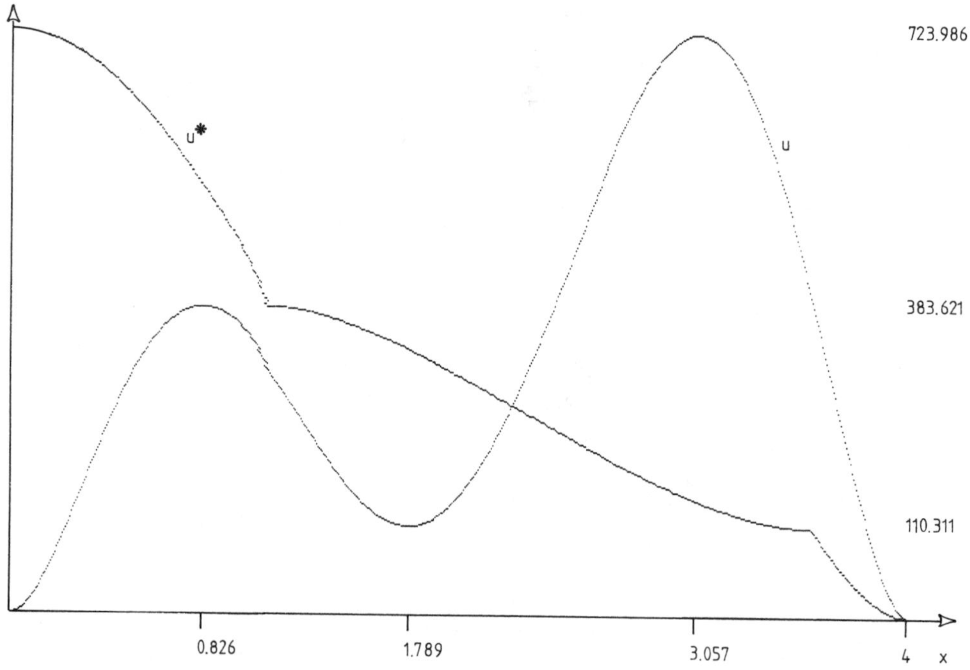

Figure 4

```
195  Clear screen : PRINT "rearranging"
200  DIM R(N)                : REM list R is used as a variable - initialized first,
then updated iteratively - and eventually hosts the decr.rearr. of LS(1),...LS(n).
210  R(1)=LS(1)
220  FOR I=2 TO N
230    Y=LS(I)
240    FOR J=I-1 TO 1 STEP -1
242    IF R(J)>=Y THEN 260
244    R(J+1)=R(J)
246    NEXT J
260  R(J+1)=Y
270  NEXT I
280  FOR I=1 TO N : PRINT I,R(I) : NEXT I     : REM display LS(1),...LS(n) in
decreasing order
290  STOP

295  Clear screen : PRINT "plotting"
300  PRINT "enter scale factors"
302  PRINT "no. of pages (enter a moderate integer or a decimal)";
304  INPUT L                 : REM plots can be split into a number
of pages,that can be displayed successively - specify such a number.
306  PRINT "height (enter a number between 1 and 100)";
308  INPUT H                 : REM specify the vertical stretching of plots
320  SX=...  : SY=...        : REM coordinates of the last (i.e. bottom right) pixel
330  L=SX*L  : H=INT(.5+SY*H/100)
340  J=0
350  Prepare screen for high resolution
360  Clear screen
370  Line between (0,H) & (SX,H)      : REM drawing x-axis
380  FOR X=0 TO SX                    : REM drawing a page
390    I=INT(1+J*N/L)
400    Y=H*(1-LS(I)/R(1)) : Y=INT(.5+Y)
410    Plot point (X,Y)               : REM graphing |stpfn|
420    Y=H*(1-R(I)/R(1))  : Y=INT (.5+Y)
430    Plot point (X,Y)               : REM graphing (stpfn)*
440    J=J+1 : IF J>=L THEN 480
450  NEXT X
460  Wait until the operator hits a key    : REM a page has just been completed
470  GOTO 360                              : REM start a new page
480  Wait until the operator hits a key    : REM plots have been completed

500  END
```

Fig.4 has been obtained via the code above. It shows the graph of function u given by

$$u(x) = 51x^6 - 592x^5 + 2461x^4 - 4328x^3 + 2768x^2$$

for $0 \le x \le 4$,and the graph of the decreasing rearrangement of u .

§5. Applications to partial differential equations.

Rearrangements of functions and partial differential equations are variously related. Interestingly enough, rearrangements à la Hardy & Littlewood are an appropriate tool for deriving a priori estimates of solutions to boundary value problems for elliptic second-order partial differential equations. Let us focus this subject.

Consider

$$-\operatorname{div} a(x,u,\operatorname{grad} u) = H(x,u) \quad , \qquad (5.1)$$

a second-order partial differential equation whose leading part has a divergence structure. Let

$$G = \text{an open subset of } \mathbb{R}^n \quad , \qquad (5.2)$$

and consider a Dirichlet problem; i.e. solutions u such that

$$u \text{ satisfies equation } (5.1) \text{ in } G \qquad (5.3)$$

and the condition

$$u = 0 \text{ on the boundary } \partial G \text{ of } G . \qquad (5.4)$$

Assume the following hypotheses:

(i) A function A exists such that

$$\begin{aligned} &]0,\infty[\ni r \longrightarrow A(r) \text{ is convex,} \\ &A(0+) = A'(0+) = 0 \ , \end{aligned} \qquad (5.5a)$$

and the principal term a satisfies

$$a(x,u,\xi)\cdot\xi \geq A(|\xi|) \quad \text{for all } (x,u,\xi) \quad . \qquad (5.5b)$$

Hypothesis (i) stands for ellipticity.

(ii) The right-hand side is small, if ellipticity is poor - i.e. $A(r)$ grows linearly as $r \longrightarrow \infty$. More precisely,

$$\|H(\cdot,0)\|_{L(n,\infty)} < n\, C_n^{1/n} \cdot \lim_{r \to \infty} B(r) \quad . \qquad (5.6)$$

Here

$$B(r) = A(r)/r \quad , \qquad (5.7)$$

C_n is the measure of the unit n-dimensional ball and $L(n,\infty)$ is a Lorentz space. Thus,

$$\left\| H(\cdot,0) \right\|_{L(n,\infty)} = \sup \left\{ m(E)^{-1+1/n} \int_E |H(\cdot,0)|\, dx \; : \; E \subset \mathbb{R}^n \right\} \quad .$$

Hypothesis (ii) stands for consistency: equation (5.1) needs not to have solutions in the large, that vanish on the boundary, in case (5.5) force $A(r)$ to grow linearly as $r \to \infty$ and inequality (5.6) is violated.

(iii) $H(x,u) \leq H(x,0) \leq H(x,-u)$ whenever $u > 0$.

(iv) $m(G)$, the measure of G, is finite.

We address ourselves to the following question. Let a set of Dirichlet boundary value problems, all having form (5.1)-(5.4) and obeying hypotheses (i)-(ii), be specified by the following parameters:

the ellipticity weight A,

the distribution function of $H(\cdot,0)$,

the measure of domain G.

Let

$\| \quad \|$

be any Luxemburg-Zaanen norm (Recall that a Luxemburg-Zaanen space is a Banach space of measurable functions, whose norm obeys: $\|u\| \leq \|v\|$ if $|u| \leq |v|$, $\|u\| = \|v\|$ if u and v are equidistributed). Does a worst problem, i.e. a problem whose solution v renders

$\|v\|$ a maximum,

exists within such a set?

The answer is the following. The worst problem is the simplest one, namely

$$-\operatorname{div} \frac{A(|Dv|)}{|Dv|^2} Dv = H(\cdot,0)^{\star} \qquad \text{in } G^{\star} , \tag{5.8}$$

$$v = 0 \qquad \text{on } \partial G^{\star} . \tag{5.9}$$

Here

$G^{\star}$ = the ball having the same measure as G,

and

$D = \operatorname{grad}$.

A more exhaustive picture is given in the next theorem.

Theorem. Suppose a solution u to problem (5.1)-(5.4) exists in an appropriate function space - i.e. u is in Sobolev-Orlicz space $W_o^{1,A}(G)$ and

$$\int_G A(|\text{grad } u|)dx < \infty \quad . \tag{5.10}$$

Define v by the following formula:

$$v(x) = \int_{C_n|x|^n}^{m(G)} B^{-1}\left(\frac{1}{nC_n^{1/n} r^{1-1/n}} \int_0^r H(\cdot,0)^*(s)\,ds\right) \frac{dr}{nC_n^{1/n} r^{1-1/n}} \quad , \tag{5.11}$$

where B^{-1} is the inverse function of B .

Claims. Function v is a solution to problem (5.8)-(5.9) - the only solution such that

$$\int_{G^\star} A(|\text{grad } v|)\,dx < \infty \quad . \tag{5.12}$$

The following inequality holds:

$$u^\star \le v \quad . \tag{5.13}$$

Moreover

$$\int_G M(|\text{grad } u|)\,dx \le \int_{G^\star} M(|\text{grad } v|)\,dx \quad , \tag{5.14}$$

provided $M(0)=0$, $M'(r) \geq 0$ and M is weaker than A ,i.e. the wronskian

$$\begin{vmatrix} A'(r) & M'(r) \\ A''(r) & M''(r) \end{vmatrix} \le 0 \quad .$$

A proof of the above theorem is based on considerations about

$$\{x \in G : |u(x)| > t\} \quad ,$$

the level sets of solution u . A crucial information on these level sets is hidden in the architecture of equation (5.1). Decoding involves three basic ingredients: Hardy & Littlewood theorem (thm A,§3),the isoperimetric inequality in $\mathbb{R}^n$,coarea formula. Decoding gives

$$nC_n^{1/n}\mu(t)^{1-1/n}\, B\left(nC_n^{1/n}\frac{\mu(t)^{1-1/n}}{-\mu'(t)}\right) \le \int_0^{\mu(t)} H(\cdot,0)^*(s)\,ds \quad ,$$

a differential inequality for the distribution function μ of u . The theorem follows. See [71] for details.

The above theorem provides us with a strategy for estimating

$$\|u\| \quad ,$$

any Luxemburg-Zaanen norm of u, or

$$\int_G M\left(|\operatorname{grad} u|\right) dx \quad .$$

The strategy consists in replacing the original problem (5.1)-(5.4) by the much simpler problem (5.8)-(5.9) and estimating a solution - that is available in the close form (5.11) - of the latter.

For example, consider the following problem :

$$-\sum_{i,k=1}^{n} \frac{\partial}{\partial x_i}\left\{a_{ik}(x)|\operatorname{grad} u|^{p-2} u_{x_k}\right\} + c(x)u = f(x) \quad \text{in } G, \tag{5.16}$$

$$u = 0 \quad \text{on } \partial G \quad . \tag{5.17}$$

Here

$$p > 1 \quad ,$$

coefficients a_{ik} and c are real-valued measurable functions such that

$$\sum_{i,k=1}^{n} a_{ik}(x)\xi_i\xi_k \geq \sum_{i=1}^{n} \xi_i^2 \qquad \text{for all } (x,\xi) \quad ,$$

$$c(x) \geq 0 \quad ,$$

the right-hand side belongs to an appropriate space. Note that, in case $p=2$, (5.16) is a standard second-order linear elliptic equation in diver= gence form; in case $a_{ik}(x) = \delta_{ik}$, the Kronecker delta, and $c(x) = 0$, (5.16) is the Euler equation of the following integral

$$\int_G \left\{\frac{1}{p}|\operatorname{grad} u|^p - f(x)u\right\} dx \quad .$$

Solutions u to problem (5.16)-(5.17), that belong to Sobolev space $W_o^{1,p}(G)$, can be estimated via inequalities (5.13) and (5.14). Here

$$v(x) = \left(nC_n^{1/n}\right)^{-\frac{p}{p-1}} \int_{C_n|x|^n}^{m(G)} \left(\frac{1}{r}\int_0^r f^*(s)\,ds\right)^{\frac{1}{p-1}} r^{-1+\frac{p}{n(p-1)}}\,dr \quad , \tag{5.18}$$

the solution in $W_o^{1,p}(G^\star)$ to the following problem

$$-\operatorname{div}\left(|Dv|^{p-2} Dv\right) = f^\star \quad \text{in } G^\star \tag{5.19}$$

$$v = 0 \quad \text{on } \partial G^\star \quad , \tag{5.20}$$

and

$$\left|\operatorname{grad} v(x)\right| = \left(\frac{1}{n C_n |x|^{n-1}} \int_0^{C_n |x|^n} f^*(s)\, ds \right)^{\frac{1}{p-1}} . \tag{5.21}$$

A sample of results is presented below. These results are derived by handling the right-hand sides of (5.18) and (5.21) via Bliss inequality [11, thm 270].

Suppose $p < n/(n-1)$, $q = n/p$ and f is in $L^q(G)$. Then u is bounded and

$$\left(n C_n^{1/n}\right)^{\frac{p}{p-1}} \frac{n-p}{n(p-1)} \sup |u| \le K \left(\int_G |f|^q dx \right)^{\frac{1}{q(p-1)}} -$$

$$[m(G)]^{-\frac{q-1}{q(p-1)}} \left(\int_G |f|\, dx \right)^{\frac{1}{p-1}} ,$$

where

$$K = \left[\frac{\Gamma\left(\frac{n}{n+p-np}\right)}{\Gamma\left(\frac{p}{n+p-np}\right) \Gamma\left(\frac{2n-np}{n+p-np}\right)} \right]^{\frac{n+p-np}{n(p-1)}} .$$

Suppose $p < n$, q is given by $1/q = 1 + 1/n - 1/p$ and f is in $L^q(G)$. Then

$$\left(n C_n^{1/n}\right)^{\frac{p}{p-1}} \frac{n-p}{n(p-1)} \int_G |\operatorname{grad} u|^p dx \le K \left(\int_G |f|^q dx \right)^{\frac{p}{q(p-1)}} -$$

$$[m(G)]^{-\frac{p(q-1)}{q(p-1)}} \left(\int_G |f|\, dx \right)^{\frac{p}{p-1}} ,$$

where

$$K = \left[\frac{\Gamma(n)}{\Gamma\left(\frac{n}{p}\right) \Gamma\left(\frac{n}{q}\right)} \right]^{\frac{q}{n-q}} .$$

References

Part one: miscellanea.

[1] T.Aubin,Problèmes isopérimetriques et espaces de Sobolev (J.Diff.Geometry,11,1976)

[2] C.Borell,The Brunn-Minkowski inequality in Gauss space (Inv.Math.,30,1974)

[3] H.J.Brascamp-E.H.Lieb-J.M.Luttinger,A general rearrangement inequality for multiple integrals (J.Functional Anal.,17,1974)

[4] J.E.Brothers-W.P.Ziemer,Minimal rearrangements of Sobolev functions (to appear)

[5] G.Chiti,Rearrangements of functions and convergence in Orlicz spaces (Applicable Anal.,9,1979)

[6] A.Ehrhard,Inégalités isopérimetriques et integrales de Dirichlet gaussiennes (Ann.Scient.Ec.Norm.Sup.,ser.4,vol.17,1984)

[7] G.Faber,Beweis dass unter allen homogenen membranen von gleicher fläche und gleicher spannung die kreisförmige den tiefsten grundton gibt (Sitzungsber.Bayer.Akad.Wiss.,Math.Naturwiss.Kl.,1923)

[8] A.Friedman-B.McLeod,Strict inequalities for integrals of decreasingly rearranged functions (Proc.Roy.Soc.Edinburgh,102A,1986)

[9] A.M.Garsia-E.Rodemich,Monotonicity of certain functionals under rearrangements (Ann.Inst.Fourier Grenoble,24,1974)

[10] L.Gross,Logarithmic Sobolev inequalities (Amer.J.Math.,97,1976)

[11] Hardy-Littlewood-Pòlya,Inequalities (Cambridge Univ.Press,1964)

[12] C.Herz,The Hardy-Littlewood maximal theorem (Symposium on harmonic analysis,Univ.of Warwick,1968)

[13] K.Hildén,Symmetrization of functions in Sobolev spaces and the isoperimetric inequality (Manuscripta Math.,18,1976)

[14] B.Kawohl,Rearrangements and convexity of level sets in PDE (Lecture Notes in Math. 1150,Springer-Verlag,1985)

[15] E.Krahn,Über eine von Rayleigh formulierte minimaleigenschaft des kreises (Math.Ann.,94,1924)

[16] E.H.Lieb,Existence and uniqueness of the minimizing solution of Choquard's nonlinear equation (Studies in Appl.Math.,57,1977)

[17] E.H.Lieb,Sharp constants in the Hardy-Littlewood-Sobolev and related inequalities (Ann.of Math.,118,1983)

[18] J.Moser,A sharp form of an inequality by N.Trudinger (Indiana Univ. Math.J.,20,1971)

[19] R.O'Neil,Convolution operators and L(p,q) spaces (Duke Math.J.,30, 1963)

[20] R.O'Neil-G.Weiss,The Hilbert transform and rearrangements of functions (Studia Math.,23,1963)

[21] G.Pòlya-G.Szegö,Isoperimetric inequalities in mathematical physics (Princeton Univ.Press,1951)

[22] F.Riesz,Sur une inégalité intégrale (J.London Math.Soc.,5,1930)

[23] C.Somigliana,Sulle funzioni reali d'una variabile,Considerazioni sulle funzioni ordinate (Rend.R.Accedemia Lincei,vol.8,1899)

[24] E.Sperner,Zur symmetrisierung für funktionen auf sphären (Math.Z., 134,1973)

[25] E.Sperner,Symmetrisierung für funktionen mehrerer reeller variablen (Manuscripta Math.,11,1974)

[26] W.Spiegel,Über die symmetrisierung stetiger funktionen im euklidishen raum (Archiv der Math.,24,1973)

[27] G.Talenti,Best constant in Sobolev inequality (Ann.Mat.Pura Appl., 110,1976)

[28] G.Talenti,Some inequalities of Sobolev type on two-dimensional spheres (General Inequalities 5,W.Walter ed.,Birkhäuser-Verlag,1987)

[29] G.Talenti,Assembling a rearrangement (Arch.Rat.Mech.Anal.,98,1987)

Part two: papers on rearrangements and PDE.

[30] A.Alvino,Formule di maggiorazione e regolarizzazione per soluzioni di equazioni ellittiche del second'ordine in un caso limite (Rend. Acc.Naz.Lincei,ser.8,vol.62,1977)

[31] A.Alvino-G.Trombetti,Equazioni ellittiche con termini di ordine inferiore e riordinamenti (Rend.Acc.Naz.Lincei,ser.8,vol.66,1979)

[32] A.Alvino-G.Trombetti,Sulle migliori costanti di maggiorazione per una classe di equazioni ellittiche degeneri (Ric.Mat.,27,1979)

[33] A.Alvino-G.Trombetti,Su una classe di equazioni ellittiche nonlineari degeneri (Ric.Mat.,29,1980)

[34] A.Alvino-G.Trombetti,Sulle migliori costanti di maggiorazione per una classe di equazioni ellittiche degeneri e non (Ric.Mat.,30,1981)

[35] A.Alvino-G.Trombetti,A lower bound for the first eigenvalue of an elliptic operator (J.Math.Anal.Applications,94,1983)

[36] A.Alvino-P.L.Lions-G.Trombetti,A remark on comparison results via symmetrization (Proc.Roy.Soc.Edinburgh,102A,1986)

[37] C.Bandle,Bounds for the solutions of boundary value problems (J. Math.Anal.Applications,54,1976)

[38] C.Bandle,On symmetrizations in parabolic equations (J.Analyse Math., 30,1976)

[39] C.Bandle,Estimates for the Green function of elliptic operators (SIAM J.Math.Anal.,9,1978)

[40] C.Bandle-J.Mossino,Application du réarrangement à une inéquation variationnelle (C.R.A.S.Paris,t.296,sér.I,1983)

[41] C.Bandle-R.P.Sperb-I.Stakgold,The single steady-state irreversible reaction (to appear)

[42] P.Buonocore,Sulla simmetrizzazione in equazioni paraboliche degeneri (Boll.UMI,3-B,1984)

[43] G.Chiti,Norme di Orlicz delle soluzioni di una classe di equazioni ellittiche (Boll.UMI,16-A,1979)

[44] G.Chiti,An isoperimetric inequality for the eigenfunctions of linear second order elliptic operators (Boll.UMI,A-1,1982)

[45] G.Chiti,A reverse Hölder inequality for the eigenfunctions of linear second order elliptic equations (Z.A.M.P.,33,1982)

[46] P.S.Crooke-R.P.Sperb,Isoperimetric inequalities in a class of nonlinear eigenvalue problems (SIAM J.Math.Anal.,9,1978)

[47] J.I.Diaz,Applications of symmetric rearrangement to certain nonlinear elliptic equations with a free boundary (Nonlinear differential equations,J.K.Hale & P.Martinez-Amores eds.,Research Notes in Math., 132,Pitman,1985)

[48] E.Giarrusso-D.Nunziante,Symmetrization in a class of first-order Hamilton-Jacobi equations (Nonlinear Analysis T.M.A.,8,1984)

[49] E.Giarrusso-D.Nunziante,Comparison theorems for a class of first-order Hamilton-Jacobi equations (to appear)

[50] P.Hess,An isoperimetric inequality for the principal eigenvalue of a periodic parabolic problem (to appear)

[51] B.Kawohl,On the isoperimetric nature of a rearrangement inequality and its consequences for some veriational problems (LCDS Report 84-4,Providence,1984)

[52] P.Laurence-E.W.Stredulinsky,A new approach to queer differential equations (to appear in Comm.Pure Appl.Math.)

[53] P.L.Lions,Quelques remarques sur la symétrisation de Schwarz (Nonlinear partial differential equations and their applications,Collège de France Seminars,vol.1,Pitman,1981)

[54] C.Maderna,Optimal problems for a certain class of nonlinear Dirichlet problems (Boll.UMI,Suppl.1,1980)

[55] C.Maderna,On level sets of Poisson integrals in disks and sectors (Boll.UMI,ser.C-2,1983)

[56] C.Maderna-S.Salsa,Symmetrization in Neumann problems (Applicable Anal.,9,1979)

[57] C.Maderna-S.Salsa,A priori bounds in nonlinear Neumann problems (Boll.UMI,16-B,1979)

[58] C.Maderna-S.Salsa,Sharp estimates for solutions to a certain type of singular elliptic boundary value problem in two dimensions (Applicable Anal.,12,1981)

[59] C.Maderna-S.Salsa,Some special properties of solutions to obstacle problems (Rend.Sem.Mat,Padova,71,1984)

[60] C.Maderna-S.Salsa,Dirichlet problem for elliptic equations with nonlinear first-order terms: a comparison result (to appear)

[61] B.McLeod,Rearrangements and extreme value of Dirichlet norms (to appear)

[62] J.Mossino,A priori estimates for a model of Grad-Mercier type in plasma confinement (Applicable Anal.,13,1982)

[63] J.Mossino,A generalization of the Payne-Rayner isoperimetric inequality (Boll.UMI,A-2,1983)

[64] J.Mossino,Inégalités isopérimetriques et applications en physique (Herman,1984)

[65] J.Mossino-J.M.Rakotoson,Isoperimetric inequalities in parabolic equations (to appear)

[66] J.Mossino-R.Temam,Directional derivative of the increasing rearrangement mapping and application to a queer differential equation in plasma physics (Duke Math.J.,48,1981)

[67] F.Pacella-M.Tricarico,Symmetrization for a class of elliptic equations with mixed boundary conditions (Atti Sem.Mat.Fis.Univ.Modena, 34,1985-86)

[68] R.P.Sperb,Maximum principles and their applications (Academic Press, 1981)

[69] E.Sperner,Spherical symmetrization and eigenvalue estimates (Math.Z.,176,1981)

[70] G.Talenti,Elliptic equations and rearrangements (Ann.Scuola Norm. Sup.Pisa,3,1976)

[71] G.Talenti,Nonlinear elliptic equations,rearrangements of functions and Orlicz spaces (Ann.Mat.Pura Appl.,120,1979)

[72] G.Talenti,Some estimates for solutions to Monge-Ampère equations in dimension two (Ann.Scuola Norm.Sup.Pisa,8,1981)

[73] G.Talenti,On the first eigenvalue of the clamped plate (Ann.Mat. Pura Appl.,129,1981)

[74] G.Talenti,A note on the Gauss curvature of harmonic and minimal surfaces (Pacific J.Math.,101,1982)

[75] G.Talenti,Linear elliptic pde's: level sets,rearrangements and a priori estimates of solutions (Boll.UMI,4-B,1985)

[76] G.Trombetti-J.L.Vazquez,A symmetrization result for elliptic equations with lower-order terms (Dip.Mat.Univ.Napoli,preprint 29,1985)

[77] J.L.Vazquez,Symétrization pour $u_t = \Delta\varphi(u)$ et applications (C.R.A.S.Paris,t.295,1982)

[78] Ch.Voas-D.Yaniro,Elliptic equations,rearrangements and functions of bounded lower oscillation (to appear)

[79] A.Weitsman,Spherical symmetrization in the theory of elliptic partial differential equations (Comm.in PDE,8(5),1983)

[80] A.Weitsman,Symmetrization and the Poincaré metric (Ann.of Math.,124, 1986)

12 Inequalities in the Theory of Function Spaces: A Tribute to Hardy, Littlewood and Pólya

Hans Triebel Universität Jena, Jena, German Democratic Republic

1. Introduction

This paper deals with selected aspects of the theory of Sobolev spaces, Hölder-Zygmund spaces, Bessel-potential spaces, Besov spaces (also denoted as Lipschitz spaces), Hardy-Sobolev spaces and related types of spaces of functions and distributions defined on the euclidean n-space $\mathbb{R}^n$. This theory is governed by more or less sophisticated inequalities in $\mathbb{R}^n$ including derivatives, weights, maximal functions, convolutions, entire analytic functions, etc. Many of these crucial n-dimensional inequalities have their origin in the work of Hardy, Littlewood and Pólya, mostly in the twenties and thirties, restricted to one dimension. The aim of this paper is to shed light on this interdependence. In this sense we collect in Section 2 some relevant classical inequalities by Hardy, Littlewood and Pólya, most of them may be found in [HLP]. This book influenced heavily the theory of function spaces in the above sense. Its original 1934 edition was translated in Russian and appeared in 1948. Hence it was available when the Russian school, after the fundamental work by Sobolev [So1-4], started to develop the theory of function spaces in a big scale at the end of the 1940s and in the 1950s. Beside Sobolev one should mention Nikol'skij and (somewhat later, around 1960) Besov and Lizorkin. This influence is well reflected in the books [Ni2] and [BIN]. On the other hand Zygmund's enormous treatise [Zy] is closely connected with [HLP] (check the "Notes" in Zygmund's book). At least partly based on the sources [HLP] and [Zy] the American school developed a somewhat different direction in the theory of function spaces. Systematic presentations including many references of this branch of the theory of function spaces may be found in [St] and [SW2]. We refer also to [Ta]. In the introduction of the latter paper one finds the sentence: "The germ of this developments is in a group of papers by Hardy and Littlewood . . .".

From Section 3 on we wish to sketch some of these developments, more or less closely connected with the inequalities from Section 2, not in a historical way but from the point of view of recent research. The flavour is more important than comprehensiveness. In other words, this paper has nothing to do with a faithful historical report. Just on the contrary we concentrate ourselves on those inequalities and related function spaces which are more or less closely connected with the author's own interests.

2. Some classical inequalities by Hardy, Littlewood and Pólya

2.1. The Hardy-Littlewood Maximal Inequality

Let $f(x)$ be a complex-valued locally Lebesgue-integrable function on the real line $\mathbb{R}$. Then

$$Mf(x) = \sup \frac{1}{|I|} \int_I |f(y)|\,dy, \quad x \in \mathbb{R} \tag{1}$$

is the Hardy-Littlewood maximal function where the supremum is taken over all finite intervals I on $\mathbb{R}$ centered at x. If $f \in L_p(\mathbb{R})$ with $1 < p \leqslant \infty$ then $Mf \in L_p(\mathbb{R})$ and

$$\|Mf \,|\, L_p(\mathbb{R})\| \leqslant c_p \|f \,|\, L_p(\mathbb{R})\|, \tag{2}$$

where c_p depends only on p. This famous maximal inequality is due to Hardy and Littlewood [HL1] and may also be found in [HLP, Theorem 398]. (The case $p = \infty$ is obvious). The extension of (2) from $n = 1$ to $n > 1$ is due to Wiener [Wi].

2.2. Sobolev Inequalities

Inequalities of this type are characterized by mapping properties of fractional integration and Riesz potentials (fractional powers of the Laplacian) between different L_p spaces. At least in the older part of the theory of function spaces they form the basis of the famous embedding theorems. The (one-dimensional) Riesz potential I_α is given by

$$I_\alpha f(x) = \int_{\mathbb{R}} |x - y|^{\alpha - 1} f(y)\,dy, \qquad 0 < \alpha < 1. \tag{1}$$

Let $1 < p < q < \infty$ and $\dfrac{1}{q} = \dfrac{1}{p} - \alpha$. Then

$$\|I_\alpha f \,|\, L_q(\mathbb{R})\| \leqslant c_{pq} \|f \,|\, L_p(\mathbb{R})\|, \tag{2}$$

where c_{pq} depends only on p and q. This assertion is due to Hardy and Littlewood, see [HL2]. We refer also to [HLP, Theorem 383]. The extension of (2) from $n = 1$ to $n > 1$ is one of the corner stones of Sobolev's theory, see [So3]. If $\alpha > 1/p$ then I_α maps $L_p(\mathbb{R})$ in a Lipschitz space of order $\alpha - 1/p$. It is surprising that even this nowadays well known fact has been observed by [HLP, 383^a].

2.3. Hardy Inequalities

The n-dimensional extensions of (2.1/2) and (2.2/2) play a decisive role in the theory of unweighted (isotropic) function spaces on $\mathbb{R}^n$. Weighted spaces attracted much attention in the last decades. Again inequalities involving different weights (besides derivatives, differences, etc.) are of crucial importance. At least some of these inequalities can be traced back to the famous Theorem 330 in [HLP] which has the following consequence. Let $1 < p < \infty$, $\epsilon \neq p - 1$ and let $f(x)$ be a function differentiable a.e. in $(0, \infty)$ such that

$$\int_0^\infty |f'(x)|^p x^\epsilon\,dx < \infty \tag{1}$$

satisfying

$$\lim_{x \downarrow 0} f(x) = 0 \qquad \text{for} \qquad \epsilon < p - 1,$$

$$\lim_{x \to \infty} f(x) = 0 \qquad \text{for} \qquad \epsilon > p - 1.$$

Then this holds:

$$\int_0^\infty |f(x)|^p x^{\epsilon - p}\, dx \leqslant \left(\frac{p}{|\epsilon - p + 1|} \right)^p \int_0^\infty |f'(x)|^p x^\epsilon dx. \tag{2}$$

This assertion is due to Hardy [Ha1]. The above formulation was taken over from [Ku] where inequalities of this type are studied systematically including n-dimensional consequences and applications to weighted Sobolev spaces.

2.4. *The Hilbert Transform*

The Hilbert transform on $\mathbb{R}$ is given by

$$Hf(x) = \int_{\mathbb{R}} \frac{f(y)}{x - y} dy, \qquad x \in \mathbb{R}, \tag{1}$$

with the usual interpretation as a singular integral. The following mapping theorem in $L_p(\mathbb{R})$ with $1 < p < \infty$ is due to M. Riesz [RiM],

$$\|Hf \,|\, L_p(\mathbb{R})\| \leqslant c_p \|f \,|\, L_p(\mathbb{R})\|, \tag{2}$$

where c_p depends only on p. This assertion is closely connected with [HLP, (9.1.2)] and [Ha2]. This theorem is a starting point of the theory of singular integrals and Fourier multipliers in spaces of type $L_p(\mathbb{R}^n)$. We refer to [St].

2.5. *Plancherel-Pólya Inequalities and Hardy Representations*

The Fourier analytical approach to function spaces of Besov-Hardy-Sobolev type on $\mathbb{R}^n$ is based on decomposition methods. The building blocks are entire analytic functions of exponential type. This means by the Paley-Wiener-Schwartz theorem that the Fourier transform of the corresponding tempered distribution has a compact support. Two types of inequalities are of interest in this connection, which go back to Plancherel-Póyla [PP] on the one hand, and to Nikol'skij on the other hand. Inequalities of Nikol'skij type will be discussed later on. As far as Plancherel-Póyla inequalities are concerned we restrict ourselves to the simplest case. Let

$$f \in S(\mathbb{R}) \qquad \text{with} \qquad \operatorname{supp} \hat{f} \subset [-\pi, \pi], \tag{1}$$

where $S(\mathbb{R})$ is the usual Schwartz space and $\hat{f}$ stands for the Fourier transform of f. Then f can be represented as

$$f(x) = \frac{\sin \pi x}{\pi} \sum_{k=-\infty}^{\infty} (-1)^k \frac{f(k)}{x - k} \tag{2}$$

This interpolation formula goes back to Hardy [Ha3]. We refer also to [Zy, XVI, (7.19)]. Furthermore let $1 < p < \infty$. Then

$$c_p \left(\sum_{k=-\infty}^{\infty} |f(k)|^p \right)^{1/p} \leqslant \|f|L_p(\mathbb{R})\| \leqslant C_p \left(\sum_{k=-\infty}^{\infty} |f(k)|^p \right)^{1/p} \tag{3}$$

where the positive constants c_p and C_p depend only on p. It is remarkable that (3) can be extended in the following way. Let $0 < p \leqslant \infty$. Then

$$c_p \left(\sum_{k=-\infty}^{\infty} |f(\kappa_p k)|^p \right)^{1/p} \leqslant \|f|L_p(\mathbb{R})\| \leqslant C_p \left(\sum_{k=-\infty}^{\infty} |f(\kappa_p k)|^p \right)^{1/p} \tag{4}$$

(with the obvious modification if $p = \infty$) where the positive constants c_p, C_p and κ_p depend only on p. (Of course if $1 < p < \infty$ then $\kappa_p = 1$ is the best possible choice). These inequalities go back to [PP], we refer also to [Bo, pp. 197–199].

2.6. *Hardy Spaces*

One of the most striking features of the recent Fourier analytical approach in the theory of function spaces is the fact that it covers not only Sobolev spaces, Bessel-potential spaces, Besov(-Lipschitz) spaces and Hölder-Zygmund spaces, which are Banach spaces, but also Hardy spaces and (with some modifications which will not be discussed here) the fashionable space BMO in a unified way. The origin of the Hardy spaces comes from complex function theory. Let $0 < p < \infty$. Then an analytic function f in the unit disk in the complex plane is said to belong to the space H_p if

$$\mu_p(r, f) = \left(\frac{1}{2\pi} \int_0^{2\pi} |f(re^{i\theta})|^r d\theta \right)^{1/p} \tag{1}$$

is bounded for $0 \leqslant r < 1$. Of special interest are the spaces H_p with $0 < p \leqslant 1$. These spaces were first considered by Hardy [Ha4]. Afterwards, F. Riesz [RiF] gave a simplified and systematic approach to these spaces. Instead of the unit disk one can use the upper half-plane. The extension from one dimension (this corresponds to the boundary behaviour of the above analytic functions in the unit disk or in the upper half-plane) to higher dimensions caused some trouble and was done by Stein and Weiss in [SW1]. A genuine real variable approach which provides the possibility to incorporate the Hardy spaces in the Fourier analytical approach of the other above-mentioned function spaces is due to Fefferman and Stein [FS2]. We return to the original Hardy spaces on the unit disk in the complex plane. We formulate an important result, also from the standpoint of recent research, which is due to Hardy and Littlewood [HL1]. We use the formulation given in [Zy, VII, (7.36)]. For any $0 \leqslant r < 1$, $0 \leqslant \theta \leqslant 2\pi$ let $\Omega_r(\theta)$ denote the domain bounded by the two tangents from the point $e^{i\theta}$ to the circle $|z| = r$ and by the larger of the two arcs of that circle between the points of contact. (For $r = 0$, the domain reduces to the radius through $e^{i\theta}$). Let $0 < p < \infty$, $f \in H_p$ and

$$N(\theta) = N_{r,f}(\theta) = \sup_{z \in \Omega_r(\theta)} |f(z)|. \tag{2}$$

Then $N(\theta) \in L_p(0, 2\pi)$ and

$$\rho \left[\int_0^{2\pi} N^p(\theta) d\theta \right]^{1/p} \leqslant C_p \lim_{\rho \to 1} \mu_p(\rho, f). \tag{3}$$

2.7. *Concluding Remark*

The above inequalities have their roots in the work of Hardy, Littlewood, and Pólya, and they are more or less directly connected with recent research in the theory of function spaces. Of course, the corresponding inequalities used nowadays are often more sophisticated. In particular the step from one dimension to higher dimensions requires often new tools which have no analogue in one dimension. In the following sections we describe some of these extensions and how to use them in the theory of function spaces.

3. Inequalities: Several Extensions

3.1. *Inequalities for Entire Analytic Functions*

Let $S(\mathbb{R}^n)$ be the usual Schwartz space on $\mathbb{R}^n$ and let Mf be the Hardy-Littlewood maximal function from (2.1/1), now extended to $\mathbb{R}^n$, i.e.,

$$Mf(x) = \sup |B|^{-1} \int_B |f(y)| dy, \qquad x \in \mathbb{R}^n, \tag{1}$$

where the supremum is taken over all balls B in $\mathbb{R}^n$ centered at x. Let Ω be a compact subset of $\mathbb{R}^n$. Let

$$f \in S(\mathbb{R}^n), \qquad \operatorname{supp} \hat{f} \subset \Omega, \tag{2}$$

where $\hat{f}$ stands for the Fourier transform of f. Let $0 < r < \infty$. Then

$$\sup_{z \in \mathbb{R}^n} \frac{|\nabla f(x - z)|}{1 + |z|^{n/r}} \leqslant c \sup_{z \in \mathbb{R}^n} \frac{|f(x - z)|}{1 + |z|^{n/r}} \leqslant C(M|f|^r(x))^{1/r} \tag{3}$$

where c and C depend only on r and Ω. Let again f be given by (2), let $0 < p \leqslant q \leqslant \infty$, and let α be an arbitrary multi-index. Then

$$\|D^\alpha f | L_q(\mathbb{R}^n)\| \leqslant c\|f | L_p(\mathbb{R}^n)\| \tag{4}$$

where c depends only on p, q, Ω and α. The inequalities (3) and (4) together with the n-dimensional versions of (2.1/2) and (2.5/4) is the backbone of a substantial part of the theory of the Besov spaces $B^s_{pq}(\mathbb{R}^n)$ which we sketch below. Inequality (4) restricted to $1 \leqslant p \leqslant q \leqslant \infty$ is the famous Nikol'skij inequality [Ni1], we refer also to [Ni2]. The maximal inequality (3) may be found in [Tr3, Theorem 1.3.1]. There is also a connection between (3) and (2.6/3), we refer to [FS2].

3.2. *Vector-Valued Inequalities*

The inequalities described in 3.1 are a sound basis to handle the spaces $B^s_{pq}(\mathbb{R}^n)$. However if one wishes to include Sobolev spaces, Bessel-potential spaces, Hardy spaces and, more general, the spaces $F^s_{pq}(\mathbb{R}^n)$, then the above inequalities are not sufficient. Roughly speaking, one needs their vector-valued extensions. This is by no means a matter of routine.

Of crucial importance is the following extension of the (n-dimensional) Hardy-Littlewood maximal inequality (2.1/2) due to Fefferman and Stein [FS1]. Let $1 < p < \infty$ and $1 < q \leqslant \infty$, and let Mf be given by (3.1/1). Then

$$\left\| \left(\sum_{k=1}^{\infty} |Mf_k|^q \right)^{1/q} |L_p(\mathbb{R}^n)\right\| \leqslant C \left\| \left(\sum_{k=1}^{\infty} |f_k|^q \right)^{1/q} |L_p(\mathbb{R}^n)\right\| \tag{1}$$

(obvious modification if $q = \infty$), where c depends only on p, q and n. The scalar case of (1), i.e. the n-dimensional extension of (2.1/2) is due to Wiener [Wi]. Next we describe the relevant extension of (3.1/3). Let $\Omega = \{\Omega_k\}_{k=1}^{\infty}$ be a sequence of compact subsets of $\mathbb{R}^n$ and let $d_k > 0$ be the diameter of Ω_k. Let

$$\{f_k\}_{k=1}^{\infty} \subset S(\mathbb{R}^n), \qquad \operatorname{supp} \hat{f}_k \subset \Omega_k, \tag{2}$$

cf. (3.1/2). Let $0 < p < \infty$, $0 < q \leqslant \infty$ and $0 < r < \min(p, q)$. Then

$$\left\| \left(\sum_{k=1}^{\infty} \sup_{y \in \mathbb{R}^n} \frac{|f_k(\cdot - y)|^q}{1 + |d_k y|^{nq/r}} \right)^{1/q} |L_p(\mathbb{R}^n)\right\| \leqslant c \left\| \left(\sum_{k=1}^{\infty} |f_k(\cdot)|^q \right)^{1/q} |L_p(\mathbb{R}^n)\right\| \tag{3}$$

(obvious modification if $q = \infty$) where c depends only on Ω, p, q and r. We refer to [Tr3, Theorem 1.6.2]. If one combines the second half of (3.1/3) with the n-dimensional version of (2.1/2) then one obtains the scalar case of (3).

3.3. *Weighted Inequalities*

Weighted generalizations of the above inequalities have been treated extensively in recent times, mostly connected with Muckenhoupt's A_p-classes. Another type of weights has been studied by the author, see the first chapter of [ST] and the reference given there. These weights avoid local singularities and the growth restrictions come from the theory of ultra-distributions. All the inequalities from 2.5, 3.1 and 3.2 have counterparts for these weighted classes. However we restrict ourselves to the simplest case, this is the weighted extension of (3.1/4). Let $\sigma(t)$ be an increasing continuous concave function on $[0, \infty]$ with

$$\sigma(0) = 0, \int_0^{\infty} \frac{\sigma(t)}{1 + t^2}\, dt < \infty, \ \sigma(t) \geqslant c + d \log(1 + t) \quad \text{if} \quad t \geqslant 0 \tag{1}$$

where c is a real number and d is a positive number. Let $\omega(x) = \sigma(|x|)$ with $x \in \mathbb{R}^n$. Then $S_\omega(\mathbb{R}^n)$ is the collection of all complex-valued integrable C^∞ functions on $\mathbb{R}^n$ with

$$\sup_{x \in \mathbb{R}^n} e^{\lambda \omega(x)} (|D^\alpha f(x)| + |D^\alpha \hat{f}(x)|) < \infty \tag{2}$$

for all multi-indices α and all $\lambda > 0$. If $\sigma(t) = \log(1 + t)$ then one has the usual Schwartz space $S(\mathbb{R}^n)$. Another typical example is $\sigma(t) = t^\beta$ with $0 < \beta < 1$ (Gevrey classes). We return to the above general function $\omega(x)$. Let $\rho(x)$ be a continuous function on $\mathbb{R}^n$ with

$$0 < \rho(x) \leqslant c \rho(y) e^{\omega(x-y)}, \ x \in \mathbb{R}^n, \ y \in \mathbb{R}^n, \tag{3}$$

where c is independent of x and y. Let

$$f \in S_\omega(\mathbb{R}^n), \ \operatorname{supp} \hat{f} \subset \Omega, \tag{4}$$

where Ω is a given compact subset of $\mathbb{R}^n$, cf. (3.1/2). Let $0 < p \leqslant q \leqslant \infty$ and let α be an arbitrary multi-index. Then

$$\|\rho D^\alpha f|L_q(\mathbb{R}^n)\| \leqslant \|\rho f|L_p(\mathbb{R}^n)\| \tag{5}$$

where c depends only on p, q, Ω, α and ρ. This is the weighted counterpart of the Nikol'skij inequality (3.1/4), see [ST, Proposition 1.4.3] and the references given there. As had been said the other inequalities from 2.5, 3.1 and 3.2 have also respective weighted counterparts in the above sense.

3.4. *Fractional and Anisotropic Inequalities*

In 3.1–3.3 we dealt mainly with inequalities for entire analytic functions of exponential type including vector-valued and weighted generalizations. The idea behind is to use these inequalities in order to construct function spaces of type B^s_{pq} and F^s_{pq} via decomposition methods in the Fourier images. Then (3.1/2) is just a typical situation for the decomposed parts. The aim of the present section is different. We wish to introduce some anisotropic function spaces via decompositions of the function itself (and not of its Fourier transform). It comes out that fractional and anisotropic extensions of Hardy's inequality (2.3/2) are of great service for this purpose.

The simplest case of a fractional Hardy inequality reads as follows. Let $1 < p < \infty$, $0 < s < 1$ and $s \neq 1/p$. Let $f(x) \in S(\mathbb{R})$ with $f(0) = 0$ in the case $sp > 1$. Then

$$\int_{\mathbb{R}} |x|^{-sp}|f(x)|^p\,dx \leqslant c\int_{\mathbb{R}\times\mathbb{R}} \frac{|f(x)-f(y)|^p}{|x-y|^{1+sp}}\,dx\,dy \tag{1}$$

where c depends only on s and p. The first proof known to the author is due to Grisvard [Gr]. However this inequality must be known around 1960 because it is connected with the "trace method" in real interpolation by J.-L. Lions. In order to generalize (1) we recall that

$$\Delta_h^m f(x) = \sum_{j=0}^{m} (-1)^{m-j}\binom{m}{j} f(x+jh), \qquad x \in \mathbb{R}, \qquad h \in \mathbb{R}, \tag{2}$$

stands for the usual differences. Let $1 < p < \infty$, $0 < s = [s]^- + \{s\}^+$ where $[s]^-$ is an integer, $0 < \{s\}^+ \leqslant 1$ and $\{s\}^+ p \neq 1$. Let k and m be integers with $0 \leqslant k \leqslant [s]^-$ and $m > s - k$. Let $f(x) \in S(\mathbb{R})$ with $f^{(j)}(0) = 0$ for $j = 0, \ldots, [s - 1/p]$. Then

$$\int_{\mathbb{R}} |x|^{-sp}|f(x)|^p dx \leqslant c\int_{\mathbb{R}} |h|^{-(s-k)p}\|\Delta_h^m f^{(k)}|L_p(\mathbb{R})\|^p \frac{dh}{|h|}. \tag{3}$$

We refer to [ST, 4.3.1].

The next goal is to extend (3) to two dimensions, the anisotropic case. Let

$$\Delta_{h,1}^m f(x) = \sum_{j=0}^{m} (-1)^{m-j}\binom{m}{j} f(x_1 + jh, x_2), \quad x = (x_1, x_2) \in \mathbb{R}^2, \quad h \in \mathbb{R}, \tag{4}$$

and similarly $\Delta_{h,2}^m$. Let $\bar{a} = (a_1, a_2)$ with $0 < a_2 \leqslant a_1 < \infty$ and $a_1 + a_2 = 2$ be a given anisotropy. Let $\bar{s} = (s_1, s_2)$ with $s_1 = s/a_1$, $s_2 = s/a_2$ and $s > 0$. Let $1 < p < \infty$. If $\bar{m} = (m_1, m_2)$ where m_1 and m_2 stand for natural numbers with $m_1 > s_1$ and $m_2 > s_2$ then

$$\|f|B_p^{\bar{s}}(\mathbb{R}^2)\|_{\bar{m}} = \|f|L_p(\mathbb{R}^2)\| +$$

$$+ \left(\int_{\mathbb{R}} [\,|h|^{-s_1 p} \|\Delta_{h,1}^{m_1} f|L_p(\mathbb{R}^2)\|^p + |h|^{-s_2 p} \|\Delta_{h,2}^{m_2} f|L_p(\mathbb{R}^2)\|^p] \frac{dh}{|h|} \right)^{1/p} \tag{5}$$

is a norm in the anisotropic Besov space $B_p^{\bar{s}}(\mathbb{R}^2)$. Let

$$|x|_{\bar{a}} = (\,|x_1|^{2/a_1} + |x_2|^{2/a_2})^{½} \tag{6}$$

be the related anisotropic distance in $\mathbb{R}^2$. Finally we need the counterpart of the above one-dimensional condition $\{s\}^+ p \neq 1$ which reads as follows: $\bar{s} = (s_1, s_2)$ is called critical if there exist non-negative integers n_1 and n_2 with

$$\frac{1}{s_1}(n_1 + \frac{1}{p}) + \frac{1}{s_2}(n_2 + \frac{1}{p}) = 1 \tag{7}$$

Otherwise $\bar{s}$ is called non-critical. Now we are ready to formulate an anisotropic Hardy inequality: Let $1 < p < \infty$ and let $\bar{s} = (s_1, s_2)$ with $s_1 = s/a_1$, $s_2 = s/a_2$ and $s > 0$ be non-critical. Let

$$f \in S(\mathbb{R}^2) \qquad \text{with} \qquad \frac{\partial^{n_1+n_2} f(0)}{\partial x_1^{n_1} \partial x_2^{n_2}} = 0$$

if

$$\frac{1}{s_1}(n_1 + \frac{1}{p}) + \frac{1}{s_2}(n_2 + \frac{1}{p}) < 1. \tag{8}$$

Then

$$\int_{\mathbb{R}^2} |x|_{\bar{a}}^{-sp} |f(x)|^p \, dx \leqslant c \|f|B_p^{\bar{s}}(\mathbb{R}^2)\|_{\bar{m}}^p \tag{9}$$

where c is independent of f. We refer to [ST, Chapter 4].

4. How to Construct Function Spaces: Decompositions

4.1. Besov spaces

In 3.1–3.4 we selected some inequalities which are on the one hand closely connected with classical inequalities by Hardy, Littlewood and Pólya, and on the other hand the backbone of some parts of the theory of function spaces. The approach we have in mind is characterized by the key word decomposition, mostly not so much for the function itself but for its Fourier transform. Seen from this standpoint the Besov spaces are especially simple. We give a description. Let $\varphi \in S(\mathbb{R}^n)$ with

$$\operatorname{supp} \varphi \subset \{y \,|\, |y| \leqslant 2\} \qquad \text{and} \qquad \varphi(x) = 1 \text{ if } |x| \leqslant 1. \tag{1}$$

Let $\varphi_0 = \varphi$ and $\varphi_j(x) = \varphi(2^{-j}x) - \varphi(2^{-j+1}x)$ if $j = 1, 2, \ldots$ Then $\sum_{j=0}^{\infty} \varphi_j(x) = 1$ is a resolution of unity. By the Paley-Wiener-Schwartz theorem

$$\varphi_j(D)f(x) = \hat{\varphi}_j * f(x), \qquad x \in \mathbb{R}^n,\ f \in S'(\mathbb{R}^n), \tag{2}$$

is an entire analytic function, where $\hat{\varphi}_j$ is the Fourier transform of φ_j and $*$ stands for the convolution. $S'(\mathbb{R}^n)$ is the dual of $S(\mathbb{R}^n)$ (tempered distributions). Let $-\infty < s < \infty$, $0 < p \leqslant \infty$ and $0 < q \leqslant \infty$. Then $B^s_{pq}(\mathbb{R}^n)$ is the collection of all $f \in S'(\mathbb{R}^n)$ such that

$$\left(\sum_{j=0}^{\infty} 2^{jsq} \| \varphi_j(D)f \,|\, L_p(\mathbb{R}^n)\|^q \right)^{1/q} \tag{3}$$

is finite. (Obvious modification of $q = \infty$). However $\varphi_j(D)f$ is essentially (besides smoothness assertions) of type (3.1/2) where Ω is a ball centered at the origin with radius 2^{j+1}. Hence instead of f one deals with the sequence $\{\varphi_j(D)f\}_{j=0}^{\infty}$ of entire analytic functions. It comes out that $B^s_{pq}(\mathbb{R}^n)$ is independent of the chosen function φ. The inequalities (3.1/3) (together with the n-dimensional version of (2.1/2)) and (3.1/4) are extremely useful (of course these inequalities can be extended by completion). We give an example. We choose $\alpha = 0$ in (3.1/4) and replace f by $\varphi_j(D)f$. By homogeneity we have $c = c' 2^{jn(1/p - 1/q)}$ where c' is independent of j. This proves the embedding theorem

$$B^{s_0}_{p_0 q}(\mathbb{R}^n) \subset B^{s_1}_{p_1 q}(\mathbb{R}^n), \qquad -\infty < s_1 \leqslant s_0 < \infty, \quad s_0 - n/p_0 = s_1 - n/p_1, \tag{4}$$

where $0 < p_0 \leqslant p_1 \leqslant \infty$ and $0 < q \leqslant \infty$. If $1 < p < \infty$, $1 \leqslant q \leqslant \infty$ and $s > 0$ then $B^s_{pq}(\mathbb{R}^n)$ are the classical Besov spaces. If $p = q = \infty$ and $s > 0$ then $\mathcal{C}^s(\mathbb{R}^n) = B^s_{\infty\infty}(\mathbb{R}^n)$ are the Hölder-Zygmund spaces. For the full range of the parameters these spaces have been treated in [Tr3]. We refer also to [Pe4], [Ni2], [BIN] and [Tr2]. The origin of the above approach to these spaces will be described below. We mention that some cases of (4) (one dimension, restricted parameters) have been known to Hardy and Littlewood [HL2].

4.2. Hardy-Sobolev Spaces and All That

Let $-\infty < s < \infty$, $0 < p < \infty$ and $0 < q \leqslant \infty$. Then $F^s_{pq}(\mathbb{R}^n)$ is the collection of all $f \in S'(\mathbb{R}^n)$ such that

$$\left\| \left(\sum_{j=0}^{\infty} 2^{jsq} |\varphi_j(D)f(\cdot)|^q \right)^{1/q} \,\Big|\, L_p(\mathbb{R}^n) \right\| \tag{1}$$

if finite (obvious modification if $q = \infty$). In other words, we simply change the order of the two involved quasi-norms l_q and $L_p(\mathbb{R}^n)$ in (4.1/3). Again one has a decomposition of f into a sequence of entire analytic functions of exponential type. But it is quite clear that the scalar inequalities from 3.1 are now not sufficient in order to handle these new spaces. One needs their vector valued counterparts, where (3.2/1) and in particular (3.2/3), with (3.2/2), are typical and important examples. The theory of the spaces $F^s_{pq}(\mathbb{R}^n)$ can at least partly be based on inequalities of this type. First we have to remark that $F^s_{pq}(\mathbb{R}^n)$ is independent of the chosen function φ. The counterpart of (4.1/4) looks as follows,

$$F^{s_0}_{p_0 q}(\mathbb{R}^n) \subset F^{s_1}_{p_1 r}(\mathbb{R}^n) \qquad -\infty < s_1 < s_0 < \infty, \quad s_0 - \frac{n}{p_0} = s_1 - \frac{n}{p_1} \tag{2}$$

where $0 < p_0 < p_1 < \infty$, $0 < q \leqslant \infty$ and $0 < r \leqslant \infty$. Some special cases are of interest.

Let $W_p^m(\mathbb{R}^n)$ be the usual Sobolev spaces on $\mathbb{R}^n$ where $1 < p < \infty$ and $m = 0, 1, 2, \ldots,$ normed by

$$\sum_{|x| \leqslant m} \|D^\alpha f | L_p(\mathbb{R}^n)\|. \tag{3}$$

Then we have

$$W_p^m(\mathbb{R}^n) = F_{p2}^m(\mathbb{R}^n), \quad 1 < p < \infty \quad \text{and} \quad m = 0, 1, 2, \ldots \tag{4}$$

This theorem of Paley-Littlewood type can be extended the Bessel-potential spaces $H_p^s(\mathbb{R}^n)$,

$$H_p^s(\mathbb{R}^n) = F_{p2}^s(\mathbb{R}^n), \quad -\infty < s < \infty, \quad 1 < p < \infty, \tag{5}$$

and to the (nonhomogeneous) Hardy spaces $H_p(\mathbb{R}^n)$ (n-dimensional real-variable version)

$$H_p(\mathbb{R}^n) = F_{p2}^0(\mathbb{R}^n), \quad 0 < p \leqslant 1. \tag{6}$$

We refer for details to [Tr3] where we developed the full theory of these spaces as it stood at the beginning of the 1980s. There one can find also the necessary historical remarks; see also [Pe4] and [Tr2]. But we wish to mention that the Fourier analytical approach to the spaces $B_{pq}^s(\mathbb{R}^n)$ and $F_{pq}^s(\mathbb{R}^n)$ described above is due to Peetre [Pe1-3], Lizorkin [Li] and the author [Tr1]. We combine (2) and (4) and obtain

$$W_p^m(\mathbb{R}^n) \subset L_q(\mathbb{R}^n), \qquad 1 < p < \infty, \qquad m - \frac{n}{p} = -\frac{n}{q} \tag{7}$$

provided that m is a natural number and $1 < q < \infty$. This is closely connected with the n-dimensional version of (2.2/2). But instead of mapping theorems for fractional integrals one uses now inequalities for entire analytic functions.

4.3. Weighted Spaces

The spaces $B_{pq}^s(\mathbb{R}^n)$ from 4.1 are closely connected with the inequalities from 3.1, and the spaces $F_{pq}^s(\mathbb{R}^n)$ from 4.2 are closely connected with the inequalities from 3.2. If one wishes to deal with weighted spaces $B_{pq}^s(\mathbb{R}^n, \rho(x))$ and $F_{pq}^s(\mathbb{R}^n, \rho(x))$ then it is quite clear now that corresponding weighted inequalities are highly desirable. As far as the scalar case is concerned first steps have been done in 3.3. There exist also weighted counterparts of the vector-valued inequalities from 3.2. We developed this theory in [ST, Chapter 1 and 5.1]. We restrict ourselves to a typical result, the counterpart of the Paley-Littlewood theorem (4.2/4). Let $\omega(x)$ and $\rho(x)$ be the same functions as in 3.3. Furthermore let $S'_\omega(\mathbb{R}^n)$ be the dual of $S_\omega(\mathbb{R}^n)$, its elements are ultra-distributions. We have always

$$S_\omega(\mathbb{R}^n) \subset S(\mathbb{R}^n) \subset S'(\mathbb{R}^n) \subset S'_\omega(\mathbb{R}^n)$$

in the sense of dense and continuous embedding. The first embedding is a consequence of (3.3/1). Let $\varphi \in S_\omega(\mathbb{R}^n)$ with (4.1/1) (there exist functions with these properties) and let φ_j be constructed as in 4.1. We put

$$\|f | L_p(\mathbb{R}^n, \rho(x))\| = \|\rho f | L_p(\mathbb{R}^n)\|, \qquad 0 < p \leqslant \infty.$$

Let $-\infty < s < \infty$, $0 < p \leqslant \infty$ and $0 < q \leqslant \infty$. Then $B^s_{pq}(\mathbb{R}^n, \rho(x))$ is the collection of all $f \in S'_\omega(\mathbb{R}^n)$ such that (4.1/3) with $L_p(\mathbb{R}^n, \rho(x))$ instead of $L_p(\mathbb{R}^n)$ is finite. In the same way one defines $F^s_{pq}(\mathbb{R}^n, \rho(x))$ where $-\infty < s < \infty$, $0 < p < \infty$ and $0 < q \leqslant \infty$. One has to replace $L_p(\mathbb{R}^n)$ in (4.2/1) by $L_p(\mathbb{R}^n, \rho(x))$. Again these spaces are independent of the chosen function φ. Let $1 < p < \infty$ and $m = 0, 1, 2, \ldots$ Then $W^m_p(\mathbb{R}^n, \rho(x))$ is the collection of all $S'_\omega(\mathbb{R}^n)$ such that

$$\sum_{|\alpha| \leqslant m} \|D^\alpha f | L_p(\mathbb{R}^n, \rho(x))\|$$

is finite. Then the Paley-Littlewood assertion (4.2/4) can be generalized by

$$W^m_p(\mathbb{R}^n, \rho(x)) = F^m_{p2}(\mathbb{R}^n, \rho(x)), \quad 1 < p < \infty, \quad m = 0, 1, 2, \ldots$$

We refer to [ST, 5.1] for further results.

4.4. *Anisotropic Spaces*

In 4.1–4.3 we described spaces closely connected with the inequalities from 3.1–3.3, respectively. In particular the underlying idea is the decomposition of functions or distributions in the sense of (4.1/2) (decompositions in the Fourier image). The anisotropic Hardy inequality from (3.4/9) is of a somewhat different character. One can use it in order to study anisotropic function spaces on $\mathbb{R}^n$ or on special domains in $\mathbb{R}^n$ which show a distinguished behaviour near isolated points. This, in turn, can be used to treat semi-elliptic differential operators of the type

$$f(x) \longrightarrow -\frac{\partial^2 f}{\partial x_1^2} + \frac{\partial^4 f}{\partial x_2^4} + \rho(x) f, \quad x = (x_1, x_2) \in \mathbb{R}^2$$

and connected boundary value problems. See [ST, Chapter 4] and the references given there, where we dealt with a model case. We give a description of some representative results, but we shall not describe in detail how the decompositions work. The interested reader is asked to consult [ST] and the references given there. Let K be the unit circle in the $x_1 - x_2$ plane. First we modify (3.4/4) and (3.4/5). Let

$$\Delta^m_{h,1}(K) f(x) = \begin{cases} \Delta^m_{h,1}(K) f(x) & \text{if } x \in K \text{ and } (x_1 + mh, x_2) \in K \\ 0 & \text{otherwise,} \end{cases} \tag{1}$$

where $m = 1, 2, 3, \ldots$ and $h \in \mathbb{R}$. Similarly one defines $\Delta^m_{h,2}(K) f(x)$. As in 3.4 let $\bar{a} = (a_1, a_2)$ with $0 < a_2 \leqslant a_1 < \infty$ and $a_1 + a_2 = 2$ be the given anisotropy, let $\bar{s} = (s_1, s_2)$ with $s_1 = s/a_1$, $s_2 = s/a_2$ and $s > 0$, and let $1 < p < \infty$. Let $\bar{m} = (m_1, m_2)$ where m_1 and m_2 stand for natural numbers with $m_1 > s_1$ and $m_2 > s_2$. Then

$$\|f | B^{\bar{s}}_p(K)\|_{\bar{m}} = \|f | L_p(K^n)\| +$$

$$+ \left(\int_0^1 [h^{-s_1 p} \|\Delta^{m_1}_{h,1}(K) f | L_p(K)\|^p + h^{-s_2 p} \|\Delta^{m_2}_{h,2}(K) f | L_p(K)\|^p] \frac{dh}{h} \right)^{1/p}. \tag{2}$$

Let $|x|_{\bar{a}}$ be given by (3.4/6). Let $x^0 = (-1, 0)$ and $x^1 = (1, 0)$ be the two points on the circumference which are distinguished in the above sense. Then we introduce

$$|x|_{\bar{a}}(K) = \min(|x - x^0|_{\bar{a}}, |x - x^1|_{\bar{a}})$$

which stands for an anisotropic distance to the points x^0 and x^1. It is convenient for us to specify in the sequel $\bar{s} = (s_1, s_2)$ by $s_2 = 2s_1$, i.e. $a_2 = a_1/2$ or $a_1 = 4/3$, $a_2 = 2/3$, and $s = (4s_1/3)$. Then the corresponding anisotropic spaces are especially well adapted to the unit circle with x^0 and x^1 as distinguished points. Now $B_p^{\bar{s}}(K)$ is the collection of all $f \in L_p(K)$ such that $\|f | B_p^{\bar{s}}(K)\|_{\bar{m}}$ is finite, and $B_p^{\bar{s}}(\mathring{K})$ is the collection of all $f \in L_p(K)$ such that

$$\|f | B_p^{\bar{s}}(K)\|_{\bar{m}} + \left(\int_{\mathring{K}} |x|_{\bar{a}}(K)^{-sp} |f(x)|^p \, dx\right)^{1/p} \tag{3}$$

is finite. Both spaces are independent of $\bar{m}$, they are Banach spaces. We describe some properties (recall that we always assume $s_2 = 2s_1$). Let $f \in B_p^{\bar{s}}(K)$ or, even more restrictive, $f \in B_p^{\bar{s}}(\mathring{K})$. Then

$$\frac{\partial^{n_1+n_2} f}{\partial x_1^{n_1} \partial x_2^{n_2}} \in C(\bar{K}) \quad \text{if} \quad \frac{1}{s_1}\left(n_1 + \frac{1}{p}\right) + \frac{1}{s_2}\left(n_2 + \frac{1}{p}\right) < 1 \tag{4}$$

where n_1 and n_2 are non-negative integers. Of course, $C(\bar{K})$ stands for the space of all continuous functions on $\bar{K}$, and (4) must be interpreted in the sense of embedding theorems. In particular, the finite co-dimensional subspace

$$B_p^{\bar{s}}(\dot{K}) = \{f \mid f \in B_p^{\bar{s}}(K), \frac{\partial^{n_1+n_2} f(x^0)}{\partial x_1^{n_1} \partial x_2^{n_2}} = \frac{\partial^{n_1+n_2} f(x^1)}{\partial x_1^{n_1} \partial x_2^{n_2}} = 0$$
$$\text{if} \quad \frac{1}{s_1}\left(n_1 + \frac{1}{p}\right) + \frac{1}{s_2}\left(n_2 + \frac{1}{p}\right) < 1\} \tag{5}$$

of $B_p^{\bar{s}}(K)$ makes sense. Let $\bar{s}$ be non-critical, i.e. (3.4/7) does not hold for any choice of the non-negative integers n_1 and n_2. Then for $f \in B_p^{\bar{s}}(\dot{K})$ we have an obvious counterpart of (3.4/9) with $|x|_{\bar{a}}(K)$ instead of $|x|_{\bar{a}}$ and K instead of $\mathbb{R}^2$. As a consequence it is almost clear that

$$B_p^{\bar{s}}(\dot{K}) = B_p^{\bar{s}}(\mathring{K}), \qquad \bar{s} \text{ non-critical} \tag{6}$$

holds. If $\bar{s}$ is critical then (6) is not true. This shows that the anisotropic Hardy inequality (3.4/9) is optimal. As we said these spaces are helpful in order to study semi-elliptic differential operators. For this purpose boundary values of $f \in B_p^{\bar{s}}(\mathring{K})$ are of interest. Again Hardy inequalities of type (3.4/3) and (3.4/9) are of crucial importance. We formulate a special result and refer for more details to [ST, Chapter 4]. For this purpose we introduce a weighted Besov space on the circumference

$$\partial K = \{(x_1, x_2) \mid x_1^2 + x_2^2 = 1\}.$$

Let λ be the arc length on ∂K such that $\lambda = 0$ corresponds to $x^1 = (1,0)$. Then $0 \leqslant \lambda < 2\pi$ and $\lambda = \pi$ corresponds to $x^0 = (-1,0)$. Let $d(\lambda)$ be the distance on ∂K of a point on ∂K to the distinguished points x^0 and x^1, i.e.

$$d(\lambda) = \begin{cases} \min(\lambda, \pi - \lambda) & \text{if} \quad 0 \leqslant \lambda \leqslant \pi \\ \min(\lambda - \pi, 2\pi - \lambda) & \text{if} \quad \pi < \lambda < 2\pi. \end{cases}$$

Let $1 < p < \infty$, $\sigma > 0$ and $\mu \in \mathbb{R}$. Let m be a natural number with $m > \sigma$. Then $B_p^\sigma(\partial K, \mu)$ is the collection of all functions $g(\lambda)$, defined on ∂K, such that

$$\left(\int_0^{2\pi} d^{-\sigma p - \mu p}(\lambda)|g(\lambda)|^p d\lambda\right)^{1/p} + \left(\int_0^{2\pi}\int_0^{2\pi} \tau^{-\sigma p}|\Delta_\tau^m(d^{-\mu}g)(\lambda)|^p d\lambda \frac{d\tau}{\tau}\right)^{1/p} \tag{7}$$

is finite. Of course Δ_τ^m are differences of type (3.4/2) taken along ∂K, and all functions are extended 2π-periodically. The space $B_p^\sigma(\partial K, \mu)$ is independent of m, it is a Banach space, and the norms in (7) are pairwise equivalent. Recall our assumptions $0 < 2s_1 = s_2$ and $1 < p < \infty$. Furthermore let $s_1 > 1/2p$, $\sigma = s_1 - 1/2p$ and $\mu = s_1 - 3/2p$. Then the trace operator $f(x) \longrightarrow f|\partial K$ (embedding on the boundary), is a continuous mapping

$$\text{from } B_p^{\bar{s}}(\mathring{K}) \qquad \text{onto} \qquad B_p^\sigma(\partial K, \mu).$$

Furthermore, there exist a linear and continuous extension operator from $B_p^\sigma(\partial K, \mu)$ into $B_p^{\bar{s}}(\mathring{K})$.

5. More Inequalities and Their Use

5.1. Singular Integrals, Fourier Multipliers, Equivalent Norms

Up to this moment we tried to give an imagination how the classical inequalities from Section 2 via their n-dimensional (maybe also vector-valued or weighted) versions are connected with the theory of function spaces. There is one exception, the Hilbert transform (2.4/1) and the inequality (2.4/2), which will be discussed now. In some sense the Hilbert transform is the simplest one-dimensional singular integral. The corresponding n-dimensional theory started in 1952 with the paper [CZ] by A. P. Calderón and Zygmund. We formulate a typical result. Let $k(x)$ be a complex-valued function, differentiable in $\mathbb{R}^n - \{0\}$ with

$$k(tx) = t^{-n}k(x) \qquad \text{if} \qquad x \in \mathbb{R}^n - \{0\}, \quad t > 0, \tag{1}$$

$$|\nabla k(x)| \leqslant c|x|^{-n-1} \qquad \text{if} \qquad x \in \mathbb{R}^n - \{0\}, \tag{2}$$

$$\int_{|y|=1} k(y)d\omega_y = 0, \tag{3}$$

where the last integral is taken over the unit sphere with $d\omega_y$ as the surface element. Of course, c in (2) is independent of x. The crucial condition is (3). Let $1 < p < \infty$. Then

$$f(x) \longrightarrow \int_{\mathbb{R}^n} k(x-y)f(y)dy \tag{4}$$

interpreted as a singular integral, is a linear and bounded operator from $L_p(\mathbb{R}^n)$ into itself. (First (4) is defined on $D(\mathbb{R}^n)$ or $S(\mathbb{R}^n)$; the rest is a matter of completion). In the case of (2.4/1) we have $k(x) = 1/x$ and the unit sphere reduces to the points 1 and -1. Let $\hat{k}$ be the Fourier transform of k. Then the above assertion can be reformulated as follows: $\hat{k}$ is a Fourier multiplier in $L_p(\mathbb{R}^n)$, where $1 < p < \infty$. Recall what is meant by a Fourier multiplier: A L_∞ function $K(x)$ is called a Fourier multiplier if

$$f \longrightarrow (K\hat{f})^\vee \tag{5}$$

is a linear and bounded operator from $L_p(\mathbb{R}^n)$ into itself, $1 < p < \infty$, (maybe after extension from $S(\mathbb{R}^n)$ to $L_p(\mathbb{R}^n)$). Here $\check{g}$ is the inverse Fourier transform of g. Closely connected with the above singular integrals and the related Fourier multipliers are assertions for equivalent norms, say, in Sobolev spaces. Let $n = 2$ then

$$\frac{x_1 x_2}{1 + x_1^2 + x_2^2}$$

is a Fourier multiplier in $L_p(\mathbb{R}^2)$ where $1 < p < \infty$. As a consequence one has the estimate

$$\left\| \frac{\partial^2 f}{\partial x_1 \partial x_2} \,|\, L_p(\mathbb{R}^2) \right\| \leqslant$$

$$c \left(\|f \,|\, L_p(\mathbb{R}^2)\| + \left\| \frac{\partial^2 f}{\partial x_1^2} \,|\, L_p(\mathbb{R}^2) \right\| + \left\| \frac{\partial^2 f}{\partial x_2^2} \,|\, L_p(\mathbb{R}^2) \right\| \right) \tag{6}$$

for some constant c which is independent of $f \in W_p^2(\mathbb{R}^2)$, $1 < p < \infty$. Even more, generalization to $\mathbb{R}^n$ and some minor modifications yield that

$$\|f \,|\, L_p(\mathbb{R}^n)\| + \|\Delta f \,|\, L_p(\mathbb{R}^n)\| \tag{7}$$

is an equivalent norm in $W_p^2(\mathbb{R}^n)$, $1 < p < \infty$, where Δ is the Laplacian on $\mathbb{R}^n$. However, (7) and the corresponding counterparts on (smooth) domains are the backbone of the theory of elliptic differential operators. Details and many references may be found, e.g. in [ST], [Tr2] or [Tr3].

5.2. *Localization Methods*

We introduced the spaces $B_{pq}^s(\mathbb{R}^n)$ and $F_{pq}^s(\mathbb{R}^n)$ in 4.1 and 4.2 via decompositions of exponential type (4.1/2), where $\varphi_j(D)f(x)$ is an entire analytic function of exponential type. Then the theory of these function spaces is based on (maybe vector-valued or weighted) inequalities for entire analytic functions or exponential type. Unfortunately this approach obscures the local nature of some main properties of these spaces completely: In order to calculate $\varphi_j(D)f(x)$ in a given point $x \in \mathbb{R}^n$ one needs a knowledge of f on whole $\mathbb{R}^n$. In the remaining parts of this paper we wish to describe a method which, on the one hand, avoids this awkward behaviour, but on the other hand, preserves the powerful inequalities on which the study of these spaces is based. In 5.2 we give the necessary modifications compared with 4.1 and 4.2, in 5.3 we describe two inequalities and in 5.4 we mention a mapping theorem for pseudodifferential operators which is just based on these inequalities. First we follow [Tr4] and the modifications described in the survey [Tr5].

Let $B = \{y|\ |y| < 1\}$ be the unit ball in $\mathbb{R}^n$ and let k_0 and k be two C^∞ functions in $\mathbb{R}^n$ with

$$\operatorname{supp} k_0 \subset B, \quad \operatorname{supp} k \subset B, \tag{1}$$

$$\hat{k}(0) \neq 0 \qquad \text{and} \qquad \hat{k}_0(y) \neq 0 \text{ for all } y \in \mathbb{R}^n. \tag{2}$$

If N is a natural number then we put $k_N(x) = \Delta^N k(x)$, where Δ stands for the Laplacian. We introduce the means

$$k_N(t, f)(x) = \int_{\mathbb{R}^n} k_N(y) f(x + ty)\, dy, \quad x \in \mathbb{R}^n, \quad t > 0, \tag{3}$$

where now $N = 0, 1, 2, \ldots$ This makes sense for any distribution f in $\mathbb{R}^n$. Let $0 < \epsilon < \infty$ and $0 < r < \infty$ be given numbers. If $0 < p \leqslant \infty$, $0 < q \leqslant \infty$, $s \in \mathbb{R}$ and $2N > \max(s, n(1/p - 1)_+)$, then

$$\|k_0(\epsilon, f)\,|L_p(\mathbb{R}^n)\| + \left(\int_0^r t^{-sq} \|k_N(t, f)\,|L_p(\mathbb{R}^n)\|^q \frac{dt}{t}\right)^{1/q} \tag{4}$$

(modification if $q = \infty$) is an equivalent quasi-norm in $B^s_{pq}(\mathbb{R}^n)$. If $0 < p < \infty$, $0 < q \leqslant \infty$, $s \in \mathbb{R}$ and $2N > \max(s, n(1/p - 1)_+)$ then

$$\|k_0(\epsilon, f)\,|L_p(\mathbb{R}^n)\| + \left\| \left(\int_0^r t^{-sq} |k_N(t, f)(\cdot)|^q \frac{dt}{t}\right)^{1/q} \Big| L_p(\mathbb{R}^n)\right\| \tag{5}$$

(modification if $q = \infty$) is an equivalent quasi-norm in $F^s_{pq}(\mathbb{R}^n)$. We compare (4) with (4.1/3) and (5) with (4.2/1). The replacement of the discrete variable j in (4.1/3) and (4.2/1) by the continuous variable t in (4) and (5) is immaterial, this can be done as one wishes. The important point is the replacement of (4.1/2) by the means (3). It makes clear why some properties of the spaces under consideration are of local nature.

5.3. *Some Maximal Inequalities*

The basic idea is to use (5.2/4) and (5.2/5) in order to prove properties for these function spaces or the continuity of mappings between these function spaces generated by (pseudo)differential operators. For this purpose one needs some sophisticated maximal inequalities. First we recall what is meant by the Hörmander class S^0_1, δ with $0 \leqslant \delta < 1$. A C^∞ function $a(x, \xi)$ in $\mathbb{R}^n \times \mathbb{R}^n$ belongs to the symbol class $S^0_{1,\delta}$ if

$$|D^\alpha_x D^\beta_\xi a(x, \xi)| \leqslant c_{\alpha,\beta}(1 + |\xi|)^{\delta|\alpha| - |\beta|}, \quad x \in \mathbb{R}^n, \quad \xi \in \mathbb{R}^n, \tag{1}$$

holds for all multi-indices α and β. The coefficients $c_{\alpha,\beta}$ are independent of x and ξ. Let

$$x \longrightarrow a(z, D)f(x) = \int_{\mathbb{R}^n} e^{ix\xi} a(z, \xi)\hat{f}(\xi) d\xi, \; x \in \mathbb{R}^n, \; z \in \mathbb{R}^n, \tag{2}$$

be a family of special pseudodifferential operators. Here $x\xi$ is the scalar product in $\mathbb{R}^n$. Because $a \in S^0_{1,\delta}$ it follows that $\xi \longrightarrow a(z,\xi)$ is for any $z \in \mathbb{R}^n$ a Fourier multiplier in $L_p(\mathbb{R}^n)$, $1 < p < \infty$. We need an extension of this assertion to $F^s_{pq}(\mathbb{R}^n)$ combined with a maximal inequality. Let $0 < p < \infty$, $0 < q \leqslant \infty$, $s \in \mathbb{R}$, $r > 0$, $\epsilon > 0$ and let N be a large natural number. Then

$$\left\| \left(\int_0^r t^{-sq} \sup_{z\in\mathbb{R}^n} |a(z,D)k_N(t,f)(\cdot)|^q \frac{dt}{t} \right)^{1/q} |L_p(\mathbb{R}^n) \right\| +$$

$$+ \| \sup_{z\in\mathbb{R}^n} |a(z,D)k_0(\epsilon,f)\| L_p(\mathbb{R}^n)\| \leqslant c\|f|F^s_{pq}(\mathbb{R}^n)\|, \tag{3}$$

where c is independent of $f\in F^s_{pq}(\mathbb{R}^n)$. Of course, k_0 and k_N have the same meaning as in 5.2. This is a combination of a vector-valued Fourier multiplier theorem and a maximal inequality. It is almost obvious that one has a corresponding assertion with $B^s_{pq}(\mathbb{R}^n)$ instead of $F^s_{pq}(\mathbb{R}^n)$. We need a second maximal inequality. Let $0 < p < \infty$, $0 < q \leqslant \infty$, $s > n/p$ and $b > 0$. Then

$$\| \sup_{|x-y|\leqslant b} \sup_{z\in\mathbb{R}^n} |a(z,D)f(y)\| L_p(\mathbb{R}^n)\| \leqslant c\|f|F^s_{pq}(\mathbb{R}^n)\|, \tag{4}$$

where c is independent of $f\in F^s_{pq}(\mathbb{R}^n)$. Of course, the $L_p(\mathbb{R}^n)$ quasi-norm in (4) is taken with respect to x. See [Tr6] for details and references.

5.4. Pseudodifferential Operators

Let again $a(x,\xi)$ be a C^∞ function on $\mathbb{R}^n \times \mathbb{R}^n$ belonging to the Hörmander class $S^0_{1,\delta}$ with $0 \leqslant \delta < 1$ introduced in 5.3. Then the corresponding pseudodifferential operator $a(x,D)f(x)$ is obtained from (5.3/2) if one identifies z with x, i.e.

$$a(x,D)f(x) = \int_{\mathbb{R}^n} e^{ix\xi} a(x,\xi)\hat{f}(\xi)d\xi, \quad x\in\mathbb{R}^n. \tag{1}$$

Pseudodifferential operators have attracted much attention in the last two decades. Mapping properties are considered first in L_2, then also in L_p with $1 < p < \infty$ and in Hölder spaces. There are now several books dealing with this subject, e.g. [Tay]. In [Pä] the following assertion is proved: If $0 < p < \infty$, $0 < q \leqslant \infty$, $s\in\mathbb{R}$ then $a(x,D)$ is a linear and continuous operator from $F^s_{pq}(\mathbb{R}^n)$ into itself. In [Tr6] we gave a new proof of this assertion which is based on the inequalities (5.3/3), (5.3/4), and on the equivalent quasi-norms (5.2/5). We give an idea how the proof runs. First by lifting arguments (algebraic properties of the considered class of pseudodifferential operators) we may assume $s > n/p$. The crucial point is to insert (1) in (5.2/3), i.e. to replace $f(x+ty)$ by $a(x+ty,D)f(x+ty)$. One has to use (1) with $x+ty$ instead of x. The critical factor is $a(x+ty,\xi)$. But this function can be expanded in a Taylor polynomial with x as the off-point. Then x and ty are separated. The derivatives of a are under control by (5.3/1) and the factors $t^{|\gamma|}y^\gamma$ improve the situation. Now (5.3/3) are just the inequalities one needs. For the remainder term in the Taylor expansion of $a(x+ty,\xi)$ one needs an inequality of type (5.3/4). For details we refer to [Tr6]. A similar assertion holds for $B^s_{pq}(\mathbb{R}^n)$.

6. References

[BIN] O. V. Besov, V. P. Il'in, S. M. Nikol'skij. Integral representations of functions and embedding theorems. (Russian) Moskva: Nauka 1975 [English translation: Scripta Series in Math., Washington: Halsted Press; New York, Toronto, London: V. H. Winston & Sons 1978/79].

[Bo] R. P. Boas. Entire functions. New York: Academic Press 1954.

[CZ] A. P. Calderón, A. Zygmund. On the existence of certain singular integrals. Acta Math. 88 (1952), 85–139.

[FS1] C. Fefferman, E. M. Stein. Some maximal inequalities. Amer. J. Math. 93 (1971), 107–115.

[FS2] C. Fefferman, E. M. Stein. H^p spaces of several variables. Acta Math. 129 (1972), 137–193.

[Gr] P. Grisvard. Espaces intermédiaires entre espaces de Sobolev avec poids. Ann. Scuola Norm.Sup. Pisa 17 (1963), 255–296.

[Ha1] G. H. Hardy. Notes on some points in the integral calculus (LXIV). Messenger of Math. 57 (1928), 12–16.

[Ha2] G. H. Hardy. Note on a theorem of Hilbert concerning series of positive terms. Proc. L.M.S. (2), 23 (1925), Records of Proc. xlv–xlvi.

[Ha3] G. H. Hardy. Notes on special systems of orthogonal functions (IV): the orthogonal functions of Whittaker's cardinal series. Proc. Camb. Phil. Soc. 37 (1941), 331–348.

[Ha4] G. H. Hardy. The mean value of the modulus of an analytic function. Proc. L.M.S. 14 (1914), 269–277.

[HL1] G. H. Hardy, J. E. Littlewood. A maximal theorem with function-theoretic applications. Acta Math. 54 (1930), 81–116.

[HL2] G. H. Hardy, J. E. Littlewood. Some properties of fractional integrals, I; II. Math. Z. 27 (1927), 565–606; 34 (1932), 403–439.

[HLP] G. H. Hardy, J. E. Littlewood, G. Pólya. Inequalities. Cambridge: University Press 1964 (reprint of the sec. ed.; first edition: Cambridge: University Press 1934).

[Ku] A. Kufner. Weighted Sobolev spaces. Teubner-Texte Math. 31. Leipzig: Teubner 1980.

[Li] P. I. Lizorkin. Properties of functions of the spaces $\Lambda^r_{p\Theta}$. (Russian) Trudy Mat. Inst. Steklov 131 (1974), 158–181.

[Ni1] S. M. Nikol'skij. Inequalities for entire functions of finite order and their application in the theory of differentiable functions of several variables. (Russian) Trudy Mat. Inst. Steklov 38 (1951), 244–278.

[Ni2] S. M. Nikol'skij. Approximation of functions of several variables and imbedding theorems. (Russian) Sec. ed., Moskva: Nauka 1977 [English translation of the first edition: Berlin, Heidelberg, New York: Springer 1975].

[Pä] L. Päivärinta. Pseudo differential operators in Hardy-Triebel spaces. Z. Anal. Anwendungen 2 (1983), 235–242.

[Pe1] J. Peetre. Sur les espaces de Besov. C. R. Acad. Sci. Paris Sér. A-B 264 (1967), 281–283.

[Pe2] J. Peetre. Remarques sur les espaces de Besov. Le cas $0 < p < 1$. C. R. Acad. Sci. Paris, Sér, A-B 277 (1973), 947–950.

[Pe3] J. Peetre. On spaces of Triebel-Lizorkin type. Ark. Mat. 13 (1975), 123–130.

[Pe4] J. Peetre. New thoughts on Besov spaces. Duke Univ. Math. Series. Durham: Duke Univ. 1976.

[PP] M. Plancherel, G. Pólya. Fonctions entières et intégrales de Fourier multiples. Math. Helv. 9 (1937), 224–248.

[RiF] F. Riesz. Über die Randwerte einer analytischen Funktion. Math. Z. 18 (1923), 87–95.

[RiM] M. Riesz. Sur les fonctions conjuguées. Math. Z. 27 (1927), 218–244.

[So1] S. L. Sobolev. The Cauchy problem in a functional space. (Russian) Dokl. Akad. Nauk SSSR 3 (1935), 291–294.

[So2] S. L. Sobolev. Méthode nouvelle à resoudre le problème de Cauchy pour les équations linéaires hyperboliques normales. Mat. Sb. 1 (1936), 39–72.

[So3] S. L. Sobolev. On a theorem of functional analysis. (Russian) Mat. Sb. 4 (1938), 471–497.

[So4] S. L. Sobolev. Some applications of functional analysis in mathematical physics. (Russian) Leningrad: University Press 1950, Novosibirsk 1962.

[St] E. M. Stein. Singular integrals and differentiability properties of functions. Princeton: University Press, 1970.

[ST] H.-J. Schmeisser, H. Triebel. Topics in Fourier analysis and function spaces. Leipzig: Akad. Verlagsgesellschaft Geest & Portig, 1987; Chichester: Wiley 1987.

[SW1] E. M. Stein, G. Weiss. On the theory of harmonic functions of several variables. I. The theory of H^p spaces. Acta Math. 103 (1960), 25–62.

[SW2] E. M.Stein, G. Weiss. Introduction to Fourier analysis on Euclidian spaces. Princeton: University Press 1971.

[Ta] M.H. Taibleson. On the theory of Lipschitz spaces of distributions on Euclidean *n*-space. I, II. J. Math. Mechanics 13 (1964), 407–479; 14 (1965), 821–839.

[Tay] M. E. Taylor. Pseudodifferential operators. Princeton: University Press 1981.

[Tr1] H. Triebel. Spaces of distributions of Besov type on Euclidian *n*-space. Duality, interpolation. Ark. Mat. 11 (1973), 13–64.

[Tr2] H. Triebel. Interpolation theory, function spaces, differential operators. Berlin: Verlag Wissenschaft 1978; Amsterdam: North-Holland 1978.

[Tr3] H. Triebel. Theory of function spaces. Leipzig: Akad. Verlagsgesellschaft Geest & Portig 1983; Boston: Birkhäuser 1983.

[Tr4] H. Triebel. Characterizations of Besov-Hardy-Sobolev spaces: a unified approach. J. Approximation Theory 52 (1988), 162–203.

[Tr5] H. Triebel. Einige neuere Entwicklungen in der Theorie der Funktionenräume. Jahresber. Deutsch. Math.-Verein 89 (1987), 149–178.

[Tr6] H. Triebel. Pseudo-differential operators in F^s_{pq}-spaces. Z. Anal. Anwendunge 6 (1987), 143–150.

[Wi] N. Wiener. The ergodic theorem. Duke Math. J. 5 (1939), 1–18.

[Zy] A. Zygmund. Trigonometric series, I, II. Cambridge: University Press 1977. (First ed.: Warsaw 1935).

13 Differential Inequalities

Wolfgang Walter Universität Karlsruhe, Karlsruhe, Federal Republic of Germany

1. Preamble

In the widest sense differential inequalities are inequalities containing derivatives of functions. More to the point is the narrower description "inequalities derived from differential equations", and one could add "and from associated equations describing initial or boundary conditions".

There is not a single differential inequality treated in Hardy-Littlewood-Pólya, and for a very simple reason. Around 1930 a theory of differential inequalities deserving this name did not exist. There were scattered results dating from the 19th century, and there was also the work of O. Perron on existence theory by differential inequality methods which led to the Perron integral (1914), a new proof of Peano's existence theorem (1915) and his famous solution of the Dirichlet problem for the Laplace equation (1923).

Most of the work in differential inequalities is associated with ordinary differential equations of the first and second order, first order partial differential equations and elliptic and parabolic differential equations of the second order, where "equations" stands for "equations and systems". The very basic results in these fields were discovered between 1925 and 1960. Shortly afterwards, textbooks on the new subject appeared. The first four were

1964 W. Walter, Differential- und Integralungleichungen. Springer Tracts in Natural Philosophy, Vol. 2. English Translation: Differential and Integral Inequalities, Springer 1970

1965 J. Szarski, Differential Inequalities. Monografie Matematyczne, Vol. 43, Warszawa

1967 M.H. Protter and H.F. Weinberger, Maximum Principles in Differential Equations. Prentice Hall

1969 V. Lakshmikantham and S. Leela, Differential and Integral Inequalities. Academic Press (2 volumes).

Among the mathematicians who contributed to the new field, one deserves special notice: Mitio Nagumo. His work on boundary value problems for second order differential equations (1937), first order partial differential inequalities (1938), parabolic differential inequalities (1939) and invariant sets (1942) initiated mathematical theories in these fields. In the last two cases he was years ahead of his time. His paper on parabolic differential inequalities written in Japanese remained unknown for more than ten years until Westphal's article (1949) aroused interest in the subject, and around 1970 his results on invariance were rediscovered by several mathematicians.

2. Introduction

In general, an initial or boundary value problem is defined by a differential operator T acting on a function u in a set G and a boundary operator B (often the identity) acting on u in a set $\Gamma \subset \partial G$. Typical examples are $Tu = u'(t) - f(t,u(t))$ or $Tu = -\Delta u + f(x,u)$. Theorems on differential inequalities come in different forms.

Nonlinear problems. Monotonicity theorem. These theorems take two forms

(M<) $Bv < Bw$ on Γ, $Tv < Tw$ in G implies $v < w$ in G ;

(M≤) $Bv \le Bw$ on Γ, $Tv \le Tw$ in G implies $v \le w$ in G .

Linear problems. Positivity theorem. This name is given to theorems of the form

(P<) $Cu > 0$ on Γ, $Su > 0$ in G implies $u > 0$ in G ;

(P≤) $Cu \ge 0$ on Γ, $Su \ge 0$ in G implies $u \ge 0$ in G ,

where S is a linear differential operator and C a linear boundary operator. Examples for S corresponding to the examples for T given above are $Su = u'(t) + c(t)u$ or $Su = -\Delta u + c(x)u$.

In many instances, a theorem of type (M) contains a theorem of type (P) as a special case. On the other hand, it is often possible to derive (M) from (P) by considering the difference $u = w - v$.

Maximum and minimum principles. A theorem which states that a function u satisfying a differential inequality takes its minimum on Γ is called a minimum principle. For example, (P≤) with $Cu = u$ is a minimum principle for solutions of $Su \geq 0$ if S vanishes for constant functions. To see this, let $Su \geq 0$ and let α be the minimum of u on Γ. Then $v = u - \alpha$ satisfies $v \geq 0$ on Γ and $Sv \geq 0$ in G, hence $v \geq 0$ or $u \geq \alpha$ in G by (P≤).

Invariance theorems. Consider a problem for a system of m differential equations

$$Tu = 0 \text{ in } G, \quad u \text{ given on } \Gamma,$$

where $u = (u^1, \dots, u^m)$. A set $M \subset \mathbb{R}^m$ is said to be invariant with respect to this problem if

(I) $$u(\Gamma) \subset M \text{ implies } u(G) \subset M.$$

For example, in the case $m = 1$, $Cu = u$, theorem (P≤) is an invariance theorem with respect to the set $M = [0, \infty)$. Indeed, the assumption $u \geq 0$ on Γ can be written as $u(\Gamma) \subset M$, and $u \geq 0$ in G is equivalent with $u(G) \subset M$.

These considerations show that there are different ways to state the main theorems. It is often a matter of taste which formulation is chosen.

Subsolutions and supersolutions. Consider a problem

$$Tu = a \text{ in } G, \quad Bu = b \text{ on } \Gamma,$$

and assume that a monotonicity theorem (M<) holds. Then, for any solution u,

(SS<) $$\begin{cases} Tv < a \text{ in } G,\ Bv < b \text{ on } \Gamma \text{ implies } v < u, \\ Tw > a \text{ in } G,\ Bw > b \text{ on } \Gamma \text{ implies } w > u. \end{cases}$$

If (M≤) is true, then the two analogous propositions (SS≤) with equality permitted hold. In both cases, v is said to be a *subsolution* (*subfunction*, *lower function*) and w is said to be a *supersolution* (*superfunction*, *upper function*) for the problem in question. The corresponding German terms *Unterfunktion*, *Oberfunktion* were introduced by Perron in 1914. The abbreviation (SS) stands for sub- and supersolution.

The theory of differential equations employs differential inequalities in many ways and for many purposes, in particular for

- Bounds and estimation,
- Qualitative behavior (extrema, monotonicity of solutions,...),
- Uniqueness and continuous dependence on data,
- Asymptotic behavior and stability,

- Existence, (i) direct by Perron's method: $u = \sup v$ or $u = \inf w$, where v is a subsolution and w a supersolution,
  (ii) indirect: a priori bounds needed for the application of fixed point theorems or monotone iterative techniques.

3. "The method of differential inequalities"

The proofs of the basic theorems on differential inequalities in the different fields have a remarkable structural similarity. They use a two-step method which in the case of a monotonicity theorem can be described as follows.

Step I: Here, the theorem (M<) with strict inequalities is treated. In the proof, a point where $v = w$ (sometimes called *Nagumo point*) is considered, and a contradiction to the differential inequality $Tv < Tw$ is derived at this point. Typically, the result is very general, and no regularity assumptions are made.

Step II: The corresponding theorem (M≤) is reduced to (M<) by a simple procedure in which a proper *auxiliary function* is crucial.

The author's philosophy about differential inequalities can be summarized as follows.

The main results on differential inequalities are simple and elementary and should be proved accordingly. Heavier machinery, in particular existence theory, should only be used when necessary.

Let us apply these principles in the following

Example: Monotone functions. Let $\phi: J \to \mathbb{R}$ be continuous, where $J \subset \mathbb{R}$ is an interval. We write $\phi \in WI$ if ϕ is weakly increasing, i.e., if $s \le t$ implies $\phi(s) \le \phi(t)$. By the mean value theorem, $\phi' \ge 0$ in J implies $\phi \in WI$. A more sophisticated version is *Scheeffer's theorem* (1884) which states that

$$D^+\phi(t) \ge 0 \text{ in } J \setminus C \text{ (C countable) implies } \phi \in WI .$$

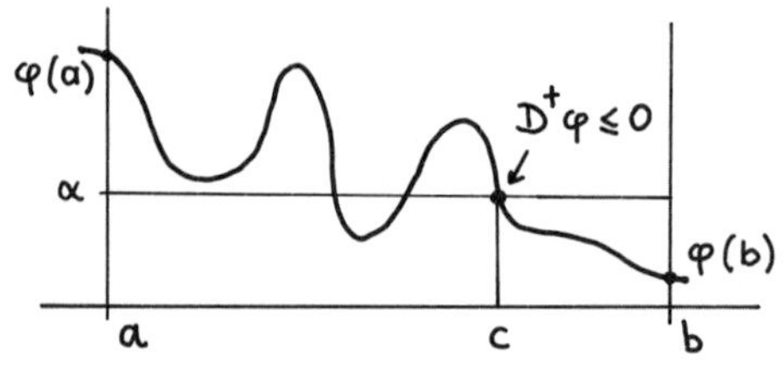

Here, D^+ denotes the right upper Dini derivative.

In step I, we show that $D^+\phi > 0$ in J implies $\phi \in WI$. If this is false, then there are points $a, b \in J$ with $a < b$ and $\phi(a) > \phi(b)$. Let

$\phi(a) > \alpha > \phi(b)$ and let c be the last (largest) point in J such that $\phi(c) = \alpha$. Since $\phi(t) < \alpha$ for $c < t < b$, right-sided difference quotients at c are negative, and hence $D^+\phi(c) \le 0$, which is a contradiction. Hence $\phi \in WI$. The same conclusion follows from the weaker assumption $D^+\phi(t) > 0$ in $J \setminus C$ (C countable), since we can vary α between $\phi(a)$ and $\phi(b)$ and obtain an uncountable set of points $c \in J$ with $D^+\phi(c) \le 0$.

In step II, we assume that $D^+\phi \ge 0$ in $J \setminus C$. The function $\phi_\varepsilon(t) = \phi(t) + \varepsilon t$ satisfies $D^+\phi \ge \varepsilon > 0$, hence $\phi_\varepsilon \in WI$ by the result of step I. Letting $\varepsilon \to 0$, the conclusion $\phi \in WI$ follows.

The reader should compare this proof (which is not far from Scheeffer's original reasoning) with the complicated proofs found in the literature (e.g. Bourbaki, I.2.2 in vol. IX).

INITIAL VALUE PROBLEMS FOR ORDINARY DIFFERENTIAL EQUATIONS

In the next four sections the problem

(IVP) $$u'(t) = f(t,u(t)) \text{ in } J = [0,T] , \quad u(0) = c$$

is considered.

4. One scalar equation (u, f real-valued)

The corresponding monotonicity theorem reads

(M<) $$v(0) < w(0),\ v' - f(t,v) < w' - f(t,w) \text{ in } J \text{ implies } v < w \text{ in } J .$$

There are no assumptions on f. The proof is very simple: If the conclusion is false and s is the first point where $v = w$, then $v(t) < w(t)$ for $t < s$ gives $v'(s) \ge w'(s)$, while the differential inequality implies $v'(s) < w'(s)$.

The corresponding theorem

(M≤) $$v(0) \le w(0),\ v' - f(t,v) \le w' - f(t,w) \text{ in } J \text{ implies } v \le w \text{ in } J$$

is not true in general, since it implies uniqueness for (IVP). If f satisfies, e.g., a Lipschitz condition with a Lipschitz constant < L, then (M<) is applicable to v and $w_\varepsilon = w + \varepsilon e^{Lt}$. Hence $v < w_\varepsilon$, and the desired inequality $v \le w$ follows by letting $\varepsilon \to 0$.

Super- and subsolutions for (IVP) are defined by

$$(SS<)\qquad \begin{cases} v' < f(t,v) \text{ in } J\,, & v(0) < c \\ w' > f(t,w) \text{ in } J\,, & w(0) > c \end{cases}$$

or (SS≤). We give four basic theorems A, B, C, D.

Theorem A. (SS<) => $v < u < w$ *for each solution* u *of* (IVP) (f *arbitrary*).

Theorem B. (SS≤) + (U) => $v \le u \le w$ *for the solution* u *of* (IVP).

Here, (U) denotes a uniqueness condition, for example a Lipschitz condition.

Theorem C. Maximal and minimal solutions. *If* f *is continuous, then there exists a maximal solution* u^* *with the properties that* $v \le u^*$ *for each continuous function* v *satisfying* $v(0) \le c$ *and* $Dv \le f(t,v)$, *(D denotes a Dini derivative) and a minimal solution* u_* *with similar properties.*

Theorem D. Perron's existence theorem. *If two continuous functions* ϕ, ψ *satisfying* $\phi \le \psi$, $D\phi \le f(t,\phi)$ *and* $D\psi \ge f(t,\psi)$ *in* J, $\phi(0) \le c \le \psi(0)$ *exist and if* f *is continuous in the set* $K = \{(t,x):\ t \in J,\ \phi(t) \le x \le \psi(t)\}$, *then there exists a solution* u *in* J *with* $\phi \le u \le \psi$.

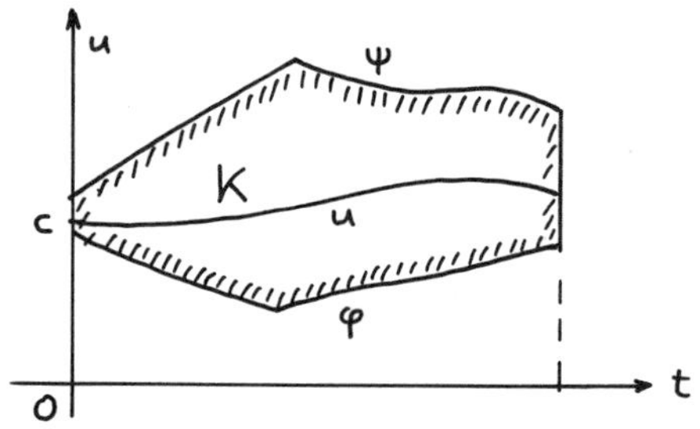

Remarks. Theorems A and B follow immediately from (M), Theorem C uses existence theory (the solutions w_n of $w_n(0) = c + \frac{1}{n}$, $w_n' = f(t,w_n) + \frac{1}{n}$ form a decreasing sequence with limit u^*), and the existence of u^* is only guaranteed in a right neighborhood of 0. In contrast, Theorem D is a global theorem, the solution u exists in J. The proof uses two new ideas, (i) a cut-off and extension procedure, by which f is defined outside K by

$$f(t,x) = \begin{cases} f(t,\psi(t)) \text{ for } x > \psi(t) \\ f(t,\phi(t)) \text{ for } x < \phi(t)\,, \end{cases}$$

and (ii) the definition $u = \sup v(t)$, where v is any "Unterfunktion", i.e., $v \in C(J)$, $v(0) \le c$, $Dv \le f(t,v)$ in J. The cut-off technique has become a standard procedure in nonlinear existence theory, e.g., for elliptic equations and systems. Let us remark in passing that Theorem B remains true for continuous functions v, w with the derivative replaced by a Dini derivative.

5. Systems of first order equations

We consider the problem (IVP), where now $u = (u_1,\dots,u_m) \in \mathbb{R}^m$ and $f(t,x) = (f_1,\dots,f_m): D \subset \mathbb{R}^{m+1} \to \mathbb{R}^m$. Inequalities in $\mathbb{R}^m$ are defined componentwise, that is

$$x \le y \iff x_i \le y_i \quad \text{for} \quad i = 1,\dots,m ,$$

$$x < y \iff x_i < y_i \quad \text{for} \quad i = 1,\dots,m .$$

Theorems A - D are now well defined propositions, and the basic problem is to find conditions on f such that they hold true.

Theorems A - D *carry over to systems if (and only if)* f(t,x) *is quasimonotone increasing in* x.

The notion of quasimonotonicity, which also governs the behavior of first order systems of partial differential equations and second order parabolic and elliptic systems, is defined as follows.

Quasimonotonicity. The function $f: D \subset \mathbb{R}^{m+1} \to \mathbb{R}^m$ is said to be quasimonotone increasing in x, if f_i is increasing in x_j for all $i \neq j$, more exactly, if

$$(t,x),(t,y) \in D,\ x \le y,\ x_i = y_i \implies f_i(t,x) \le f_i(t,y) .$$

Results close to Theorems A, B and D for systems with a quasimonotone right-hand side were discovered by Max Müller (1926, 1927), while Theorem C was proved by E. Kamke (1932). The term quasimonotone was first used in the book by Walter (1964). We have put the "only if" case in brackets, because the construction of a counterexample uses existence theory. If f is such that (IVP) has a unique solution, then (M≤) holds if and only if f is quasimonotone increasing in x.

M. Müller showed (1927) that in the general case, where a monotonicity theorem is not available, upper and lower bounds for the solution can still be described by inequalities. His theorem uses a condition (SSM) (sub- and supersolutions of the Müller type) which is stronger than the earlier condition (SS), since (i) it uses a subsolution v and a supersolution w simultaneously, while in the earlier theorems the conditions on v and w were independent of each other, and (ii) there is a sharper requirement in the differential inequality. The condition reads

$$
(\mathrm{SSM}<) \qquad \begin{aligned} & v_i' < f_i(t,x) \quad \text{for} \quad v(t) \le x \le w(t),\ x_i = v_i(t) \ , \\ & w_i' > f_i(t,x) \quad \text{for} \quad v(t) \le x \le w(t),\ x_i = w_i(t) \ , \\ & v_i(0) \le c_i \le w_i(0) \quad (i = 1, \dots, m) \ . \end{aligned}
$$

In the case $m = 2$ the set described by $v(t) \le x \le w(t)$ (t fixed) is a rectangle with lower left corner $v(t)$ and upper right corner $w(t)$, and in each of the four differential inequalities x varies on a side (and not in the interior) of this rectangle.

M. Müller's results can be summarized in the statement:

Theorems A, B *and* D *remain true for arbitrary systems when the conditions* (SSM<) *and* (SSM≤) *are used.*

Theorem C cannot be transferred to the general case since maximal and minimal solutions do not exist.

The proofs are similar to the earlier ones. For example, if the assertion "$v < u < w$ in J" in Theorem A is not true, then there is a first point s where one of the $2m$ inequalities is violated, i.e., $v(t) < u(t) < w(t)$ for $0 \le t < s$, $v_i(s) = u_i(s)$ or $u_i(s) = w_i(s)$ for some index i. Then a contradiction is obtained as in Section 4.

The conditions in (SSM) become simpler when f has monotonicity properties. We give an

Example. The problem

$$u'' + u = g(t) \ , \qquad u(0) = c_1 \ , \qquad u'(0) = c_2$$

is, when transformed into a first order system $u_1' = u_2$, $u_2' = -u_1 + g(t)$ not quasimonotone. Conditions for a subfunction v and a superfunction w are given by

$$
(\mathrm{SSM}\le) \qquad \begin{cases} v'' + w \le g(t) \ , & v(0) \le c_1 \le w(0) \ , \\ w'' + v \ge g(t) \ , & v'(0) \le c_2 \le w'(0) \ , \end{cases}
$$

and the conclusion is

$$v \le u \le w \quad \text{and} \quad v' \le u' \le w' \quad \text{in } J \ .$$

Theorems of the M. Müller type also hold for elliptic and parabolic systems, as we shall see in later sections.

6. Ordinary differential equations in ordered Banach spaces

Infinite systems of ordinary differential equations were considered by H. von Koch as early as 1899 and later by E.H. Moore in 1906 and others.

For an infinite system $u' = f(t,u)$, more explicitly (I index set)

$$u_i'(t) = f_i(t,u)\ , \quad u_i(0) = c_i \quad (i \in I,\ \text{card}\ I = \infty)\ ,$$

there is a natural order relation ($x \le y$ iff $x_i \le y_i$ for all $i \in I$) and a corresponding definition of quasimonotonicity ($f(t,x)$ is quasimonotone increasing in x if $x \le y$, $x_i = y_i$ implies $f_i(t,x) \le f_i(t,y)$). In this framework, Mlak and Olech (1963) have proved a theorem C on maximal and minimal solutions, and Walter (1969) has generalized Theorems A and B to the space $\ell_\infty(I)$.

An interesting example is obtained when the Cauchy problem for a parabolic differential equation

$$u_t = u_{xx} + f(t,x,u) \quad \text{in } J \times \mathbb{R}\ , \qquad u(0,x) = \phi(x) \quad \text{for } x \in \mathbb{R}$$

is discretized by the so-called longitudinal line method, where only the space variable x is discretized. Let $h > 0$ be the mesh size and $x_i = hi$ for $i \in \mathbf{Z}$. When the derivative u_{xx} is replaced by the corresponding central difference quotient, the following initial value problem for the functions $u_i(t) \approx u(t,x_i)$

$$u_i'(t) = \frac{1}{h^2}(u_{i+1} + u_{i-1} - 2u_i) + f(t,x_i,u_i),\ u_i(0) = \phi(x_i) \quad (i \in \mathbb{Z})$$

is obtained. This system is quasimonotone increasing, and corresponding theorems A - D have been established in various spaces. In this way, results on the boundedness and convergence as $h \to 0$ and furthermore existence theorems for nonlinear parabolic equations based on the line methods have been obtained by several authors. In particular, an existence proof for the boundary layer equations in fluid dynamics has been found using these ideas. See [W; Sections 35 and 36], Walter (1970, 1974), Lemmert (1977) for more details and further literature.

Now we turn to the general case where B is a real Banach space and C a closed cone in B, i.e., a closed subset of B with the property that $x,y \in C$ implies $x+y \in C$ and $\lambda x \in C$ for $\lambda \ge 0$. The set C induces an order relation in B,

$$x \le y \quad \Longleftrightarrow \quad y-x \in C\ ,$$

which has the usual properties ($x \le y$, $y \le z$ implies $x \le z$; $x \le y$ implies $\lambda x \le \lambda y$ for $\lambda \ge 0$; $x_n \le y$, $x = \lim x_n$ implies $x \le y$). The set C is sometimes called the positive cone since it contains all elements $x \ge 0$ and only these. If C has a nonempty interior C^o, then strict inequalities can be defined by

$$x < y \quad \Longleftrightarrow \quad x-y \in C^o\ .$$

Monotonicity theorems use the following concept of

Quasimonotonicity. The function $f(t,x)\colon D \subset J \times B \to B$ is said to be quasimonotone increasing in x if $(t,x),(t,y) \in D$,

$$\phi \in B',\ \phi(C) \geq 0,\ x \leq y,\ \phi(x) = \phi(y) \text{ implies } \phi(f(t,x)) \leq \phi(f(t,y)) .$$

Here, B' is the dual space of B, hence ϕ is a continuous, positive linear functional. In the case of $\mathbb{R}^m$ with the natural order introduced in the previous section, the positive linear functionals are given by $\phi = \sum \alpha_i \phi_i$ with $\alpha_i \geq 0$ and $\phi_i(x) = x_i$. The condition $x \leq y$, $\phi(y-x) = 0$ then becomes $x_j \leq y_j$ for all j and $x_i = y_i$ for all i with $\alpha_i > 0$, and it is easily seen that the above condition reduces to the quasimonotone condition of Section 5.

The general quasimonotonicity condition was discovered by P. Volkmann (1972). Different versions of Theorems A - D and of corresponding theorems of the M. Müller type were given by Volkmann, K. Deimling, R. Lemmert and others. Results prior to 1977 can be found in the book by Deimling (1977). A simple and elementary approach in the case where C^o is nonempty is found in Redheffer and Walter (1986).

7. Uniqueness and continuous dependence of the solution

We consider the general case of a differential equation $u' = f(t,u)$ in a Banach space B with norm $|\cdot|$. Usually uniqueness is derived from a Lipschitz condition or a more general condition

$$\text{(N)} \qquad |f(t,x)-f(t,y)| \leq \omega(t,|x-y|) ,$$

where $\omega(t,z)$ is a real-valued function (defined for $t \in J$ and $z \geq 0$) with certain properties. It is important that instead of the norm condition (N) a much weaker *condition of dissipative type*

$$\text{(D)} \qquad (x-y, f(t,x)-f(t,y)) \leq |x-y|\omega(t,|x-y|)$$

suffices. Here, $(\cdot,\cdot)$ is the inner product when B is a Hilbert space (in particular in the case $B = \mathbb{R}^m$) and the generalized inner product $(\cdot,\cdot)_+$ when B is a Banach space; see Deimling (1977; § 3.1). We recall that a linear operator A in a Hilbert space is called dissipative if $(Ax,x) \leq 0$. Hence a function $f(t,x) = A(t)x + b(t)$ with linear dissipative operators A(t) satisfies (D) with $\omega = 0$ (if A(t) is unbounded, an estimate (N) does not hold). In the real-valued case ($m = 1$), (D) is equivalent with a one-sided condition

$$f(t,x)-f(t,y) \leq \omega(t,x-y) \quad \text{for} \quad x > y .$$

For example, it holds with $\omega = 0$ whenever $f(t,x)$ is weakly decreasing in x.

<u>Estimation theorem.</u> *Assume that* u *is a solution of* (IVP) *and* v *an approximate solution satisfying*

$$|v'-f(t,v)| \le \delta(t) \text{ in } J \text{ and } |v(0)-c| \le \varepsilon .$$

If $\rho(t): J \to \mathbb{R}$ *has the properties*

$$\rho' > \omega(t,\rho) + \delta(t) \text{ in } J , \quad \rho(0) > \varepsilon ,$$

then

$$|u - v| < \rho \quad \text{in } J .$$

There are no assumption on ω and δ. The proof uses the fact that the function $\sigma = |u-v|$ satisfies $\sigma\sigma' = (u-v,u'-v')$, whence $\sigma' \le \omega(t,\sigma) + \delta$ and $\sigma(0) \le \varepsilon$ follows. Theorem A of Section 4 (applied to $\omega+\delta$ instead of f) gives the desired result $\sigma < \rho$.

A function ω belongs to the class UCD, $\omega \in$ UCD, if to each $\varepsilon > 0$ there corresponds a number $\delta > 0$ and a differentiable function ρ such that

$$\rho' > \omega(t,\rho) + \delta \quad \text{and} \quad 0 < \delta < \rho < \varepsilon \quad \text{in } J .$$

An inequality (D) (or (N)) with $\omega \in$ UCD is called a uniqueness condition. The letters UCD stand for uniqueness and continuous dependence. For example, the Lipschitz function $\omega(t,z) = Lz$ belongs to UCD. More generally, $\omega \in$ UCD if ω is continuous and nonnegative in $J \times [0,\infty)$, $\omega(t,0) \equiv 0$ and if $\sigma(t) \equiv 0$ is the only solution of $\sigma' = \omega(t,\sigma)$ in J, $\sigma(0) = 0$.

<u>Theorem on uniqueness and continuous dependence.</u> *If* f *satisfies the uniqueness condition* (D) *with* $\omega \in$ UCD, *then* (IVP) *has at most one solution* u, *and it depends continuously on* c *and* f, *i.e., to* $\varepsilon > 0$ *there exists* $\delta > 0$ *such that for every differentiable function* v *with the properties* $|v(0)-c| < \delta$ *and* $|v'-f(t,v)| < \delta$ *the inequality* $|v-u| < \varepsilon$ *in* J *holds.*

This follows immediately from the above estimation theorem.

The idea of using instead of an explicit uniqueness condition such as a Lipschitz or Osgood condition a more general condition which is characterized by certain properties of the "Ersatzproblem" $\sigma' = \omega(t,\sigma)$, was first introduced by Bompiani (1925) and Perron (1926). Our condition is essentially Perron's condition.

There are more general uniqueness conditions than our condition with $\omega \in$ UCD, for example Kamke's condition (1930; Section 78). But we emphasize that with respect to applications uniqueness alone is not sufficient. Each mathematical model contains data which are only approximately known (material constants, measurements,...). In order to make the model meaningful we must therefore require that small changes in the data affect the solution only slightly, i.e. that the solution depends continuously on the data. This property follows from Kamke's condition only when f is continuous.

INTERMEZZO

Around 1932 the theory of differential inequalities related to the system $u' = f(t,u)$ was rather complete. As we know today, the extension of this theory to parabolic equations and systems is quite natural. Yet it took time and effort to discover this fact. In the historical development, first order partial differential equations came first, serving as kind of an intermediate station between ordinary and parabolic equations.

The following estimate was given by A. Haar (1928):

$$|u_t| \le A \sum |u_{x_i}| + B|u| + C \text{ in } G , \qquad |u(0,x)| \le D$$

implies

$$|u(t,x)| \le \frac{C}{B}(e^{Bt}-1) + De^{Bt} ,$$

where G is a truncated pyramid with a slope determined by A and $u = u(t,x) = u(t,x_1,\dots,x_n)$.

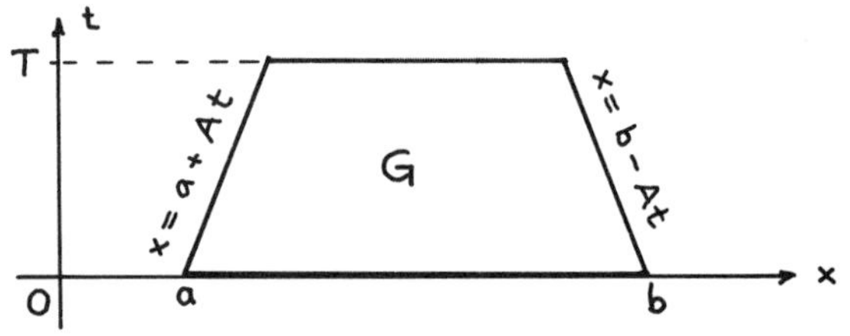

More in the spirit of differential inequalities is Nagumo's monotonicity theorem (1938):

$$v_t - f(t,x,v,v_x) < w_t - f(t,x,w,w_x) \text{ in } G , \quad v(0,x) < w(0,x)$$

implies

$$v < w \quad \text{in } G ,$$

if

$$|f(t,x,z,p)-f(t,x,z,p')| \le A \sum |p_i - p_i'| .$$

It is quite obvious that this result and the underlying method of proof led Nagumo to his famous lemma for nonlinear parabolic inequalities (1939).

A similar development occurred in parabolic systems. J. Szarski (1948) extended Nagumo's monotonicity theorem to weakly coupled first order systems

$$u_t^i = f_i(t,x,u,u_x^i) \qquad (i = 1,\dots,m) ,$$

where $f(t,x,z,p)$ is assumed to be quasimonotone increasing in z. Several years later he and other Polish mathematicians established the theory of systems of parabolic differential inequalities.

PARABOLIC EQUATIONS AND SYSTEMS

8. Preliminaries and notations

The set of real symmetric $n \times n$ matrices $a = (a_{ij})$, $r = (r_{ij}),\dots$ is denoted by S^n. If a is positively semidefinite, we write $a \geq 0$,

$$a \geq 0 \iff \xi^T a \xi \equiv \sum a_{ij}\xi_i\xi_j \geq 0 \quad \text{for all} \quad \xi \in \mathbb{R}^n .$$

Here and below, the indices i, j run from 1 to n.

This notion generates an order relation in S^n:

For $r,s \in S^n$ we write $r \leq s$ iff $s-r \geq 0$.

For example (I_n denotes the $n \times n$ unit matrix),

$$a \geq \alpha I_n \iff \xi^T a \xi = \sum a_{ij}\xi_i\xi_j \geq \alpha \sum \xi_i^2 .$$

If this inequality holds with $\alpha > 0$, then a is positive definite. We remark that S^n is an ordered Banach space of dimension $n(n+1)/2$ in the sense of Section 6. The positive cone consists of all elements $a \geq 0$, and $a > 0$ iff $a \in C^o$.

A function $f: S^n \to \mathbb{R}$ is said to be (weakly) increasing if

$r,s \in S^n$, $r \leq s$ implies $f(r) \leq f(s)$.

In particular,

(a) $f(r) = \sum a_{ij}r_{ij}$ is increasing whenever $a \geq 0$.

This is a consequence of

Schur's lemma. $a,b \in S^n$, $a \geq 0$, $b \geq 0$ implies $\sum a_{ij}b_{ij} \geq 0$.

Parabolic differential equations. Let $t \in \mathbb{R}$, $x = (x_1, \ldots, x_n) \in \mathbb{R}^n$ and let $u(t,x)$ be a real-valued function. The gradient and the Hessian of u with respect to x are denoted by u_x and u_{xx},

$$u_x = (u_{x_1}, \ldots, u_{x_n}) , \qquad u_{xx} = (u_{x_i x_j}) \in S^n .$$

The differential equation

$$\text{(b)} \qquad F(t,x,u,u_t,u_x,u_{xx}) = 0$$

is called *parabolic* if $F(t,x,z,q,p,r)$ (with $p \in \mathbb{R}^n$, $r \in S^n$) is decreasing in q and increasing in r, i.e., if $q \geq q'$, $r \leq r'$ implies

$$F(t,x,z,q,p,r) \leq F(t,x,z,q',p,r') .$$

In particular, the explicit equation

$$\text{(c)} \qquad u_t = f(t,x,u,u_x,u_{xx})$$

is parabolic if $f(t,x,z,p,r)$ is increasing in r in the sense described above. For the linear equation we use the notation

$$\text{(d)} \qquad u_t = au_{xx} + bu_x + cu ,$$

where

$$au_{xx} := \sum a_{ij} u_{x_i x_j} , \qquad bu := \sum b_i u_{x_i} .$$

The coefficients a, b, c are functions of t and x, and it is always assumed that $a \in S^n$ and $a \geq 0$. Under this condition, the equation is parabolic due to (a). The linear equation is called *strongly parabolic* in a region G if there exists $\alpha > 0$ such that $a(t,x) \geq \alpha I_n$ for all $(t,x) \in G$.

Parabolic interior and parabolic boundary. A lower half-neighborhood U_- of $(\bar{t},\bar{x}) \in \mathbb{R}^{m+1}$ is the set of all (t,x) satisfying $0 < \bar{t}-t < \varepsilon$, $|x-\bar{x}| < \varepsilon$ ($\varepsilon > 0$). For a closed set $\bar{G} \subset \mathbb{R}^{m+1}$, the parabolic interior G consists of all points of $\bar{G}$ which have a lower half-neighborhood $U_- \subset \bar{G}$. The parabolic boundary Γ is the set $\bar{G} \setminus G$.

We assume that $\bar{G}$ is bounded below by 0 in the t-direction. If $\bar{G}$ is a cylinder $[0,T] \times \bar{D}$, where D is open in $\mathbb{R}^m$, then

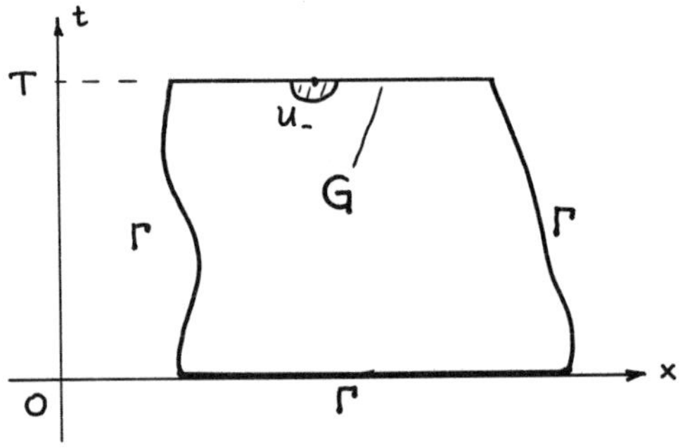

$$G = (0,T] \times D , \qquad \Gamma = \{0\} \times \bar{D} \cup (0,T] \times \partial D .$$

The function $u(t,x): \bar{G} \to \mathbb{R}$ is said to be *admissible*, $u \in Z = Z(G)$ if u is continuous in $\bar{G}$ and u_t, u_x, u_{xx} are continuous in G. The functions u, v, w,.. in our theorems are always assumed to be admissible.

(e) *Minimum with respect to* x. Assume that, for fixed $t = t_o$, u has a minimum with respect to x at x_o, i.e., $u(t_o,x) \geq u(t_o,x_o)$ with $(t_o,x_o) \in G$. Then $u_x = 0$ and $u_{xx} \geq 0$ and hence $au_{xx} + bu_x \geq 0$ at (t_o,x_o). Here, Schur's lemma is used.

9. Nonlinear parabolic differential inequalities

The first landmark result in this field is

Nagumo's lemma (1939, 1951). Assume that F is parabolic, G bounded and $v,w \in Z$. Then $F(t,x,v,v_t,v_x,v_{xx}) < F(t,x,w,w_t,w_x,w_{xx})$ in G, $v < w$ on Γ implies $v < w$ in $\bar{G}$.

Actually, Nagumo proved a more general statement for bounded or unbounded domains. For proof, assume that $u = w-v$ is not positive in $\bar{G}$. Then there is a maximal t_o such that $u > 0$ for $t < t_o$, and a point x_o with $u(t_o,x_o) = 0$. Since $v < w$ on Γ, we have $(t_o,x_o) \in G$ and $u(t_o,x) \geq 0$ for $(t_o,x) \in G$. By 8 (e) we have $u_x = 0$ and $u_{xx} \geq 0$ at (t_o,x_o). Since $u(t,x_o)$ is positive for $t < t_o$, we also have $u_t(t_o,x_o) \leq 0$. These inequalities, which are equivalent to $v = w$, $v_x = w_x$, $v_{xx} \leq w_{xx}$, $v_t \geq w_t$, are incompatible with the differential inequality; see the definition of parabolicity.

We consider the first boundary value problem for equation 8 (c)

$$(P_c) \qquad u_t = f(t,x,u,u_x,u_{xx}) \text{ in } G , \quad u = \phi \text{ on } \Gamma$$

and assume that G is bounded and $t \in J = [0,T]$ for $(t,x) \in G$. Sub- and supersolutions are defined by

$$(SS<) \qquad \begin{cases} v_t < f(t,x,v,v_x,v_{xx}) \text{ in } G , & v < \phi \text{ on } \Gamma, \\ w_t > f(t,x,w,w_x,w_{xx}) \text{ in } G , & w > \phi \text{ on } \Gamma. \end{cases}$$

Since Nagumo's lemma applies to (P_c), we get the same theorems as in Section 4:

Theorem A. (SS<) => $v < u < w$ *in* $\bar{G}$.

Theorem B. (SS≤) + (U) => $v \leq u \leq w$ *in* $\bar{G}$.

Here, (U) is a uniqueness condition given in the next theorem.

A special case of Theorem B was proved by Westphal (1949). Mlak (1960) has given a theorem of type C on maximal and minimal solutions.

The theorem of Section 7 has the following counterpart for parabolic equations.

Estimation theorem. *Let* $u,v \in Z$, *where* u *is a solution of* (P_c) *and*

$$|v_t - f(t,x,v,v_x,v_{xx})| \le \delta(t) \text{ in } G, \quad |v-\phi| \le \varepsilon(t) \text{ on } \Gamma,$$

(U) $$f(t,x,z,v_x,v_{xx}) - f(t,x,z',v_x,v_{xx}) \le \omega(t,|z-z'|) \quad \text{for} \quad z > z'.$$

Then

$$\rho'(t) > \omega(t,\rho) + \delta(t) \text{ and } \rho(t) > \varepsilon(t) \text{ in } J$$

implies

$$|u(t,x) - v(t,x)| < \rho(t) \quad \text{in } \bar{G}.$$

This theorem can be proved by showing that $v+\rho$ is a supersolution and $v-\rho$ a subsolution for (P_c) and applying Theorem A.

Corollary. *If* (U) *holds with* $\omega \in$ UCD *(see Section 7), then problem* (P_c) *has at most one solution, and it depends continuously on* ϕ *and* f.

10. The linear case

We consider the linear equation 8 (d), which we write in the form

$$u_t = Lu + cu, \quad \text{where} \quad Lu = au_{xx} + bu_x.$$

A short outline on methods of proof will be given. There are, apart from $a \ge 0$, no conditions on the coefficients unless explicitly stated. We start with two positivity theorems for *bounded regions* G.

(P<) $$w > 0 \text{ on } \Gamma, \; w_t > Lw + cw \text{ in } G \Rightarrow w > 0 \text{ in } G.$$

(P≤) $$w \ge 0 \text{ on } \Gamma, \; w_t \ge Lw + cw \text{ in } G \Rightarrow w \ge 0 \text{ in } G, \text{ if } \sup c < \infty.$$

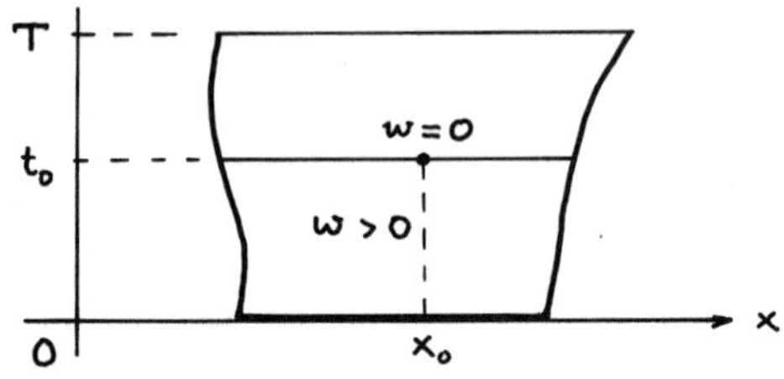

For proof, we assume that w is not positive in G. Then there exists a point $(t_o,x_o) \in G$ with the properties $w(t_o,x_o) = 0$ and $w > 0$ for $t < t_o$. At the point (t_o,x_o) we have $Lw \ge 0$ by 8 (e)

and $cw = 0$, hence $u_t(t_o, x_o) > 0$. But this contradicts the fact that $w(t,x_o) > 0$ for $t < t_o$ and $w(t_o, x_o) = 0$. (P≤) is reduced to (P<) by considering $w_\varepsilon = w + \varepsilon e^{\alpha t}$, where $\alpha > c(t,x)$.

For unbounded domains G (lying in the half-space $t \geq 0$), in particular for the Cauchy problem, a positivity theorem holds only in certain growth classes and with restrictions on the coefficients which are tied to the growth class. We consider the case where w grows not faster than $\exp(Kx^2)$ for $|x| \to \infty$. Here and below we use the notation $x^2 := \sum x_i^2$ and also $|a|$, $|b|$ for norms in S^n or $\mathbb{R}^n$.

(P≤, G *unbounded*) *If, for some constant* K,

$$|a| \leq K, \quad |b| \leq K(1+|x|), \quad c \leq K(1+x^2), \quad |w| \leq K\exp(Kx^2),$$

then propostion (P≤) *given above is true.*

The proof by reduction to (P≤) uses the auxiliary functions $w_\varepsilon = w + \varepsilon\rho$, $\rho(t,x) = \exp(At + (B+Ct)x^2)$. By choosing $B > K$, the set where $w_\varepsilon \leq 0$ becomes bounded due to the growth condition. The differential inequality holds for w_ε in strips $t_o \leq t \leq t_o + \delta$, $\delta = K/C$, if A and C are properly chosen, and one treats the strips $0 \leq t \leq \delta$, $\delta \leq t \leq 2\delta, \ldots$ successively.

Preliminary results of this kind were given by Holmgren in 1924, Picone in 1929, and there is a famous counterexample by Tychonoff (1935) which shows that the Cauchy problem for the heat equation $u_t = u_{xx}$ with zero initial data has nonvanishing solutions (which do not belong to the growth class given above).

<u>Boundary minimum principle.</u> Assume that $w \in Z(S)$,

$$w_t \geq Lw + cw \quad \text{and} \quad w \geq 0 \quad \text{in } S,$$

where S is a truncated paraboloid given by the inequalities $(R, B > 0)$

$$S: \ |x-x_o|^2 - R^2 < B(t-t_o) \leq 0,$$

a, b, c are bounded and $a \geq \alpha I_n$ in S $(\alpha > 0)$. If $w(t_o, x_o) > 0$, there is $\varepsilon > 0$ such that

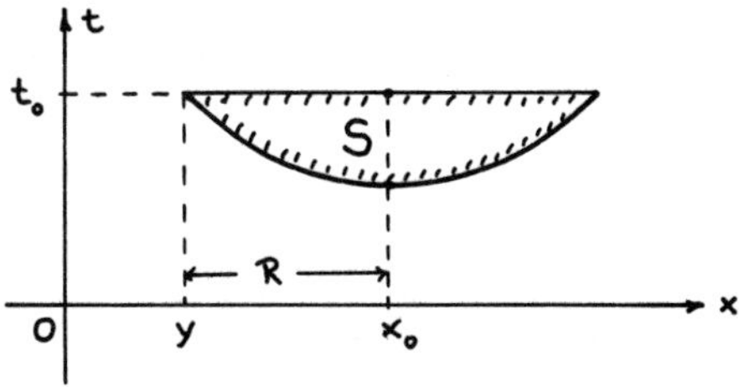

$$w(t_o, x) \geq \varepsilon(R^2 - |x-x_o|^2) \quad \text{for} \quad |x-x_o| \leq R.$$

Hence, if w vanishes at a point (t_o, y) with $|y-x_o| = R$, then

$$\frac{\partial w}{\partial \nu}(t_o, y) < 0, \quad \text{where} \quad \nu = (y-x_o)/R$$

is the outer normal at (t_o, y) (if the derivative exists).

The proof uses the function $v = \exp A(R^2 - x^2 + Bt) - 1$ (we assume $(t_o, x_o) = 0$) which is positive in S and vanishes on the parabolic boundary of S. The function v satisfies $v_t \leq Lv + cv$ for $|x| \geq \delta > 0$, if A and B are properly chosen, and (P≤), applied to the set of those points in S with $|x| \geq \delta$, gives $\varepsilon v \leq w$ in S for small $\varepsilon > 0$.

This theorem, which is the parabolic analogue of Hopf's well-known boundary minimum principle for elliptic equations, was discovered by Vyborny (1957) and Friedman (1958). It is a very useful tool and seems not to be generally known.

Strong minimum principle. Assume that $w \in Z(G)$, G bounded or unbounded,

$$w_t \geq Lw + cw \quad \text{and} \quad w \geq 0 \quad \text{in } G ,$$

where a, b, c are locally bounded and $a \geq \alpha I_n$ locally in G ($\alpha > 0$).

If w vanishes at a point $(t_o, x_o) \in G$, then $w \equiv 0$ below t_o, i.e. at all points $(t,x) \in G$ which can be connected with (t_o, x_o) by a polygonal line lying in G along which t increases. In the case of a cylinder $G = (0,T] \times \Omega$, the statement reads

$$w(t_o, x_o) = 0 \Longrightarrow w \equiv 0 \quad \text{in} \quad [0, t_o] \times \bar{\Omega} .$$

Nirenberg (1953) has proved this important principle for a more general type of equations (with several time variables). A simple proof is described in Walter (1985). An even simpler proof uses the boundary minimum principle.

Normal derivatives. We consider boundary value problems, where on part of the parabolic boundary, say, on $\Gamma_o \subset \Gamma$, the values of the solution are given, while on the rest $\Gamma_\nu = \Gamma \setminus \Gamma_o$ the normal derivative $w_\nu \equiv \partial w/\partial \nu$ with respect to an *outer* normal ν or a combination $w_\nu + dw$ is prescribed. In the case of a cylinder $G = (0,T] \times \Omega$, one has often (but not always) $\Gamma_o = \{0\} \times \bar{\Omega}$ and $\Gamma_\nu = (0,T] \times \partial\Omega$.

The earlier positivity theorems carry over to this case. For example, (P<) remains valid if the assumption $w > 0$ on Γ is replaced by $w > 0$ on Γ_o, $w_\nu + dw > 0$ on Γ_ν. There is no assumption on $d = d(t,x)$. The same is true for (P≤), if a and b are bounded, $\sup c < \infty$, $\inf d > -\infty$, and if Γ_ν is *regular*, which means that there exists a smooth function $\phi(x)$ such that

$$|\phi| + |\phi_x| + |\phi_{xx}| < K \text{ in } G , \quad \frac{\partial \phi}{\partial \nu} \geq 1 \text{ on } \Gamma_\nu .$$

This is also true for unbounded domains G if the earlier conditions are imposed and if Γ_ν is bounded.

The earlier proof of (P<) has to be amended. If $w(t_o, x_o) = 0$, where $(t_o, x_o) \in \Gamma_\nu$, then $w_\nu(t_o, x_o) > 0$ in contradiction to the inequality $w(t_o, x) \geq 0$.

Mildly nonlinear equations. Consider the nonlinear problem

(P_{nl}) $\quad u_t = Lu + g(t,x,u)$ in G , $\quad u = \phi$ on Γ_o , $\quad u_\nu + \theta(t,x,u) = 0$ on Γ_ν .

Using the operators

$$Pu = u_t - Lu - g(t,x,u) \quad \text{and} \quad Ru = \begin{cases} u - \phi & \text{on } \Gamma \\ u_\nu + \theta(t,x,u) & \text{on } \Gamma_\nu \end{cases},$$

problem (P_{nl}) can be written $Pu = 0$ in G, $Ru = 0$ on Γ.

We have the following

Monotonicity theorem. *Assume that* G *is bounded,* Γ_ν *regular,* $g(t,x,u)$ *and* $\theta(t,x,u)$ *are locally Lipschitz continuous in* u, a *and* b *are bounded. Then*

$$Pv \le Pw \text{ in } G, \quad Rv \le Rw \text{ on } \Gamma \quad \text{implies} \quad v \le w \text{ in } \bar{G} .$$

This is also true for unbounded G *if* $|v|, |w| \le K \exp(Kx^2)$.

This is proved by applying (P≤) to the difference $u = w - v$. From this theorem one derives easily theorems on uniqueness, sub- and super-functions, continuous dependence and also maximum or minimum principles. For example,

$$u_t = Lu + g(u) \text{ in } G, \quad u \le \alpha \text{ on } \Gamma_o, \quad u_\nu = 0 \text{ on } \Gamma_\nu, \quad g(\alpha) = 0$$

implies $u \le \alpha$ in G and, if $u(t_o,x_o) = \alpha$ with $(t_o,x_o) \in G \cup \Gamma_\nu$, then $u \equiv \alpha$ below t_o (in the sense defined in the strong minimum principle).

11. Weakly coupled parabolic systems

We consider the boundary value problem for $u(t,x) = (u^1, \dots, u^m)$,

$$\left.\begin{aligned} u^i_t &= f_i(t,x,u,u^i_x,u^i_{xx}) \quad \text{in } G , \\ u^i &= \phi^i \quad \text{on } \Gamma_o , \quad u^i_\nu + \theta_i(t,x,u^i) = 0 \quad \text{on } \Gamma_\nu \end{aligned}\right\} \quad (i = 1, \dots, m) ,$$

where $f_i(t,x,z,p,r)$ is weakly increasing in $r \in S^n$ in the sense described in Section 8. Γ_o, Γ_ν and $u_\nu = \partial u/\partial\nu$ have the same meaning as in the previous section, and for the sake of simpler notation the components of u are denoted by u^i instead of u_i. The system is weakly coupled, i.e., the i-th equation contains $u^1, \dots, u^m$ but derivatives only of u^i. The first paper by Szarski (1955) contains an estimation theorem similar to the theorem in Section 10. Two years later, Mlak (1957) showed that a monotonicity theorem holds when $f(t,x,z,p,r) = (f_1,\dots,f_m)$ is quasimonotone increasing in $z \in \mathbb{R}^m$. He obtained theorems of the type A and B, where assumption (SS<) now reads

$$(SS<) \qquad \begin{cases} v_t^i < f_i(t,x,v,v_x^i,v_{xx}^i) & \text{in } G, \quad v^i < \phi^i \quad \text{on } \Gamma_o, \dots, \\ w_t^i > f_i(t,x,w,w_x^i,w_{xx}^i) & \text{in } G, \quad w^i > \phi^i \quad \text{on } \Gamma_o, \dots . \end{cases}$$

The corresponding theorem of the M. Müller type for arbitrary systems, which uses the assumption

$$(SSM<) \qquad \begin{cases} v_t^i < f_i(t,x,z,v_x^i,v_{xx}^i) & \text{for } v \le z \le w, \quad z_i = v^i, \dots \\ w_t^i > f_i(t,x,z,w_x^i,w_{xx}^i) & \text{for } v \le z \le w, \quad z_i = w^i, \dots \end{cases}$$

was proved by Walter (1964); see [W; 32.V]. It is quite remarkable that these basic theorems on parabolic systems have almost the same structure as the corresponding theorems for systems of ordinary differential equations.

The theory sketched above is presented in the books [LL], [S] and [W]; many historical remarks are given in [W]. We shall describe in detail

Mildly nonlinear systems. Let us consider the problem

$$(P_{sy}) \qquad u_t^i = L^i u^i + g_i(t,x,u) \text{ in } G, \quad B^i u^i = 0 \text{ on } \Gamma \quad (i = 1,\dots,m),$$

where ($h(t,x)$ real-valued)

$$L^i h = a^i h_{xx} + b^i h_x, \quad a^i \in S^n, \quad a^i \ge 0, \quad b^i = (b_1^i,\dots,b_n^i),$$

$$B^i h = \begin{cases} h - \phi^i & \text{on } \Gamma_o, \\ h_\nu + \theta_i(t,x,h) & \text{on } \Gamma_\nu . \end{cases}$$

The differential operator T is defined by

$$T^i u^i = u_t^i - L^i u^i - g_i(t,x,u) .$$

A monotonicity theorem

$$(M<) \qquad Tv < Tw \text{ in } G, \ Bv < Bw \text{ on } \Gamma \implies v < w \text{ in } \bar{G}$$

holds for bounded G if $g(t,x,z)$ is quasimonotone increasing in z (no assumptions on g, θ and L), while the corresponding (M≤) is true if g and θ are locally Lipschitz continuous in z, Γ_ν is regular and the coefficients a^i, b^i are bounded. Theorem (M≤) holds also for unbounded G (Γ_ν bounded) if v and w satisfy a growth condition $|v|, |w| \le K \exp(Kx^2)$.

This result generates corresponding theorems A and B which we formulate in the special case that v and w depend only on t and that $\theta(t,x,z) \equiv 0$ (the latter assumption implies that $v(t)$ and $w(t)$ satisfy the boundary condition on Γ_ν). Under these conditions

$$(SS<) \quad \begin{cases} v'(t) < g(t,x,v) \text{ in } G, & v < \phi \text{ on } \Gamma_o, \\ w'(t) > g(t,x,w) \text{ in } G, & w > \phi \text{ on } \Gamma_o \end{cases}$$

implies $v < u < w$ for every solution u of (P_{sy}), and

$$(SS\le) \text{ implies } v \le u \le w \text{ in } \bar{G}$$

under the assumptions given for $(M\le)$.

For arbitrary $g(t,x,z)$ a theorem of the M. Müller type holds. The assumptions

$$(SSM<) \quad \begin{cases} v_i' < g_i(t,x,z) & \text{for } v(t) \le z \le w(t), \; z_i = v_i(t), \; v_i < \phi^i \text{ on } \Gamma_o, \\ w_i' > g_i(t,x,z) & \text{for } v(t) \le z \le w(t), \; z_i = w_i(t), \; w_i > \phi^i \text{ on } \Gamma_o \end{cases}$$

imply $v < u < w$ in G (similarly with $\le$).

Estimation theorem. *Let* u *be a solution of* (P_{sy}), *and let* $v(t,x)$ *satisfy*

$$|Tv| \le \delta(t) \text{ in } G, \quad |v-\phi| \le \varepsilon(t) \text{ on } \Gamma_o, \quad v_\nu^i + \theta_i(t,x,v^i) = 0 \text{ on } \Gamma_\nu,$$

where $|\cdot|$ *is the maximum norm in* $\mathbb{R}^m$. *If* $g(t,x,z)$ *and* $\rho(t)$ *satisfy*

$$|g(t,x,z)-g(t,x,z')| \le \omega(t,|z-z'|),$$

$$\rho' > \omega(t,\rho) + \delta(t) \text{ and } \rho(t) > \varepsilon(t),$$

then

$$|u-v| < \rho \quad \text{in} \quad \bar{G}.$$

Theorems on uniqueness and continuous dependence follow in an obvious way.

Mildly nonlinear parabolic systems appear in a great variety of applications, such as nonlinear diffusion, flame propagation, superconductive liquids, conduction of nerve impulses, predator-prey and competing species models. The above theorems have been used (and sometimes rediscovered) by many authors. The case where v and w are constant is known under the name "invariant rectangles".

INVARIANCE

12. Invariant sets

The notion of invariant sets with respect to initial or boundary value problems has been explained in Section 1. We treat two cases, systems of ordinary differential equations and parabolic systems.

Ordinary differential equations. A closed set $M \subset \mathbb{R}^m$ is said to be (positively) invariant with respect to the system $u' = f(t,u)$, where $u = (u_1,\dots,u_m)$, $f = (f_1,\dots,f_m)$, if

$$u(t_o) \in M \quad \text{implies} \quad u(t) \in M \quad \text{for} \quad t > t_o$$

(as long as the solution exists).

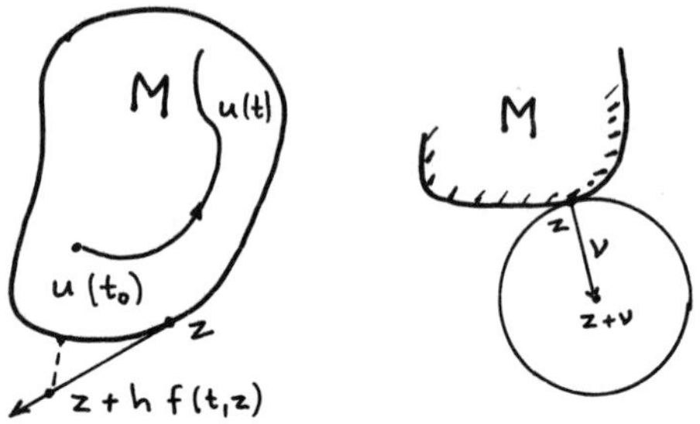

It is intuitively clear that the set M can only be invariant if at a boundary point $z \in \partial M$ the vector $f(t,z)$ points into the interior of M or is tangent to M. There are two ways to give this vague idea a precise meaning without assuming that a tangent plane exists at z. The first such *tangential condition* (dist (x,M) is the distance from x to M)

$$(T_n) \qquad \lim_{h \to 0+} \frac{1}{h} \operatorname{dist}(z + hf(t,z), M) = 0 \quad (z \in M)$$

was given by Nagumo (1942), the second one, due to Bony (1969), uses the inner product $(\cdot,\cdot)$ in $\mathbb{R}^m$ and outer normals ν_z at z,

$$(T_i) \qquad (\nu_z, f(t,z)) \le 0 \text{ for } z \in \partial M \text{ and all outer normals } \nu_z .$$

Here, $\nu = \nu_z \neq 0$ is said to be an outer normal to M at z, if the open ball with center at $z+\nu$ and radius $|\nu|$ has no point in common with M.

It has been shown by Redheffer (1972) that (T_n) implies (T_i) and by Crandall (1972) that, when f is continuous, (T_i) implies (T_n) (even uniformly on compact sets). We give two basic results.

Invariance. $(T_i) + (U)$ implies (positive) invariance of M.

Existence. If (T_n) holds and if f is continuous in $J \times M$ ($J = [0,T]$), then for every $c \in M$ the initial value problem $u' = f(t,u)$, $u(0) = c$ has a solution u which remains in M. The solution exists in J if M is bounded.

The existence theorem was proved by Nagumo (1942) and rediscovered around 1970 by several authors. The invariance theorem (without a continuity assumption) is found in Redheffer (1972). Condition (U) is a Lipschitz condition or the weaker condition (D) of Section 7 with $\omega \in UCD$.

A simple proof of the invariance theorem runs as follows. Let u(t) be a solution of $u' = f(t,u)$, let $u(0) \in M$ and $\rho(t) = \operatorname{dist}(u(t),M)$. Assume that $u(s) \notin M$, i.e., $\rho(s) > 0$, and let $y \in \partial M$ be such that $\rho(s) = |u(s)-y|$. The function $\sigma(t) = |u(t)-y|$ satisfies $\rho(t) \le \sigma(t)$, $\rho(s) = \sigma(s)$, hence $D^+\rho(s) \le \sigma'(s)$. Note that σ is differentiable near s and

$$\tfrac{1}{2}(\sigma^2)' = \sigma\sigma' = (u(t)-y, u'(t)) = (u(t)-y, f(t,u(t))) .$$

As $t = s$ we have, using (T_i) and (D), since $\nu_y = u(s)-y$ is an outer normal to M at y,

$$\sigma\sigma' = (\nu_y, f(s,u(s)) \leq (\nu_y, f(s,u)-f(s,y)) \leq \rho\omega(t,\rho) .$$

Hence $D^+\rho \leq \omega(t,\rho)$. Since s was arbitrary, we have $D^+\rho \leq \omega(t,\rho)$ at all points in J where $\rho > 0$. For $\varepsilon > 0$ there exists a function $\tau(t)$ satisfying $\tau' > \omega(t,\tau)$ and $\tau < \varepsilon$ in J. By Theorem A of Section 4, $\rho < \tau$, hence $\rho \equiv 0$.

The above results have been extended to differential equations in Hilbert and Banach spaces. Condition (T_n) uses the distance and is defined in any Banach space, while (T_i) employs the inner product and carries over to Hilbert spaces. Results of this nature and historical remarks are found in Deimling's book (1977; § 5). The following relation between monotonicity theorems and invariance theorems is important from a methodological point of view. A monotonicity theorem for the operator $Tu = u'-f(t,u)$ in the Banach space B with positive cone C

$$v(0) \leq w(0),\ Tv \leq Tw \text{ in } J \Rightarrow v \leq w \text{ in } J$$

can be reformulated as an invariance theorem for the difference $u = w-v$ with respect to the set C

$$u(0) \in C,\ u' = g(t,u) \text{ in } J \Rightarrow u(t) \in C \text{ for } t \in J ,$$

where

$$g(t,z) = f(t,v(t)+z) - f(t,v(t)) + Tw - Tv .$$

One can show that the tangential condition for g with respect to C is equivalent to the quasimonotonicity condition for f given in Section 6. From this point of view, which was described in Redheffer and Walter (1975) (see also Deimling (1977; § 5)), the monotonicity theorem is a corollary of an invariance theorem.

Parabolic systems. We restrict consideration to systems of the form

$$u^i_t = Lu^i + g_i(t,x,u) \quad (i = 1,\ldots,m)$$

with one and the same elliptic operator $L\phi = a\phi_{xx} + b\phi_x$ in all equations. Invariance of a set $M \subset \mathbb{R}^m$ is defined in the usual way by $u(\Gamma) \subset M \Rightarrow u(G) \subset M$. Using the same tangential condition as before,

(T_i) $\qquad (\nu_z, g(t,x,z)) \leq 0$ for $z \in \partial M$, ν_z outer normal at z ,

we get the following

Invariance theorem. *Assume that* G *is bounded, that* M *is closed and convex and that* (T_i) *and the uniqueness condition* (U) *of Section* 10 *holds. Then* M *is invariant.*

This carries over to solutions u which satisfy a condition

$$u_\nu^i + \theta_i(t,x,u^i) = 0 \quad \text{on } \Gamma_\nu ,$$

if $\theta = (\theta^1,\dots,\theta^m)$ satisfies also a tangential condition

$$(T_\nu) \qquad (\nu_z, \theta(t,x,z)) \le 0 .$$

Under these conditions

$$(T_i),\ (T_\nu),\ (U),\ u(\Gamma_o) \in M \text{ implies } u(G) \in M ,$$

if a, b are bounded, Γ_ν is regular and θ is Lipschitz continuous in z.

These results can be extended to the Cauchy problem and other problems with unbounded G.

The first invariance theorem of this type was proved by Weinberger (1975). Shortly afterwards, Amann (1978), Bebernes and Schmitt (1977) and others have extended this result. An approach which uses no smoothness assumptions on g (and hence no existence theory in the proofs) is described in Redheffer and Walter (1977).

Invariance theorems for parabolic systems differ from those for systems of ordinary differential equations in an essential point: The set M has to be convex. This additional assumption is necessary in the parabolic case as the following example shows.

Let $M \subset \mathbb{R}^m$ be an invariant set, let $z_o, z_1 \in M$ and let z_o, z_1 be connected by a polygonal line lying in M which has a parameter representation $z = \phi(x)$ for $0 \le x \le 1$, where $\phi(0) = z_o$ and $\phi(1) = z_1$. The problem

$$u_t = u_{xx} \text{ in } (0,\infty)\times(0,1) ,\quad u(t,0) = z_o ,\quad u(t,1) = z_1 ,\quad u(0,x) = \phi(x)$$

for $u = (u^1,\dots,u^m)$ (which consists of m uncoupled heat equations) has a solution u(t,x) which tends to the stationary solution $\bar{u}(x) = z_o + x(z_1 - z_o)$ as $t \to \infty$. The tangential condition (T_i) holds since $g = 0$. Hence the points on the line segment which connects z_o and z_1 belong to M, i.e., M is convex.

ELLIPTIC EQUATIONS AND SYSTEMS

Maximum principles for ellliptic equations and related results are presented in many textbooks. For this reason we shall give only a short outline of basic results. We begin with the one-dimensional case.

13. Ordinary differential equations of the second order

We consider the boundary value problem

(BVP) $$u''(x) = f(x,u,u') \quad \text{in} \quad J = [a,b] \,, \quad u(a) = c \,, \quad u(b) = d \,.$$

Conditions for sub- and superfunctions read (note that the differential inequalities for v and w are reversed)

(SS<) $$\begin{cases} v'' > f(x,v,v') & \text{in} \quad J \,, \quad v(a) < c \,, \quad v(b) < d \,, \\ w'' < f(x,w,w') & \text{in} \quad J \,, \quad w(a) > c \,, \quad w(b) > d \,. \end{cases}$$

In the case of one scalar equation we have four Theorems A - D of the type given in Section 4.

Theorem A. (SS<) + (I) => $v < u < w$ *in* J.

Theorem B. (SS≤) + (SI) => $v \le u \le w$ *in* J.

Here, (I) or (SI) signifies that $f = f(x,z,p)$ is increasing or strictly increasing in z, respectively. The counterexample $u'' + u = 0$ in $[0,\pi]$, $u(0) = u(\pi) = 0$ with the solutions $u(x) = C \sin x$ shows that without a monotonicity condition these theorems do not hold. More generally, Theorem B is certainly not true for the homogeneous problem

(*) $$u'' + \lambda p(x)u = 0 \quad \text{in } J \,, \quad u(a) = u(b) = 0 \,, \qquad (p > 0 \text{ in } J) \,,$$

when a nonvanishing solution exists, i.e., when λ is an eigenvalue. Yet it remains true not only for $\lambda < 0$ (as stated above), but for $\lambda < \lambda_1$, where λ_1 is the first eigenvalue of (*). This is contained in the following version of Theorem B.

Theorem. *Assume that* f *satisfies*

$$f(x,z,p) - f(x,z',p') \le L(x)(z'-z) + M(x)|p-p'| \quad \textit{for} \quad z' \ge z$$

and that there exists a function ρ *with the properties*

$$\rho'' + L(x)\rho + M(x)\rho' < 0 \ \textit{and} \ \rho > 0 \ \textit{in} \ J \,.$$

Then (SS≤) *implies* $v \le u \le w$ *in* J. *In particular, uniqueness for* (BVP) *follows from these assumptions.*

Theorem D (if v, w satisfy (SS≤) and $v \le w$ and if f is continuous, then there exists a solution of (BVP) between v and w) and a corresponding Theorem C (there exists a maximal solution between v and w) hold without a monotonicity assumption if f is independent of u'. In the general case a growth condition in p such as $|f(x,z,p)| \le L(1+p^2)$ is required, and the following result is needed.

(a) If $|u(x)| \le M$ and $|u''(x)| \le L(1+|u'(x)|^2)$ in J, then there exists a constant $C = C(L,M,b-a)$ such that $|u'(x)| \le C$ in J.

Most of these results are contained in a pioneering article by Nagumo (1937). In the sixties, the theory was perfected and extended to weakly coupled systems

$$u_i'' = f_i(x,u,u_i') \qquad (i = 1,\dots,m) ,$$

where the i-th equation depends on $u = (u_1,\dots,u_m)$, but on the derivative of u_i only. Much of this work was accomplished by L. Jackson and his school; see Jackson (1968). For example, it was shown by Heimes (1966) that Theorem A remains true if $f(x,z,p) = (f_1,\dots,f_m)$ is quasimonotone decreasing in z and if (I) holds in the form that $f_i(x,z_1+t,\dots,z_m+t,p)$ is increasing in t $(i = 1,\dots,m)$. In the linear case

$$f_i(x,z,p) = \sum_{j=1}^{m} a_{ij}(x)z_j + b_i(x)p ,$$

these assumptions are satisfied iff $a_{ij} \le 0$ for $i \neq j$ and $\sum_j a_{ij} \ge 0$.

Summarizing the results regarding problem (BVP) for weakly coupled systems, we discover a remarkable similarity with the results for (IVP) in Section 5: The basic results for scalar equations $(m = 1)$ carry over (with a natural extension of monotonicity assumptions) to those systems for which f is quasimonotone decreasing in z, while in the general case the differential inequalities (SS) have to be replaced by stronger conditions of the M. Müller type, e.g.,

$$(\text{SSM}\le)\quad \begin{cases} v_i'' \ge f_i(x,z,v_i') & \text{for } v \le z \le w ,\quad z_i = v_i \quad (x \in J) ,\\ w_i'' \le f_i(x,z,w_i') & \text{for } v \le z \le w ,\quad z_i = w_i \quad (x \in J) ,\\ v(a) \le c \le w(a) , & v(b) \le c \le w(b) . \end{cases}$$

14. Elliptic equations

In what follows, L is a linear elliptic operator as defined in Sections 8 and 10,

$$Lu = au_{xx} + bu_x \quad \text{with} \quad a \geq 0 \, ,$$

where a(x) and b(x) are defined in a bounded open set $D \subset \mathbb{R}^m$ with boundary $\partial D = \Gamma$. We note that a formally self-adjoint operator

$$Lu = \sum_{i,j=1}^{n} (a_{ij}(x)u_{x_i})_{x_j}$$

can be reduced to that form if a is of class C^1. It is always assumed that the functions u, v, ... belong to $Z = Z(D) = C^2(D) \cap C^0(\bar{D})$.

It was discovered as early as 1839 by Gauß and at about the same time by Earnshaw that a function u satisfying $\Delta u > 0$ in D cannot have a maximum in D. The following positivity theorems, which are essentially due to Paraf (1892), generalize this statement ($c = c(x)$ is defined in D).

(P<) $c \leq 0$, $Lu + cu < 0$ *in* D, $u \geq 0$ *on* Γ *implies* $u > 0$ *in* D.

(P≤) $c \leq 0$, $Lu + cu \leq 0$ *in* D, $u \geq 0$ *on* Γ *implies* $u \geq 0$ *in* D *if* $c < 0$ *or if there exists* $h \in Z$ *with* $Lh > 0$ *in* D.

The second theorem holds, e.g., if $a \not\equiv 0$ and $b = 0$ in D (take $h = x^2$) or if $a_{11} \geq \delta > 0$ and $|b_1| \leq K$ (take $h = \exp(\alpha x_1)$).

The proof of (P<) is simple. If u is not positive in D, then there exists $x_o \in D$ such that min u = $u(x_o) \leq 0$, hence $Lu \geq 0$ by 8 (e) and $cu \geq 0$ at x_o, which is a contradiction. (P≤) is reduced to (P<) by considering $u_\varepsilon = u + \varepsilon(A-h)$, where $A > \max h$.

It follows immediately from (P≤) that the maximum-minimum principle holds for solutions of $Lu = 0$. A stronger statement was proved by E. Hopf (1927):

<u>Strong minimum principle.</u> Assume that $Lu + cu \leq 0$ in D and $u \geq 0$ on Γ, where $c \geq 0$ and the coefficients a, b, c are locally bounded and $a(x) \geq \delta > 0$ locally in D. Then $u \equiv 0$ in D or $u > 0$ in D.

This principle follows easily from another principle, which was discovered by Hopf (1952) and (independently) by Oleinik (1952).

<u>Boundary minimum principle.</u> Let $Lu + cu \leq 0$ in the ball $B = B_R(x_o)$, $u \geq 0$ on ∂B and $u(x_o) > 0$, where a, b and $c \leq 0$ are bounded and $(x-x_o)^T a(x)(x-x_o) \geq \alpha |x-x_o|^2$ ($\alpha > 0$) in B. Then there exists $\delta > 0$ such that

$$u(x) \geq \delta(R^2 - |x-x_o|^2) \quad \text{in} \quad B \, .$$

Hence, if u vanishes at a point $\bar{x} \in \partial B$, then the outer normal derivative at $\bar{x}$ is negative.

The name 'principle' points to the fact that both theorems are indispensable tools in the theory of elliptic equations. We were careful in stating the necessary assumptions, because there are important examples of elliptic equations which degenerate at the boundary. Take Tricomi's equation $yu_{xx} + u_{yy} = 0$, for example. If D is lying in the upper half-plane and Γ has points in common with the x-axis, then the strong minimum principle applies. Likewise, for the ball B: $x^2+(y-1)^2 < 1$, the ellipticity condition for the matrix $a = \mathrm{diag}\,(y,1)$ is satisfied, but an inequality $a \geq \alpha I_2$ in B does not hold with $\alpha > 0$.

The above positivity theorems can be used to derive monotonicity and uniqueness theorems for the nonlinear Dirichlet problem

$$\text{(E)} \qquad Lu = f(x,u) \text{ in } D\,, \quad u = \phi \text{ on } \Gamma\,.$$

Corresponding conditions for sub- and superfunctions are now given by

$$\text{(SS<)} \qquad \begin{cases} Lv > f(x,v) \text{ in } D\,, & v < \phi \text{ on } \Gamma\,, \\ Lw < f(x,w) \text{ in } D\,, & w > \phi \text{ on } \Gamma\,. \end{cases}$$

All four Theorems A - D of the preceding section are valid for problem (E) under the conditions on f given there, i.e., f(x,z) increasing in z in Theorem A, strictly increasing in z in Theorem B, continuous in Theorems C and D.

A new type of monotonicity theorems which uses a family of upper solutions instead of just one upper solution was discovered by McNabb (1961). It has proved useful and is sometimes called Serrin's sweeping principle.

Theorem. *Assume that* $v, w_\lambda \in Z$ $(\alpha \leq \lambda \leq \beta)$ *and that the family* (w_λ) *is continuous in* (x,λ) *and increasing in* λ. *If*

$$Lv + f(x,v) > Lw_\lambda + f(x,w_\lambda) \text{ in } D \quad \text{and} \quad v < w_\lambda \text{ on } \Gamma \quad \text{for} \quad \alpha < \lambda \leq \beta\,,$$

then $v \leq w_\beta$ *in* D *implies* $v \leq w_\alpha$ *in* D.

A similar theorem holds with equality permitted. Among others, this theorem can be used to generalize (P) by requiring only $c(x) < \lambda_1 p(x)$ instead of $c \leq 0$, where λ_1 is the first eigenvalue of the eigenvalue problem $Lu + \lambda p(x)u = 0$ in D, $u = 0$ on Γ $(p(x) > 0$ in D).

For proof, one assumes that the conclusion is false and considers the minimal index μ such that $v \leq w_\mu$ in D. Then there is $x_0 \in D$ with $v(x_0) = w_\mu(x_0)$, and one obtains a contradiction to the differential inequality by applying 8 (e) to the difference $w_\mu - v$.

As in the case $n = 1$ discussed in Section 13, the theorems carry over to weakly coupled systems of the form

$$(E_{Sy})\qquad L^i u \equiv a^i u^i_{xx} + b^i u^i_x = f^i(x,u) \quad (i = 1,\ldots,m;\ a^i \geq 0)$$

in a way which is by now familiar. Whenever f is quasimonotone decreasing in z, then Theorems A - D remain true with the same interpretation of condition (I) as in Section 13, and McNabb's theorem on families of upper solutions is also true as it stands. In the general case, conditions of Müller type for upper and lower solutions

$$(SSM\leq)\qquad \begin{cases} L^i v^i \geq f_i(x,z) & \text{for } v(x) \leq z \leq w(x),\quad z_i = v^i(x), \\ L^i w^i \leq f_i(x,z) & \text{for } v(x) \leq z \leq w(x),\quad z_i = w^i(x), \\ v \leq \phi \leq w & \text{on } \Gamma \end{cases}$$

come into use. For example, in Theorem D it is assumed that $v,w \in Z(D)$ satisfies (SSM≤) and $v \leq w$ in D and that f is continuous in the set $K \subset \mathbb{R}^{n+m}$ of points (x,z) satisfying $x \in \bar{D}$ and $v(x) \leq z \leq w(x)$. Then, under proper regularity assumptions regarding L^i, ϕ and Γ, there exists a solution u of (E_{Sy}) satisfying $v \leq u \leq w$ in D and $u = \phi$ on Γ.

There exists a similar theory of elliptic differential inequalities with respect to boundary value problems of the second and third kind, where on (part of) the boundary the normal derivative u_ν or a combination $u_\nu + h(x,u)$ is prescribed. For these and other matters regarding elliptic equations we refer to the book by Protter and Weinberger cited in the preamble and to [W; p.304-319].

EPILOG

This article represents an attempt to cover a large body of methods and results from different areas of differential inequalities in the limited space available. The main problem was how to condense the material without destroying the flavor of the subject. It was clear from the beginning that only the very basic results could be considered. Functional differential equations and methods based on Lyapunov functions and functionals were left out, and existence theory was treated rather cursorily. When looking for unifying points of view, the subject turned out to be obliging to an amazing degree. This applies both to methods of proof and statements of results. Among the various conclusions to be drawn from a monotonicity theorem (M) we singled out four types of theorems: A (upper and lower bounds using strict inequalities), B (the same with weak inequalities), C (maximal and minimal solutions), and D (existence of a solution between an upper and lower solution). Such a restriction has also its benefits. For example, it becomes manifest that Perron's version (1915) of Peano's existence theorem and the existence theorem for the Dirichlet problem $\Delta u = f(x,u)$ in the presence of an upper solution w and a lower solution v, which goes back (at least) to Nagumo (1954), are just different cases of Theorem D and use the same method of proof. Once this interconnection has been recognized, it becomes a matter of routine to formulate and prove a similar theorem for elliptic systems along the lines of M. Müller's (1927) version of Theorem D for systems of ordinary differential equations. Let us remark here that in the setting of Theorem D the Leray-Schauder degree with respect to the set of functions between v and w is +1. In some cases this gives additional information on the number of solutions between v and w, which cannot be obtained directly from Theorem D; see, e.g., McKenna and Walter (1986).

We conclude with some remarks about existence and successive approximation which shed new light on quasimonotonicity, the basic concept for systems of equations. Let us start with the initial value problem

$$\text{(IVP)} \qquad u'(t) = f(t,u) \quad \text{in} \quad J = [0,T] \, , \quad u(0) = 0 \, .$$

If $f(t,z)$ (real-valued) is Lipschitz continuous in z, then there exists $\alpha > 0$ such that $F(t,z) = f(t,z) + \alpha z$ is increasing in z. In order to transform the problem, which now reads $u' + \alpha u = F(t,u)$, $u(0) = 0$, into an integral inequation, we solve the problem $u' + \alpha u = g(t)$, $u(0) = 0$. Since the differential equation is equivalent to $(e^{\alpha t}u)' = e^{\alpha t}g$, we get (with $D = d/dt$)

$$(D+\alpha)u = g \, , \quad u(0) = 0 \quad \Longleftrightarrow \quad u(t) = e^{-\alpha t}\int_0^t e^{\alpha s}g(s)ds =: (Rg)(t) \, .$$

The operator R is the inverse of the map $D+\alpha$, which maps the space $C_o^1(J) = \{u \in C^1(J): u(0) = 0\}$ bijectively onto $C(J)$. In this way, (IVP) is transformed into a fixed point equation in the Banach space $X = C(J)$

$$u = RF(t,u) =: Tu \, , \quad \text{where} \quad R = (D+\alpha)^{-1} \, .$$

The operator R is linear, compact and positive ($u \geq 0 \Rightarrow Ru \geq 0$), and if we assume that f is continuous in (t,z) and Lipschitz continuous in z, then T is compact (in the present case even a contraction, when a weighted maximum norm is used in X) and monotone ($u \leq v \Rightarrow Tu \leq Tv$) in X. A subsolution v in the sense of Section 4 satisfies $v \leq Tv$. If we start successive approximation $u_{n+1} = Tu_n$ with $u_o = v$, then an increasing sequence (u_n) is produced which converges to the unique solution $u = \lim u_n$ of (IVP). Since $v = u_o \leq u_n$, we get $v \leq u$, which is Theorem B of Section 4.

This line of reasoning remains true for systems if $f(t,z)$ has the property that $F(t,z) = f(t,z) + \alpha z$ is increasing in z ($z \leq z' \Rightarrow F(t,z) \leq$

$F(t,z')$, where $z,z' \in \mathbb{R}^m$) for some $\alpha > 0$. It turns out that this property is basically quasimonotonicity. More precisely:

A Lipschitz continuous function $g: M \subset \mathbb{R}^m \to \mathbb{R}^m$ *is quasimonotone increasing if and only if there exists* $\alpha > 0$ *such that* $G(z) = g(z) + \alpha z$ *is increasing.*

In this way, Theorem B is established for quasimonotone systems. It has already been pointed out in the introduction that such an approach, which uses existence theory from the beginning, is not recommended because it requires regularity assumptions. Nevertheless, it is useful in proving Theorem C. We shall explain this point in the Dirichlet problem

$$\Delta u = f(x,u) \text{ in } D\,, \quad u = 0 \text{ on } \Gamma$$

Let $F(x,z) = \alpha z - f(x,z)$ and let $R = (\alpha - \Delta)^{-1}$ be defined by

$$\alpha u - \Delta u = g(x) \text{ in } D\,, \quad u = 0 \text{ on } \Gamma \quad \Longleftrightarrow \quad u = Rg\,.$$

Then our problem becomes $u = RF(x,u) =: Tu$ as before (in a proper Banach space, e.g., $C^\alpha(\bar{D})$). The operator R is compact, and Theorem (P≤) of Section 14 implies that R is a positive operator; hence T is compact and monotone. Now consider the situation of Theorem D, where we have an upper solution w, a solution u and a lower solution v with $v \le u \le w$ in D. Since $v \le Tv$ and $w \ge Tw$ and since T is monotone, we get $v \le Tv = v_1 \le u \le Tw = w_1 \le w$. Applying this procedure n times, the inequalities $v_n \le u \le w_n$ are obtained, where v_n or w_n is the n-th successive approximation, starting from v or w, resp. In this way a version of Theorem C is proved: There exists a maximal solution $u^* = \lim w_n$ and a minimal solution $u_* = \lim v_n$ with the property that every solution between v and w is between u_* and u^*.

LITERATURE

Our intention to present only basic results is reflected in the selection of bibliographical data. The more recent literature reports on many important and beautiful new results which are of a more special nature and had to be omitted in this survey article.

The four books cited in the preamble are abbreviated (in alphabetical order) [LL], [PW], [S], [W] (English edition), while the German edition of the author's book is cited as Walter (1964).

Amann, H., Invariant sets and existence theorems for semilinear parabolic and elliptic systems. J. Math. Anal. Appl. 65 (1978), 432-467.

Bebernes, I.W., and K. Schmitt, Invariant sets and the Hukuhara-Kneser property for systems of parabolic partial differential equations. Rocky Mountain J. Math. 7 (1977), 557-567.

Bompiani, E., Un teorema di confronto ed un teorema di unicità per l'equazione differenziale y' = f(x,y). Atti Accad. Naz. Lincei. Rend. Classe Sci. Fis. Mat. Nat. (6)1 (1925), 298-302.

Bony, J.-M., Principe du maximum, inégalité de Harnack et unicité du problème de Cauchy pour les opérateurs elliptiques dégénérés. Ann. Inst. Fourier Grenoble 19 (1969), 277-304.

Crandall, M.G., A generalization of Peano's theorem and flow invariance. Proc. AMS 37 (1972), 151-155.

Deimling, K., Ordinary differential equations in Banach spaces. Lecture Notes in Math., Vol. 596. Springer 1977.

Friedman, A., Remarks on the maximum principle for parabolic equations and its applications. Pacific J. Math. 8 (1958), 201-211.

Gauß, C.F., Allgemeine Lehrsätze in Beziehung auf die im verkehrten Verhältnisse des Quadrats der Entfernung wirkenden Anziehungs- und Abstoßungskräfte. Beob. d. magn. Vereins f. 1839. Ges. Werke 5, p.200-242 (Lehrsatz 20 (p.222) is the mean value theorem for harmonic functions, and Lehrsatz 21 (p.223) can be interpreted as a version of the strong maximum principle).

Haar, A., Über die Eindeutigkeit und Analyzität der Lösungen partieller Differentialgleichungen. Atti del Congresso Internazionale dei Matematici Bologna, Bd.3, 1928 (p.5-10).

Heimes, K.A., Boundary value problems for ordinary nonlinear second order systems. J. Diff. Eqs. 2 (1966), 449-463.

Holmgren, E., Sur les solutions quasianalytiques de l'équation de la chaleur. Ark. för Mat., Astron. och Fys. 18 (1924) Nr. 9 (9 p.) (Jb. Fortschritte d. Math. 50, p.337).

Hopf, E., Elementare Bemerkungen über die Lösungen partieller Differentialgleichungen zweiter Ordnung vom elliptischen Typus. Sitz.-ber. Preuss. Akad. Wiss. 19 (1927), 147-152.

Hopf, E., A remark on linear elliptic differential equations of second order. Proc. Amer. Math. Soc. 3 (1952), 791-793.

Jackson, L., Subfunctions and boundary value problems for second order ordinary differential equations. Advances in Math. 2 (1968), 307-363.

Kamke, E., Differentialgleichungen reeller Funktionen. Akadem. Verlagsges., Leipzig 1930.

Kamke, E., Zur Theorie der Systeme gewöhnlicher Differentialgleichungen II. Acta Math. 58 (1932), 57-85.

v.Koch, H., Sur les systèmes d'ordre infini d'équations différentielles. Stockh. Öfv. 56 (1899), 395-411 (Jb. Fortschritte d. Math. 30, p.310).

Lemmert, R., Existenz- und Konvergenzsätze für die Prandtlschen Grenzschichtdifferentialgleichungen unter Benutzung der Linienmethode. ZAMM 57 (1977), 207-220.

McKenna, P.J., and W. Walter, On the Dirichlet problem for elliptic systems. Applicable Analysis 21 (1986), 207-224.

McNabb, A., Strong comparison theorems for elliptic equations of second order. J. Math. Mech. 10 (1961), 431-440.

Mlak, W., Differential inequalities of parabolic type. Ann. Polon. Math. 3 (1957), 349-354.

Mlak, W., Parabolic differential inequalities and Chaplygin's method. Ann. Polon. Math. 8 (1960), 139-153. (The critical remarks in Math. Reviews 22, No. 6930, are not justified, since a Hölder condition and not a Lipschitz condition is assumed.)

Mlak, W., and C. Olech, Integration of Infinite Systems of Differential Inequalities. Ann. Polon. Math. 13 (1963), 105-112.

Müller, M., Über das Fundamentaltheorem in der Theorie der gewöhnlichen Differentialgleichungen. Math. Z. 26 (1926), 619-645.

Müller, M., Über die Eindeutigkeit der Integrale eines Systems gewöhnlicher Differentialgleichungen und die Konvergenz einer Gattung von Verfahren zur Approximation dieser Integrale. Sitz.-ber. Heidelberg, Akad. Wiss., Math.-Naturw. Kl., 9. Abh. (1927).

Nagumo, M., Über die Differentialgleichung $y'' = f(x,y,y')$. Proc. Phys. Math. Soc. Japan 19 (1937), 861-866.

Nagumo, M., Über die Ungleichung $\frac{\partial u}{\partial x} > f\left(x,y,u,\frac{\partial u}{\partial y}\right)$. Japan. J. Math. 15 (1938), 51-56.

Nagumo, M., Note in "Kansū-Hōteisiki" No. 15 (1939). (Japanese. Reference given in Nagumo-Simoda (1951).)

Nagumo, M., Über die Lage der Integralkurven gewöhnlicher Differentialgleichungen. Proc. Phys. Math. Soc. Japan 24 (1942), 551-559.

Nagumo, M., and S. Simoda, Note sur l'inégalité différentielle concernant les équations du type parabolique. Proc. Japan. Acad. 27 (1951), 536-539.

Nagumo, M., On principally linear elliptic differential equations of the second order. Osaka Math. J. 6 (1954), 207-229.

Nirenberg, L., A strong maximum principle for parabolic equations. Commun. Pure Appl. Math. 6 (1953), 167-177.

Oleinik, O.A., On properties of solutions of certain boundary problems for equations of elliptic type. Mat. Sbornik, N.S., 30 (1952), 695-702.

Paraf, M.A., Sur le problème de Dirichlet et son extension au cas de l'équation linéaire générale du second ordre. Ann. Fac. Sci. Toulouse 6, Fasc. 47 - Fasc. 54 (1892).

Perron, O., Über den Integralbegriff. Sitz.-ber. Heidelberg. Akad. Wiss. Abt. A, 14. Abhandl. (1914).

Perron, O., Ein neuer Existenzbeweis für die Integrale der Differentialgleichung $y' = f(x,y)$. Math. Ann. 76 (1915), 471-484.

Perron, O., Eine neue Behandlung des ersten Randwertproblems für $\Delta u = 0$. Math. Z. 18 (1923), 42-54.

Perron, O., Über Ein- und Mehrdeutigkeit des Integrals eines Systems von Differentialgleichungen. Math. Ann. 95 (1926), 98-101.

Picone, M., Sulla problema della propagazione del calore in un mezzo privo di frontiera, conduttore, isotropo e omogeneo. Math. Ann. 101 (1929), 701-712.

Redheffer, R.M., The theorems of Bony and Brezis on flow-invariant sets. Amer. Math. Monthly 79 (1972), 740-747.

Redheffer, R.M., and W. Walter, Flow-invariant Sets and Differential Inequalities in Normed Spaces. Applicable Analysis 5 (1975), 149-161.

Redheffer, R.M., and W. Walter, Invariant Sets for Systems of Partial Differential Equations. I. Parabolic Equations. Arch. Rat. Mech. Anal. 67 (1977), 41-52.

Redheffer, R.M., and W. Walter, Remarks on ordinary differential equations in ordered Banach spaces. Monatshefte Math. 102 (1986), 237-249.

Scheeffer, L., Zur Theorie der stetigen Funktionen einer reellen Veränderlichen. Acta math. 5 (1884), 279-296.

Szarski, J., Sur certains systèmes d'inégalités différentielles aux dérivées partielles du premier ordre. Ann. Soc. Polon. Math. 21 (1948), 7-25.

Szarski, J., Sur la limitation et l'unicité des solutions d'un système non-linéaire d'équations paraboliques aux dérivées partielles du second ordre. Ann. Polon. Math. 2 (1955), 237-249.

Tychonov, A., Théorèmes d'unicité pour l'équation de la chaleur. Math. Sb. 42 (1935), 199-215.

Volkmann, P., Gewöhnliche Differentialgleichungen mit quasimonoton wachsenden Funktionen in topologischen Vektorräumen. Math. Z. 127 (1972), 157-164.

Vyborny, R., The properties of the solutions of certain boundary problems for parabolic equations. Dokl. Akad. Nauk SSSR 117 (1957), 563-565.

Walter, W., Gewöhnliche Differential-Ungleichungen im Banachraum. Arch. Math. 20 (1969), 36-47.

Walter, W., Existence and convergence theorems for the boundary layer equations based on the line method. Arch. Rat. Mech. Anal. 39 (1970), 169-188.

Walter, W., The line method for parabolic differential equations. Problems in boundary layer theory and existence of periodic solutions. Lecture Notes in Math., Vol. 430 "Constructive and Computational Methods for Differential and Integral Equations" (p.395-413). Springer Verlag 1974.

Walter, W., On the strong maximum principle for parabolic differential equations. Proc. Edinburgh Math. Soc. 29 (1986), 93-96.

Weinberger, H., Invariant sets for weakly coupled parabolic and elliptic systems. Rend. Mat. 8, Ser. VI (1975), 295-310.

Westphal, H., Zur Abschätzung der Lösungen nichtlinearer parabolischer Differentialgleichungen. Math. Z. 51 (1949), 690-695.